Atlas of Diagnostic Cytology

Atlas of Diagnostic Cytology

Claude Gompel, M.D., F.I.A.C.

Professor and Chairman
Department of Pathology
Institut Jules Bordet
Université Libre de Bruxelles

With the Assistance of
J. F. SCARFO, B.A., M.A.

Chapter on Cytogenetics by
A. VERHEST, M.D.
Department of Pathology
Institut Jules Bordet
Université Libre de Bruxelles

A Wiley Medical Publication
John Wiley & Sons
New York • Chichester • Brisbane • Toronto

Library of Congress Cataloging in Publication Data:

Gompel, C
    Atlas of diagnostic cytology.

    (A Wiley medical publication)
    Includes index.
    1. Diagnosis, Cytologic — Atlases. I. Scarfo, J.F.,
joint author. II. Verhest, A., joint author.
III. Title. [DNLM: 1. Cytodiagnosis — Atlases. QY17
G643a]
RB43.G65   616.07′582   77-27068
ISBN 0-471-02278-0

Printed in the United States of America

10 9 8 7 6 5 4 3 2 1

Preface

My ambition in writing this book was to realize a relatively concise treatment of the different facets of modern clinical cytology and to provide a reliable morphologic reference source for the student and the established cytopathologist.

It was the work of Papanicolaou that first revived the interest of clinicians and morphologists in a cytologic approach to diagnostic problems. Since that time, numerous methods of sampling, fixing, and staining have been devised and perfected, and the cytologic images provided by such technologic advances have become more and more refined. Progress will continue and, I feel, will make the cytologic approach to diagnosis more and more important in the everyday routine of the pathology laboratory as well as in research.

However, the reader should be cautioned (and this comment is directed particularly to the student pathologist) that the images provided by the various technologic advances can be interpreted correctly *only* in the light of a sound understanding of the structure, function, and physiologic mechanisms of normal cells. I furthermore emphasize that the study of isolated cells can make the problems of differential diagnosis all the more difficult, simply because the histologic context is missing.

It is, therefore, impossible to make competent cytologically based pathologic interpretations without a firm knowledge of the tissular structure from which isolated cells arise.

I have chosen a number of plates to illustrate the text, preferring the quality of good color reproductions to a greater quantity of black and white ones. Also, I have restricted the illustrations to the more common lesions in the belief that, at any rate, the rare and highly unusual case cannot be interpreted by the simple examination of one photograph. (The color plates appear between pages 52 and 53.)

This atlas could not have been prepared without the years of experience acquired at the Institut Gustave Roussy in Villejuif, France, at the Memorial Hospital for Cancer and Allied Diseases in New York, and at the Centre des Tumeurs of the Institut Jules Bordet at the Université Libre de Bruxelles. I am much indebted to all my colleagues at these institutions, and I am especially grateful to several individuals whose experience and enthusiasm inspired in me the desire to study and to teach cytopathology: Dr. C. Oberling, Dr. A. Claude, Dr. P. Dustin, Dr. F. W. Stewart, Dr. F. W. Foote, and Dr. J. de Brux. Special thanks also go to Dr. L. G. Koss, Dr. A. Meisels, and Dr. S. S. Silverberg, friends whose encouragement and advice were of inestimable value in this work.

I am deeply grateful to my secretary, Mrs. I. Chorowitz, for her continual and skillful effort in the manuscript preparation, and to Mrs. N. Delbrassine who also gave much appreciated assistance; to Mr. R. Fauconier, our medical school artist; Mr. G. Snel, chief laboratory technician; Mrs. A. Spetchinsky, chief cytology technician; and Miss F. Van Schoubroeck, cytogenetics technical assistant. Also, I greatly appreciate the collaboration of the members of my staff, Dr. R. Heimann, Dr. B. Van den Heule,

Dr. Z. Horanyi, and especially Dr. A. Verhest, who contributed the chapter on cytogenetics (Chapter 4). Lastly, I am greatly indebted to Mr. J. F. Scarfo, a medical student at the Université Libre de Bruxelles, for his constant and efficient assistance in the preparation of the text.

C. G.

Contents

Atlas of Diagnostic Cytology

1
Clinical Cytology: Definition, Perspective, and Principles of the Method

DEFINITION AND PERSPECTIVE

New ideas and new techniques must stand the test of time before their dissemination becomes general and their value established. Clinical cytology has certainly been no exception to this rule. Conceived in the nineteenth century [7, 42, 49] and since described by numerous authors [3, 29, 46] for increasingly numerous purposes, clinical cytology came into its own only with the now classic work of Papanicolaou [38] (Figure 1.1). First applied with success in gynecology, its use was progressively adapted to other disciplines, whether for the detection and identification of neoplastic cells, of infections, or of pathologic hormonal manifestations.

Cytopathology has thus become an important addition to histopathology, and the results that it furnishes are valid ones — as long as the method's guidelines are observed. Accordingly, certain principles must be respected:

- Sampling, fixation, and staining techniques must be used correctly to avoid the creation of unnecessary artifacts; these may arise from delayed or insufficient fixation, from uneven smearing resulting in clumping, or from defective staining due to old or poor quality products. Cellular structural modifications are difficult enough to appreciate: they should not be obscured by faulty preparation.

- The data of clinical cytology must be interpreted by those with firm histopathologic training; a knowledge of both the cytologic and histologic aspects of normal and pathologic tissues is indispensable since the absence of a cytohistologic correlation can lead to rather serious diagnostic mistakes.

- Each organ exhibits its own particular cytology; accordingly the specificity and sensitivity of the cytologic method vary from one organ to another; the cytopathologist therefore must be aware of both the method's advantages and its limitations, and he should advise the clinician of both.

- Modesty and honesty are essential qualities: by recognizing errors, the cytopathologist further promotes the cytologic method.

1

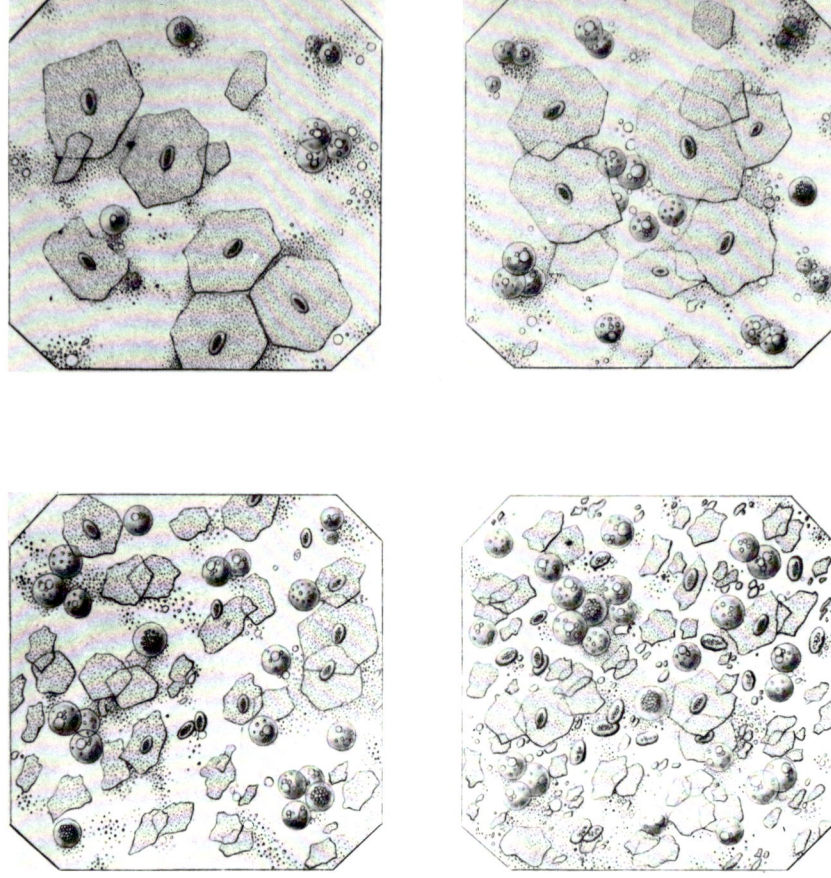

Figure 1.1. Reproduction of the original drawings published by Pouchet in 1847. The different types of squamous epithelium are represented.

PRINCIPLES OF THE METHOD

Clinical cytology is the study of the normal and abnormal morphologic characteristics of human cells. The main objective of the cytologic examination is therefore the recognition of structural changes in the nucleus and the cytoplasm. Thus, there is a fundamental difference between clinical cytology and histopathology. As its name implies, histopathology deals with the form and structure of the tissues which themselves represent complex interrelations among cells.

Whereas a histologic evaluation usually begins with a biopsy, the cytologic method requires only cells. Because these cells may come from different organs and lesions, they are collected in different ways:

- from normal desquamation products (e.g., bronchial, vaginal, cervical, and endometrial secretions)
- by superficial scraping of the lesion or organ in question (e.g., cervical scraping direct imprint on a tumor)

- by catheter or needle aspiration of an organ, a pathologic mass, or an anatomic cavity (e.g., pleural fluid, lymph node puncture)

Once collected, the cells are spread onto slides, fixed, and stained. When the cytopathologist looks at the specimen under the microscope, he observes a smear population that may be abundant or scarce, homogeneous or heterogeneous, apportioned in sheets or as isolated elements, and, generally speaking, normal or abnormal. More specifically, when performing his cytological evaluation the pathologist is looking for the following:

- modifications of normal nuclear and cytoplasmic structure
- indications of cell maturation
- the quantity of cells and their mode of desquamation
- the abnormal presence of leukocytes, histiocytes, red blood cells, etc.

Correctly interpreted, these criteria can provide a good deal of information. Among the possibilities of the cytologic examination are:

- the evaluation of gonadal hormonal activity
- the detection of inflammatory lesions and their causes
- the detection of precancerous cellular modifications
- the identification of neoplastic cells in primary, metastatic, or recurrent tumors
- the follow-up of cancer cases treated by radiation or chemotherapy
- the identification of the sex chromatin

We shall consider each of these topics in subsequent chapters. For the moment, let us mention some of the practical advantages of the cytologic method:

- the relative ease with which samples may be obtained, as opposed to the longer and more traumatic biopsy procedure of histopathology
- the possibility of multiple follow-up examinations
- the reproducibility, ease, and modest cost of the fixation and staining processes
- the excellent appreciation afforded of cellular structural modifications

It must be emphasized that the cytologic method and histopathology are different but complementary approaches to diagnosis. The results obtained by one often will be verified and completed by use of the other.

Finally, let us recall that a pathologist is a physician and not a technician. When he makes his diagnosis, he must take into consideration the relevant portions of a patient's clinical history: the precise anatomic site of a lesion, when it was first noticed, and its possible cause or causes are all criteria that may prove as important to diagnosis as the pathological cells themselves.

BIBLIOGRAPHY

1. Andrews G.S., *Exfoliative Cytology*. Thomas, Springfield, Ill., 1971.
2. Ayre J.E., *Cancer Cytology of the Uterus*. Grune and Stratton, New York, 1951.
3. Babes A., Diagnostic du cancer du col utérin par les frottis. *Presse Med.* 36:451–454, 1928.
4. Bamforth J., *Cytological Diagnosis in Medical Practice*. Little, Brown, Boston, 1966.
5. Boschann H.W., *Praktische Zytologie*. W. de Gruyter, Berlin, 1960.

6. Bourg R., Gompel C., Pundel J.P., *Diagnostic cytologique du cancer génital chez la femme*. Desoer Liège, Masson, Paris, 1954.

7. Chuquet, *Du carcinome généralisé du péritoine*. Thesis, Paris, 1879.

8. Cox H.S., *Medical Cyto-Technology*. Butterworth, London, 1968.

9. De Allende I.C.L., Orias O., *Cytology of the Human Vagina*. Hoeber, New York, 1950.

10. de Brux J., *Histopathologie gynecologique*. Masson, Paris, 1974.

11. De Neef J., *Clinical Endocrine Cytology*. Harper & Row, New York, 1965.

12. Dupre-Froment, J., *Cytologie gynécologique abdomino-Pelvienne et Mammaire*. Flammarion, Paris, 1974.

13. Frost J.K., Gynecologic and obstetric cytopathology. In Novak E.R., Woodruff J.D. (Eds.), *Novak's Gynecologic and Obstetric Pathology*. 6th Ed. Saunders, Baltimore, 1967.

14. Frost J.K., *The Cell in Health and Disease*. Karger, Basel, 1969.

15. Gates O., Warren S., *A Handbook for the Diagnosis of Cancer of the Uterus by the Use of Vaginal Smears*. Harvard University Press, Cambridge Mass., 1948.

16. Gibbs D.D., *Exfoliative Cytology of the Stomach*. Appleton-Century-Crofts, New York, 1968.

17. Gompel C., Silverberg S.G., *Pathology in Gynecology and Obstetrics*. 2nd Ed. Lippincott, Philadelphia, 1977.

18. Graham R.M., *The Cytologic Diagnosis of Cancer*. 2nd Ed. Saunders, Philadelphia, 1972.

19. Hajdu S.I., Hajdu E.O., *Cytopathology of Sarcomas and Other Nonepithelial Malignant Tumors*. Saunders, Philadelphia, 1976.

20. Henning N., Witte S., *Gastroenterologische Cytodiagnostik*. Hoffman-LaRoche, Basel, 1961.

21. Hopman B.C., *Clinical Cytology and Cytologic Research*. United States Public Health Service, Washington, D.C. 1960.

22. Hughes A., *A History of Cytology*. Abelard-Schuman, New York 1959.

23. Hughes H.E., Dodds T.C., *Handbook of Diagnostic Cytology*. Livingstone, Edinburgh, 1969.

24. Ishizuka Y., *Practical Cytodiagnosis*. Urban Schwarzenberg, Munich. Vienna Ikagu Shoin, Tokyo, 1972.

25. Kolmel H.W., *Atlas of Cerebrospinal Fluid Cells*. Springer-Verlag, Berlin, 1976.

26. Koss L.G., *Diagnostic Cytology and its Histopathologic Bases*. 2nd Ed. Lippincott, Philadelphia, 1968.

27. Lencioni L.J., (Ed.), *L'urocytogramme*. S.A. Maloine, Paris, 1975.

28. Liu W., *An Introduction to Respiratory Cytology*. Thomas, Springfield, Ill. 1964.

29. Loeper M., Binet E., Cytodiagnostic des maladies de l'estomac. *C. R. Soc. Med. Hôp.* 31:563, 1911.

30. *Manual of Cytotechnology*. 3rd Ed. National Committee for Careers in Medical Technology, Washington, 1967.

31. Marsan A., Le Capon J. (Eds.). *Atlas de Cytologie. Gynécologie*. Vol. 1. 2nd Ed. (coll. A. Sicard, C. Marsan) Varia, Paris, 1973.

32. Medak H., McGrew E.A., Burlakow P., Tiecke R. W. *Atlas of Oral Cytology*. Public Health Service Publication No. 1949, Washington, 1970.

33. Mestwerdt G., *Atlas de colposcopie*. 2nd Ed. Masson, Paris, 1955.

34. Mouriquard J., *Cytodiagnostic Mammaire*. Hoffmann-La Roche, Basel, 1962.

35. Naib Z.M., *Exfoliative Cytopathology*. Little, Brown, Boston, 1970.

36. Nieburgs H.E., *Diagnostic Cell Pathology in Tissue and Smears*. Grune and Stratton, New York, 1967.

37. Oehmichen M., *Cerebrospinal Fluid Cytology. An Introduction and Atlas*. Thieme, Stuttgart, 1976.

38. Papanicolaou, G.N., The sexual cycle in the human female as revealed by vaginal smears. *Am. J. Anat.* 52:519, 1933.

39. Papanicolaou G.N., A survey of actualities and potentialities of exfoliative cytology in cancer diagnosis. *Amer. Int. Med.* 31:661–674, 1949.

40. Papanicolaou G.N., *Atlas of Exfoliative Cytology*. The Commonwealth Fund by Harvard University Press, Cambridge, Mass., 1954.

41. Patten S.F., *Diagnostic Cytology of the Uterine Cervix*. Karger, Basel, 1969.

42. Pouchet F.A., *Théorie positive de l'ovulation spontanée et de la fécondation des mammifères et de l'espèce humaine basée sur l'observation de toute la série. Atlas.* Paris, Baillière, 1847.

43. Pundel J.P., *Précis de colpocytologie hormonale.* Masson, Paris, 1966.

44. Pundel J.P., Van Meensel F., with the collaboration of Jaworski Z. Masson, Paris, 1951.

45. Prolla J.C., Kirsner J.B., *Handbook and Atlas of Gastrointestinal Exfoliative Cytology.* University of Chicago Press, Chicago, 1972.

46. Quensel V., Zur Frage der Zytodiagnostik der Ergüsse serösen Höhlen. *Acta Med. Scand.* 68:427, 1-28.

47. Reagan J.W., Ng A.B.P., *The Cells of Uterine Adenocarcinoma.* S. Karger, Basel and New York, 1965.

48. Riotton G., Christopherson W.M., Cytology of the female genital tract. WHO Series No. 8, 1973.

49. Sanders W.R., Cancer of the bladder. *Edinburgh M. J.* 1:273, 1864.

50. Sani G., Citti U., Caramazza G., *Fluorescence Microscopy in the Cytodiagnosis of Cancer.* Thomas, Springfield, Ill., 1964.

51. Schade R.O.K., *Gastric Cytology.* Arnold, London, 1960.

52. Schiødt T., *Breast Carcinoma. A Histologic and Prognostic Study of 650 Followed-up Cases.* Munksgaard, Copenhagen, 1966.

53. Sicard A., Marsan C. (Eds.). *Atlas de cytologie.* Vol. 2. Varia, Paris, 1964.

54. Siebert S., Besancon D., *Atlas of colpocytologie. Atlas of colpocytology.* Simep (Ed.). Villeurbanne, 1971.

55. Smolka H., Soost H.J., *An Outline and Atlas of Gynaecological Cytodiagnosis.* 2nd Ed. Arnold, London, 1965.

56. Spriggs A.I., Boddington M.M., *The Cytology of Effusions.* 2nd Ed. Grune and Stratton, New York, 1968.

57. Stoll, P., *Gynäkologische Vital Cytologie in der Praxis.* Springer-Verlag, Berlin, 1969.

58. Takahashi M., *Color Atlas of Cancer Cytology.* Igaku Shoin, Tokyo, 1971. Lippincott Co., Philadelphia, 1971.

59. Taylor R.G.W., *Practical Cytology.* Academic Press, New York, 1967.

60. Tweeddale D.N., Dublilier L.D., *Cyto-pathology of Female Genital Tract Neoplasms.* Year Book Medical Publishers, Chicago, 1972.

61. Vassilakos P., *Cytopathologie des cancers broncho-pulmonaires.* Masson, Paris, 1976.

62. von Haam E., *Cytology Examination Review Book.* Vol. 2. Medical Examination Publishing Co., Flushing, N.Y., 1975.

63. Wachtel E.G., *Exfoliative Cytology in Gynaecological Practice.* 2nd Ed. Butterworth, London, 1969.

64. Wied G.L., Koss L.G., Reagan J.W., *Compendium on Diagnostic Cytology.* 4th Ed. Editors' Tutorials of Cytology, Chicago, 1976.

65. Zajicek J., *Aspiration Biopsy Cytology.* Karger, Basel, 1974.

66. Zuiger H.K., *Die Zytodiagnostik in der Gynäkologie.* Fischer, Jena, 1957.

2
Constituents
of the Cell

The cell, the unit of life, exhibits two major parts: the nucleus or nucleoplasm and the cytoplasm.[3] These two parts include a series of structures termed organelles, as well as inclusions of different types (Figure 2.1). Whatever the size of a cell, from the mycoplasms which are the smallest (0.1d) to the ostrich egg, the same organelles will be found.

The nucleus of the cell is enveloped by a double membrane whose layers enclose the perinuclear space.[29, 44, 45, 47] This membrane is interrupted by numerous, regularly spaced pores[18] (Figures 2.2 and 2.12). The nuclear substance itself is rich in deoxyribonucleic acid (DNA) which combines with proteins to form chromatin.[8, 10, 31, 43] Thus, metaphasic chromosomes consist of chromatin which contains DNA,[37] proteins, and a small amount of RNA.[5] The nucleus also contains nuclear sap and one or more nucleoli. Nuclear sap is the colloidal suspension in which the chromatin bathes. In contrast to the chromatin and nucleoli, the nuclear sap does not stain very intensely. Thus in cytologic examinations, chromatin appears on a relatively clear background.

The chromatin often appears granular: this image corresponds to the DNA filaments whose exact prefixation structure is never perfectly preserved (Figure 2.3).

The cytopathologist often is faced with the problem of evaluating normal and abnormal chromatin distributions (Figure 2.3). This task, as we shall see, is not an easy one. For example, variations in the intensity of nuclear staining do not necessarily reflect changes in DNA content. Variations of chromatin staining may result from modifications in nuclear size: an equal amount of DNA will look paler in a larger nucleus. Such variations also may be due to the quality of the staining method itself: methods that may incidentally liberate large quantities of DNA from its protein linkage will result in a denser nuclear stain.

These remarks should be borne in mind when delicate evaluations of chromatin structure or DNA content must be made, as in the case of cytologic cancer detection. The DNA content will vary according to the cell type; normal and benign cells have a diploid DNA content while atypical or malignant cells will reveal an aneuploid DNA content.[43]

In certain cells, one may observe a localized thickening of the nuclear membrane. This is due to the condensation of the sex chromatin, i.e., and X-chromosome that retains its visible, coiled structure during interphase.

The *nucleolus* is seen as a round, basophilic, intranuclear formation. It is rich in

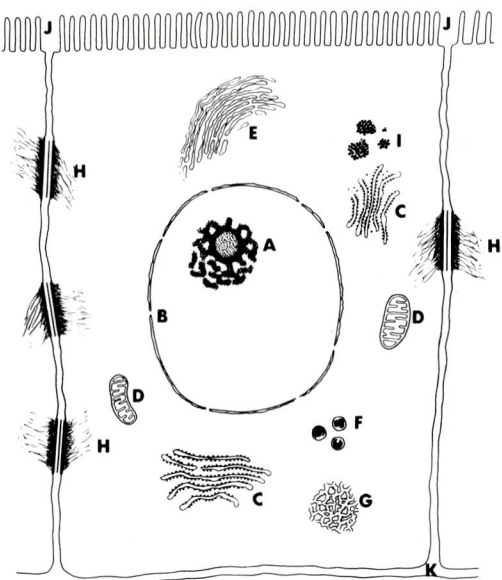

Figure 2.1a. Schematic representation of cellular organelles. *A*, nucleolus; *B*, nuclear envelope with a double membrane and nuclear pores; *C*, rough endoplasmic reticulum and ribosomes; *D*, mitochondrion; *E*, Golgi apparatus; *F*, lysosomes; *G*, smooth endoplasmic reticulum; *H*, desmosome and microfilaments; *I*, glycogen granules; *J*, microvilli; *K*, basal membrane.

ribonucleic acid (RNA). The ultrastructure of the nucleolus includes granules (15 Å), filaments (50μ diameter), and a less dense matrix. The granular and filamentous constituents are rich in RNA and play a role in the transfer of RNA to the cytoplasm[21, 26, 28, 41] (Figure 2.3). The number and size of the nucleoli vary, depending on the functional activity of the cell and the number of chromosomes. Large nucleoli are seen in cells actively engaged in protein synthesis.[26] This is why metabolically active cells such as cancerous elements frequently exhibit numerous and voluminous nucleoli. Interestingly enough, it has been shown that antibiotics have an effect on nucleolar structure: this is probably due to the blocking of protein synthesis.

The *cell membrane* (plasmalemma or plasma membrane), is approximately 80 A thick and encloses the cytoplasm.[40] Electron microscopy shows that the membrane is formed by two layers, each about 25 Å thick and containing proteins, which are separated by a less dense zone of 30 Å, containing lipids. The membrane has its own characteristic properties of permeability and conductibility and is the site of many specific enzymatic activities.[22] These characteristics show that the plasma membrane is by no means a simple container for the cytoplasm but rather a highly active and dynamic contributor to the myriad cellular processes necessary to life.

The cytoplasm includes a ground matrix and numerous organelles.[9] The matrix is a homogeneous, unstructured fluid medium in which the organelles bathe. Among the

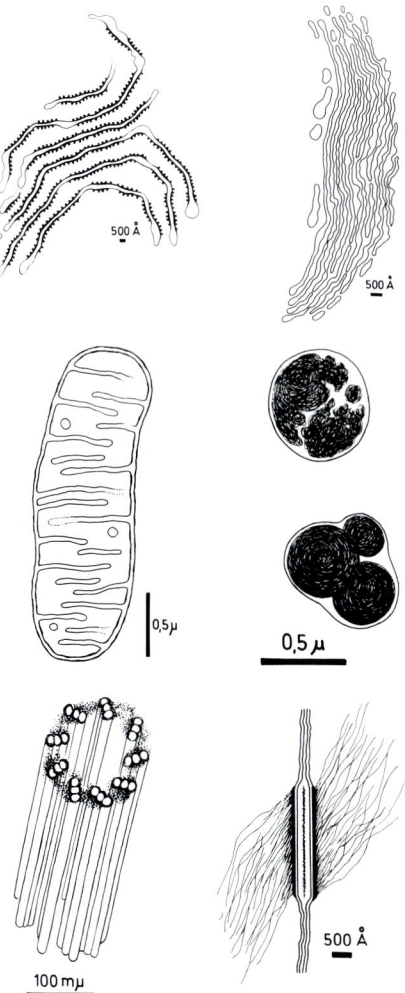

Figure 2.1b. Details of schematic representation of cellular organelles. *Left to right, top:* rough endoplasmic reticulum, Golgi zone; *middle:* mitochondrion, lysosomes; *bottom:* centriole, desmosome and microfilaments.

latter is the *endoplasmic reticulum,*[29] which is a membranous, extensively branched canal system whose volume varies with the type of cell and its activity (Figure 2.4). The membrane of the endoplasmic reticulum is continuous with the nuclear envelope.

We distinguish a rough *endoplasmic reticulum* (rER) and a *smooth endoplasmic reticulum* (sER).[32, 38] The rER is so-named because it is studded with particles ranging from 120–150 Å in diameter; these are the ribosomes, which are essential to protein synthesis. The rER is the site of synthesis of those proteins destined for secretion. The sER is not covered with ribosomes and is thought to play a role in steroid hormone synthesis and glycogen metabolism.

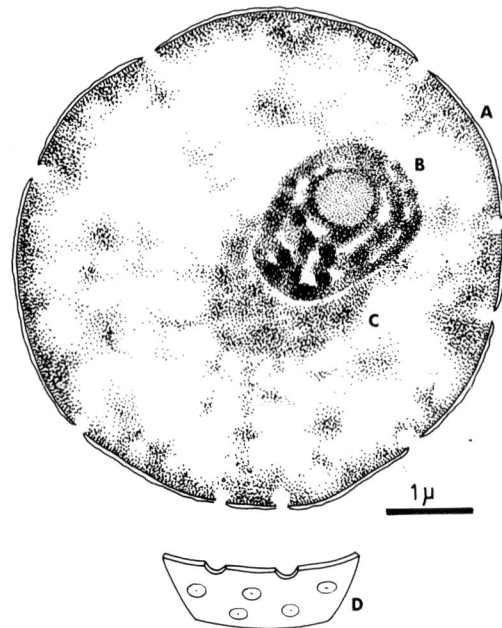

Figure 2.2. Schematic representation of the nucleus. *A*, nuclear envelope; *B*, nucleolus; *C*, nuclear substance: chromatin and nuclear sap; *D*, detail of the nuclear membranes and pores.

Ribosomes also may be found in the cytoplasm. They are rich in protein and RNA and are thus responsible for cytoplasmic basophilia. The ribosomes may form chains consisting of several units (polysomes), linked by a filament about 15 Å in diameter. This filament is a molecule of RNA, more specifically messenger RNA, which transmits genetic information from the nucleus to the ribosomes. This information directs the synthesis of the polypeptide chains of proteins, whether they are produced on the rER for secretion or by polysomes for the cell's own needs.

The acidophilia of the cytoplasm may be increased by substances having a great affinity for acid stains, such as keratin and its precursors in keratinized squamous cells.

Mitochondria, discovered by Altmann[1] in 1890, are seen with phase contrast microscopy and by using vital stains such as Janus green. These organelles, which measure from 1 to 4 μ in length, exhibit a complex structure; they possess a double membrane whose internal layer invaginates into the mitochondrial matrix to form cristae [21, 31] (Figure 2.5). There is thus a separation between the internal and external membranes called the intermembranous space. Within the mitochondrial matrix, dense granules, ranging from 300 to 500 Å in diameter, are active in ionic equilibrium; other granules are located along the crests and possess the enzyme AT Pase; finally we also find filaments of deoxyribonucleic acid (DNA) and granules of RNA. The mitochondrial enzymatic equipment is fundamental to cellular energy production, protein synthesis, and lipid metabolism.

It is the abundance of mitochondria in certain cell types that gives their cytoplasm a

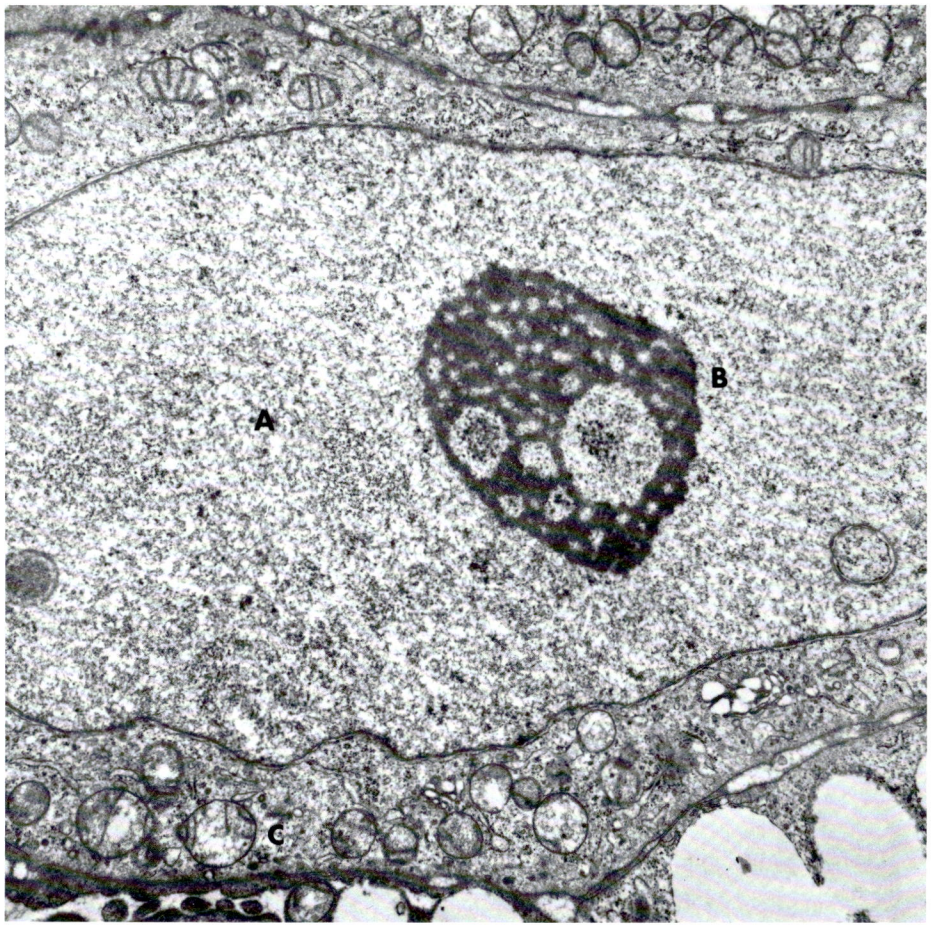

Figure 2.3. Epithelial cell: detail of the nucleus showing the granular structure of the chromatin (*A*) and a large nucleolus (*B*). The granules and the matrix of the nucleolus are visible. Mitochondria (*C*) (× 10,260).

characteristic acidophilia (e.g., the oncocytic cells of the salivary glands, certain gastric cells, etc.).

Lysosomes appear in electron microscopy as dense organelles measuring 0.2 to 0.5 μ in diameter. They are membranous structures and contain numerous hydrolytic enzymes whose role is the digestion of certain intracellular substances such as proteins, polysaccharides, nucleic acids, lipoproteins, and other constituents of fairly high molecular weight. [11,13, 14, 46]

The existence of the *Golgi apparatus* was discovered as early as 1898.[3, 20, 25, 30, 35] However, it required electron microscopy to show that it is a membranous structure exhibiting a group of vesicles, cisternae, and tubules (Figure 2.6). The Golgi ap-

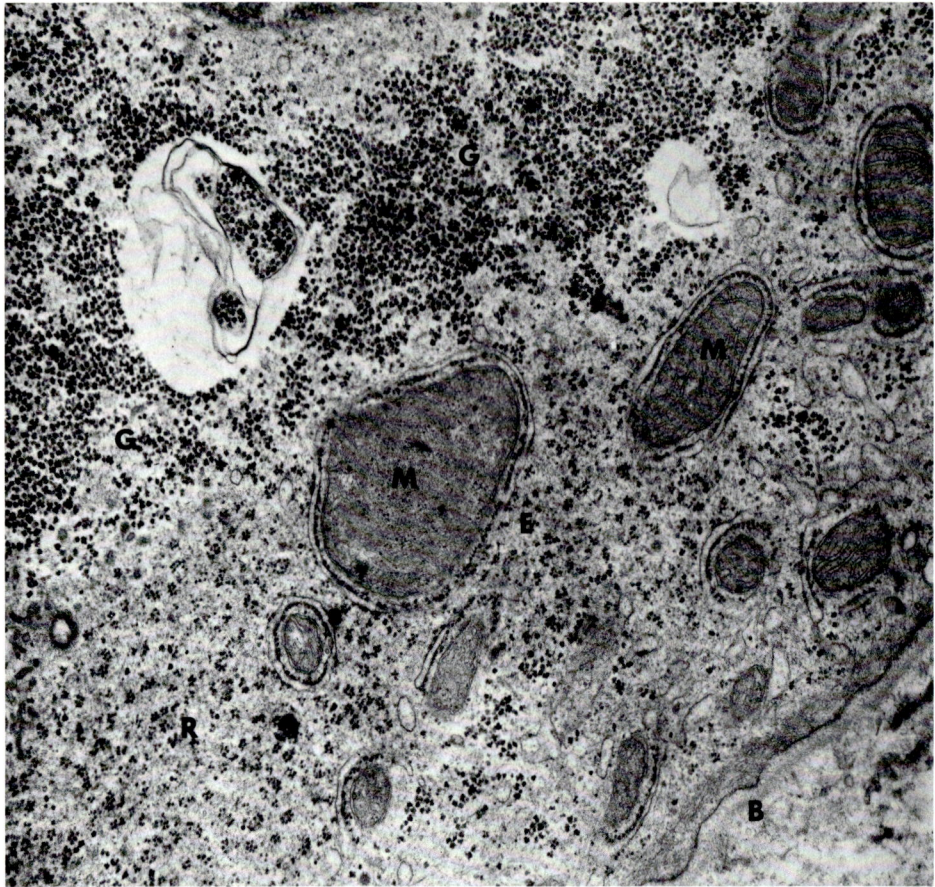

Figure 2.4. Epithelial cell: detail of cytoplasm showing the endoplasmic reticulum (*E*) with ribosomes (*R*) and glycogen granules (*G*), mitochondria (*M*); basal lamina (*B*) (× 21,870).

paratus intervenes in the synthesis, concentration, and transport of secretion products (e.g., proteins and polysaccharides).[2, 20]

Annulate lamellae, which arise from the nuclear envelope, form an assembly of parallel double membranes limiting a space which is interrupted by pores. These structures were revealed by the electron microscope, but their function is yet poorly understood. They are found especially in ciliated bronchial epithelium, in the endometrium, and in certain cancer cells (Figure 2.7).

Centrioles are hollow cylindrical structures the walls of which are formed by nine groups of tubule-triplets (Figure 2.8). They measure 150 μ in diameter and 500 μ in length. These highly complex organelles participate in the organization of the mitotic spindle as well as in the structure of cilia.[2, 15]

Microfilaments, whose diameter is at least 20 Å, are not to be confused with the mitotic fibrils (250 Å in diameter) found only during mitosis or necrosis.

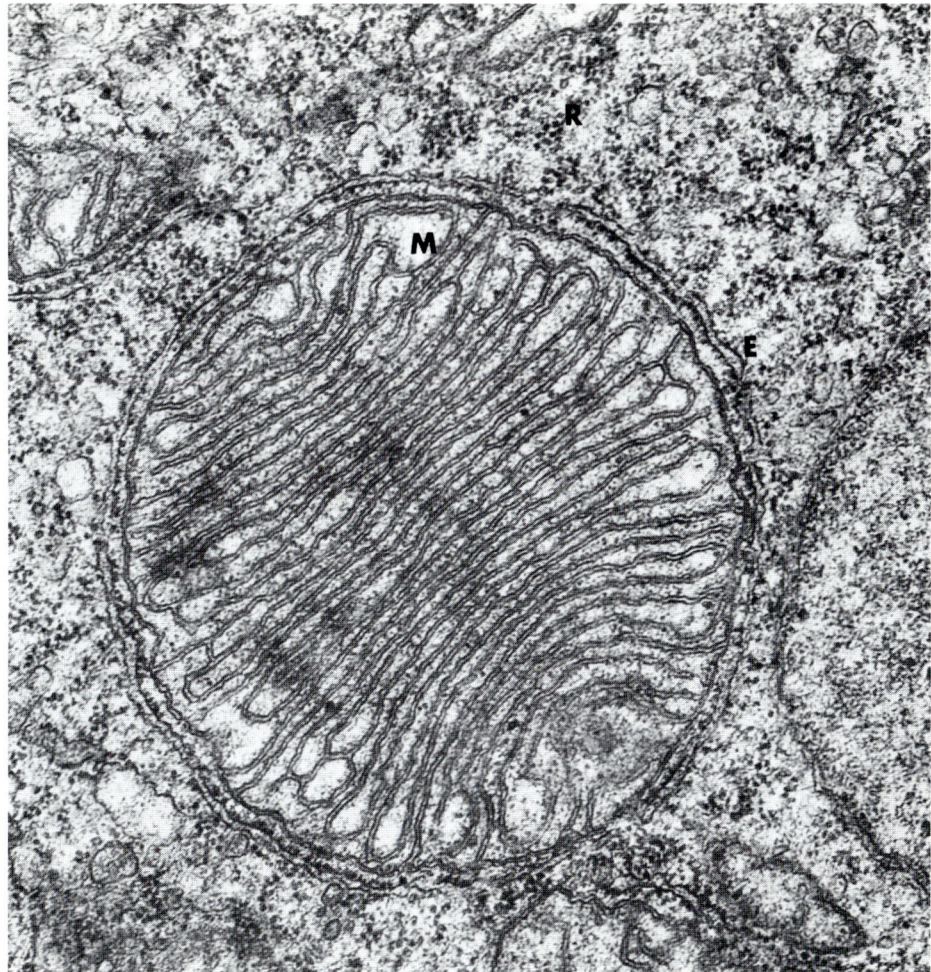

Figure 2.5. Epithelial cell: detail of the mitochondrial structure. Note the endoplasmic reticulum (*E*) surrounding the mitochondria (*M*) and numerous free ribosomes (*R*).

Microfilaments are found in most cells but are best represented in epidermal cells, where they may anchor themselves to desmosomes (Figure 2.9). Formed by chains of globular protein molecules, they also participate in cellular motility.

Microvilli are cell processes measuring up to 1μ in length and $90\,\mu$ in diameter (Figure 2.8). They border the apical pole of the cell and represent the brush or striated border. These structural devices increase the exchange surface area of the apical cell membrane. Variations in size and shape of microvilli are correlated with the metabolic cellular activity.

The *cilium* is composed of two central microtubules or fibrils and nine peripheral double microtubules. Each fibril is composed of two subfibrils. At the base of the cilium is a dense granule (basal body). Cilia have a more complex structure and are larger than microvilli (Figure 2.10).

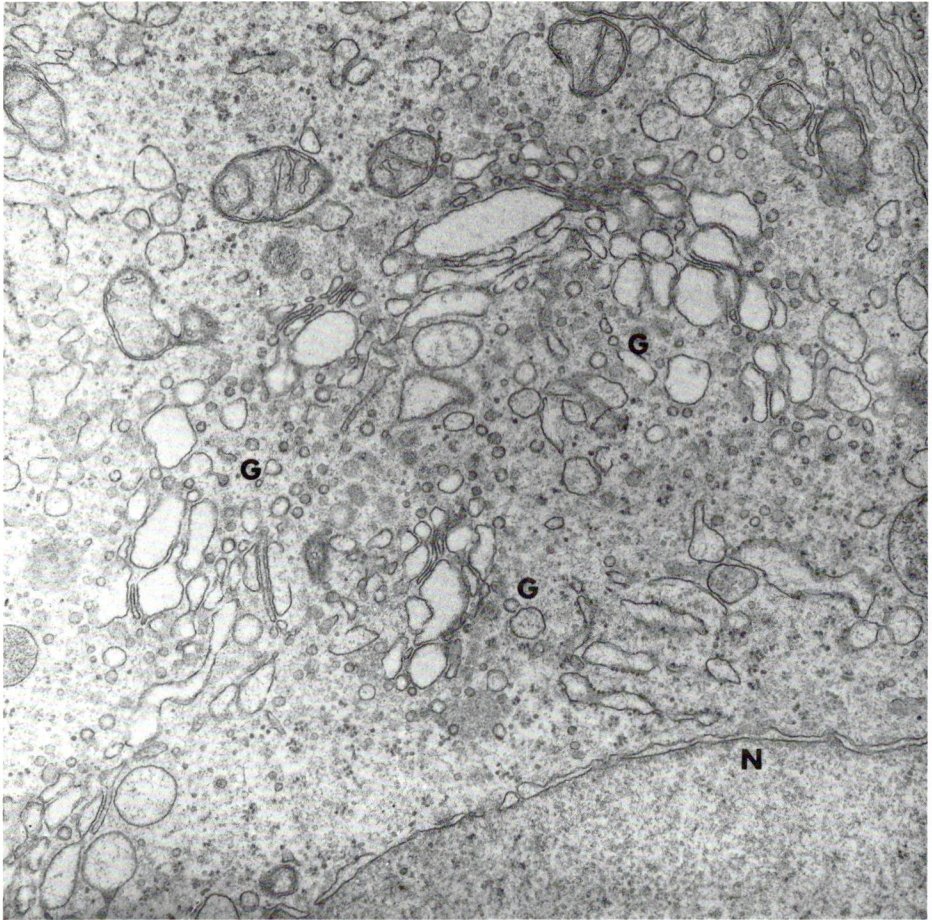

Figure 2.6. Epithelial cell: nuclear membrane (*N*); Golgi zone with tubules, cysternae, and vesicles (*G*) (× 21,870).

Microtubules have a cylindrical structure formed by 12 to 13 rows of longitudinally disposed globular proteins. This structure gives them properties different from those of the tonofibrils: both resist forces of extension; however, the microtubules are also resistant to lateral distorsion.

Desmosomes help to maintain a certain degree of cohesion among neighboring cells. Desmosomes are observed with the electron microscope as a localized thickening of adjacent cell membranes, separated by intercellular cement (Figure 2.11).

Besides these organelles, the cytoplasm may contain various inclusions, such as *lipids* (Figure 2.12), *glycogen*[16] (Figure 2.4), and *pigment granules* — melanin (melanocytes of the skin and retina), lipofuscin, and hemosiderin (phagocytes).

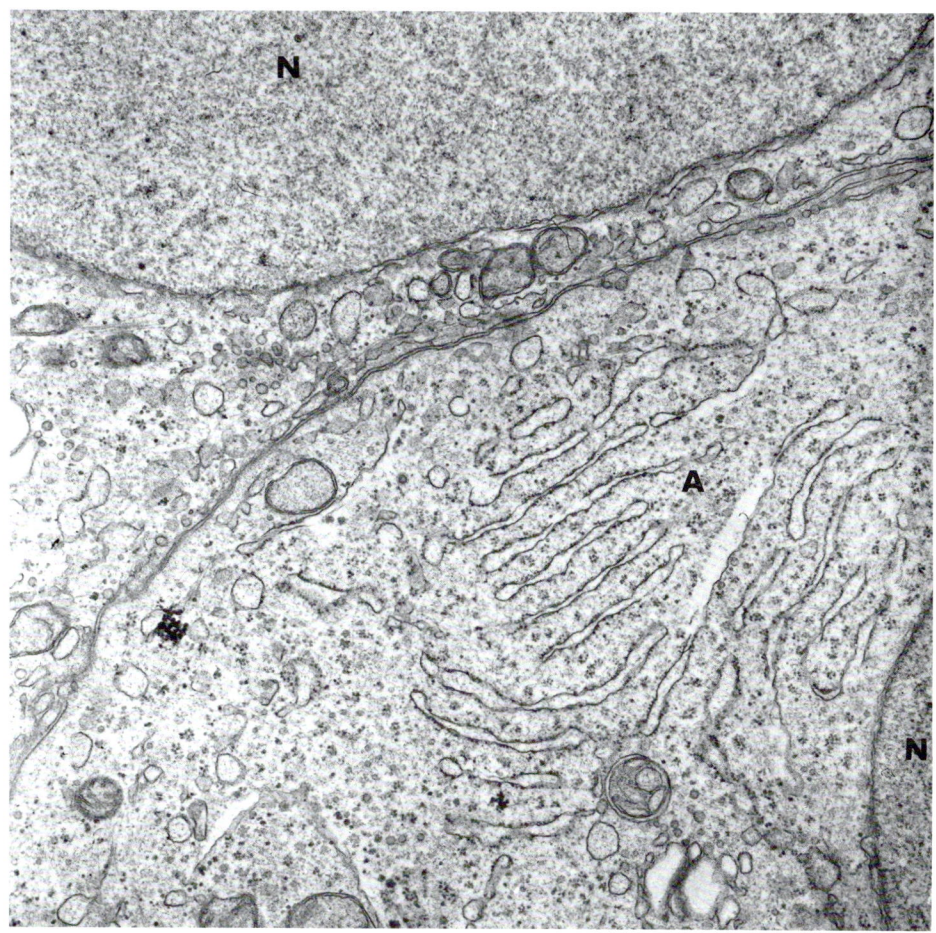

Figure 2.7. Epithelial cell: annulate lamellae (*A*); nuclei (*N*) (× 17,315).

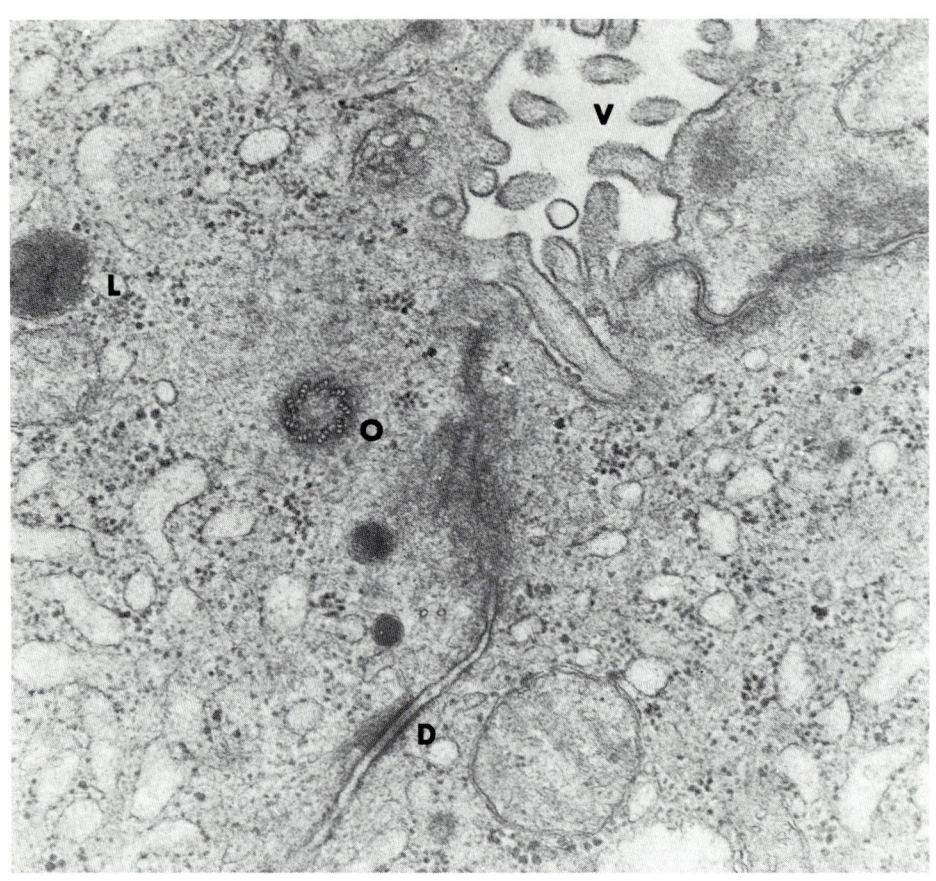

Figure 2.8. Epithelial cell: centriole (*O*); microvilli (*V*); lysosome (*L*); desmosome (*D*) (× 44,175).

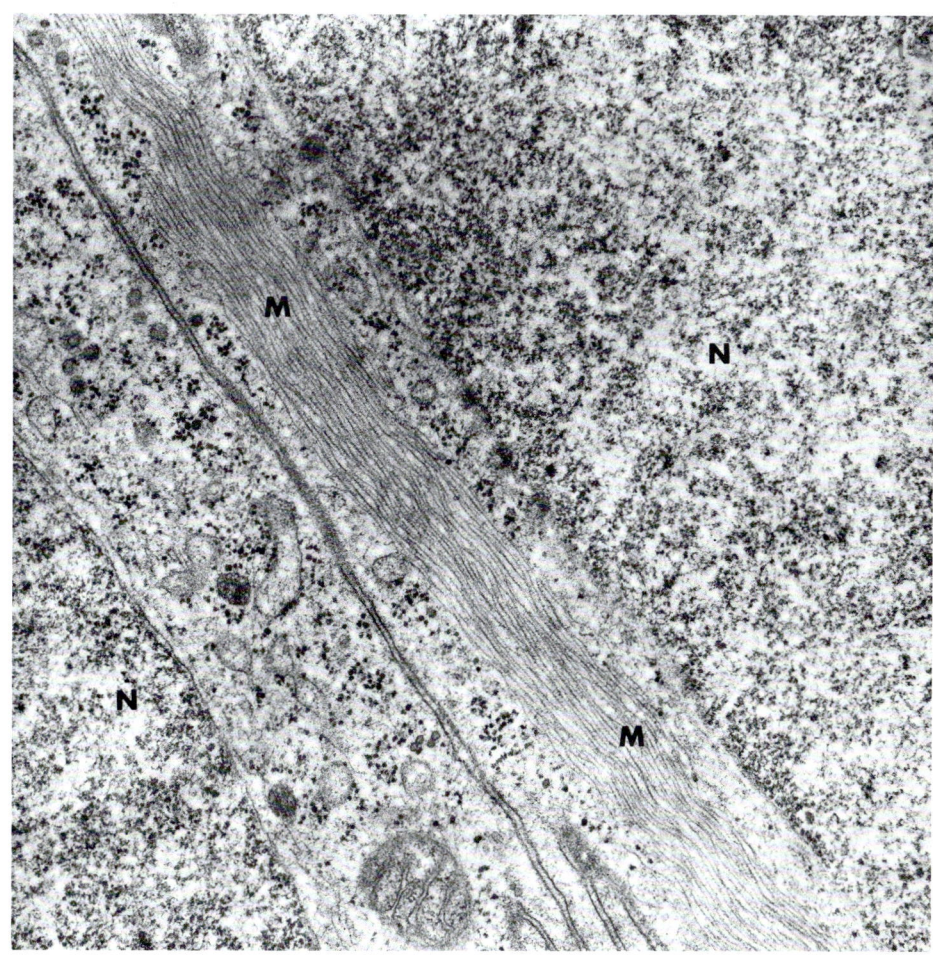

Figure 2.9. Epithelial cell: microfilaments (*M*); nuclei (*N*) (× 28,215).

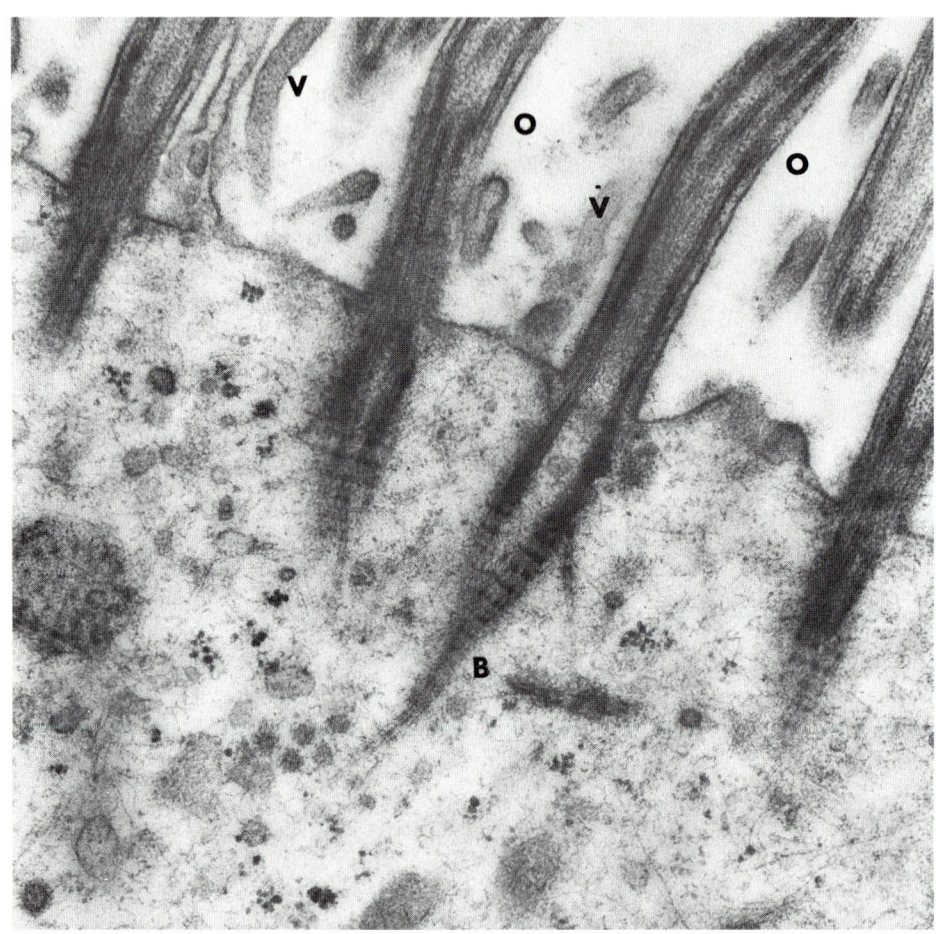

Figure 2.10. Apical pole of an epithelial cell: cilia (*O*); microvilli (*V*); basal body (*B*) (×40,279).

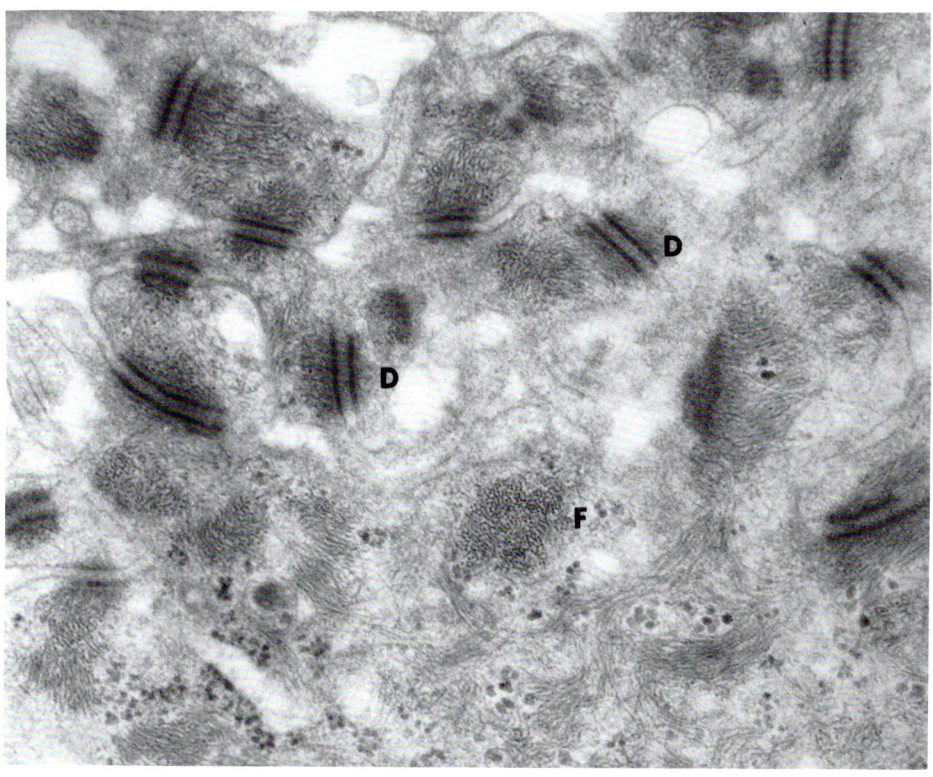

Figure 2.11. Epidermoid cell: presence of numerous desmosomes (*D*); dense fibrillar system (*F*) (× 44,175).

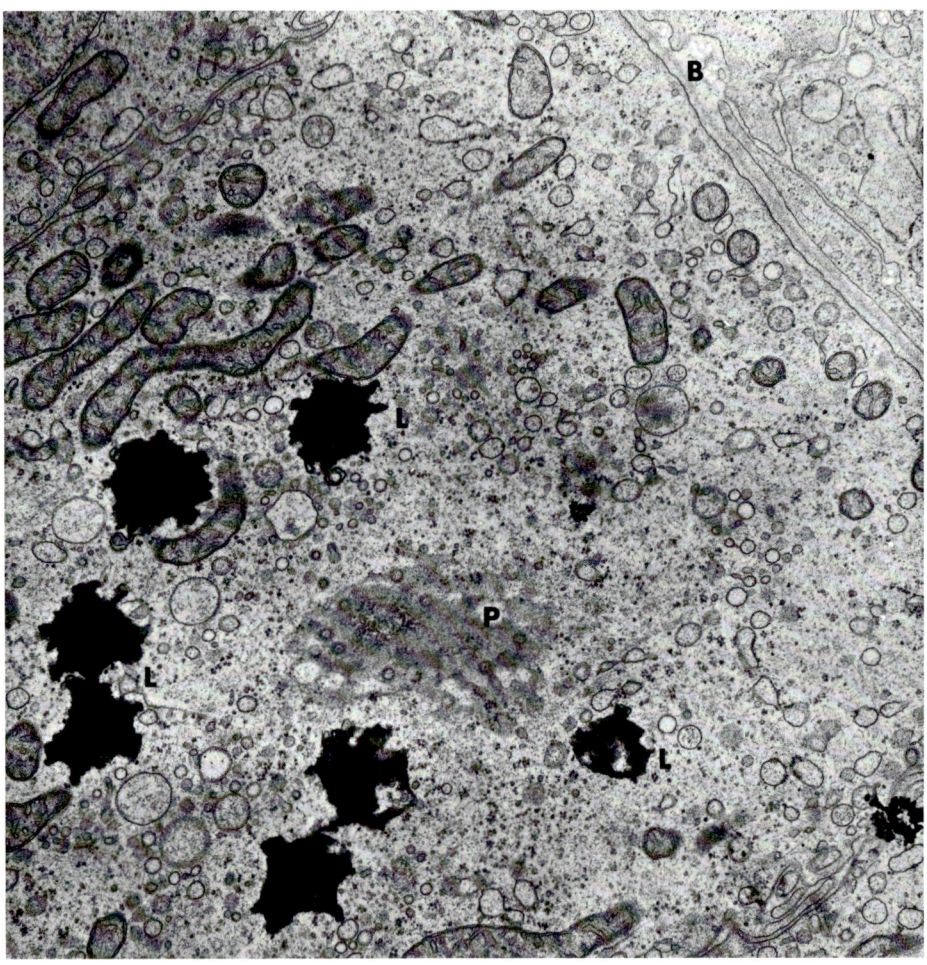

Figure 2.12. Epithelial cell: tangential section of the nucleus showing the pores (*P*); lipid inclusions (*L*); basal lamina (*B*) (× 16,065).

BIBLIOGRAPHY

1. Altmann R., *Die Elementarorganismen und ihre Beziehungen zu den Zellen.* Veit, Leipzig, 1890.

2. André J., Le centriole et la région centrosomienne. *J. Microscopie* 3:23, 1964.

3. Beams H.W., Kessel R.G., The Golgi apparatus: structure and function. *Intern. Rev. Cytol.* 23:209, 1968.

4. Bittar E.E. (Ed.). *Cell Biology in Medicine.* John Wiley & Sons, New York, 1973.

5. Brachet J., The role of nucleus and cytoplasma in synthesis and morphogenesis. *Symp. Soc. Exptl. Biol.* 6:173, 1952.

6. Brachet J., *Biochemical Cytology.* Academic Press, New York, 1957.

7. Burke J.D., *Cell Biology.* Williams & Wilkins, Baltimore, 1970.

8. Chèvremont, M., Baeckeland E., Chèvremont-Comhaire S., Contribution à l'étude du métabolisme et de la synthèse cytoplasmique d'acides deoxyribonucléiques en culture de tissus. *Bull. Acad. Roy. Med. Belg.* 15:349, 1960.

9. Claude A., Particulate components of the cytoplasm. *Cold Spring Harbor Symp. Quant. Biol.* 9:263, 1941.

10. Claude A., Distribution of nucleic acids in the cell and the morphological constitution of cytoplasm. *Biol. Symp.* 10:111, 1943.

11. Claude A., "Microbodies" et lysosomes: une étude au microscope électronique. *Arch. Int. Physiol.* 68:672, 1960.

12. Dalton A.J., Electronmicroscopy of tissue sections. *Intern. Rev. Cytol.* 2:403, 1953.

13. De Duve C., Lysosomes. Ciba Foundation Symposium. Churchill, London, 1963.

14. De Duve C., Wattiaux, R., Function of lysosomes. *Ann. Rev. Physiol.* 28:435, 1966.

15. De Harven E., Bernard W., Étude au microscope électronique de l'ultrastructure du centriole chez les vertébrés. *Zeitschr. f. Zellforsch.* 45:387, 1956.

16. Drochmans P., Morphologie du glycogène. *J. Ultrastruct. Res.* 6:141, 1962.

17. Editorial. How lymphocytes kill tumor cells. *New Engl. J. Med.* 295:165, 1976.

18. Feldherr C.M., The effect of the electron opaque material on exchanges through the nuclear annuli. *J. Cell. Biol.* 25:43, 1965.

19. Gey G., Some aspects on the constitution and behavior of normal and malignant cells maintained in continuous culture. *Harvey Lectures* 50:154, 1954/55.

20. Golgi C., Sur la structure des cellules nerveuses. *Arch. Ital. Biol.* 30:298, 1898.

21. Gompel, C., Structure fine des mitochondries de la cellule glandulaire endométriale au cours du cycle menstruel. *J. Microscopie* 35:427, 1964.

22. Holter H., Passage of particles and macromolecules through cell membranes. *Symp. Soc. Gen. Microbiol.* 15:89, 1965.

23. Kautz J., De Marsh Q.B., Fine structure of the nuclear membrane in cells from the chick embryo: on the nature of the so-called "pores" in the nuclear membrane. *Exptl. Cell. Res.* 8:394, 1955.

24. Kessel R.G., Annulate lamellae. *J. Ultrastruct. Res.* Suppl. 10, 1968, p. 5.

25. Lacy D., Challice C.E., The structure of the Golgi apparatus in vertebrate cells examined by light and electronmicroscopy. *Symp. Soc. Exptl. Biol.* 10:62, 1957.

26. La Cour L.F., The internal structure of nucleoli. *Chromosomes Today* 1:150, 1966.

27. Leuchtenberger C., A cytochemical study of pycnotic nuclear degeneration. *Chromosoma* 3:449, 1950.

28. Marinozzi, V., Cytochimie ultrastructurale du nucléole. RNA et protéines intra-nucléolaires. *J. Ultrastruct. Res.* 10:443, 1964.

29. Merriam R.W., On the fine structure and composition of the nuclear envelope. *J. Biophys. Biochem. Cytol.* 11:559, 1961.

30. Northcote D.H., The Golgi apparatus. *Endeavour* 30:26, 1971.

31. Palade G.E., An electron microscopic study of the mitochondrial structure. *J. Histochem. Cytochem.* 1:188, 1953.

32. Palade G.E., The endoplasmic reticulum. *J. Biophys. Biochem. Cytol.* Suppl. 2, 1956, p. 85.

33. Perry R.P., Hell A., Errera M., The role of the nucleolus in RNA and protein synthesis. *Biochem. Biophys. Acta* 49:47, 1961.

34. Pfitzer P., Pape H.D., Characterization of Tumor Cell Populations by DNA-measurements. *Acta Cytol.* 17:19, 1973.

35. Pollister A.W., Pollister P.F., The structure of the Golgi apparatus. *Int. Rev. Cytol.* 6:85, 1957.

36. Porter K.R., The sarcoplasmic reticulum: its recent history and present status. *J. Biophys. Biochem. Cytol.* 10:219, 1961.

37. Robbins E., Gonata S.N.K., The ultrastructure of a mammalian cell during the mitotic cycle. *J. Cell Biol.* 21:429, 1964.

38. Roels H., Metabolic DNA: A cytochemical study. *Intern. Rev. Cytol.* 19:1, 1966.

39. Ris H., Mirsky A.E., The state of chromosomes in the interphase nucleus. *J. Gen. Physiol.* 32:489, 1949.

40. Robertson J.D., The structure of biological membranes. *Arch. Int. Med.* 129:202, 1972.

41. Simard R., Bernhard W., Le phénomène de la ségrégation nucléolaire: spécificité d'action de certains antimétabolites. *Intern. J. Cancer* 1:463, 1966.

42. Unanue E. R., Secretory function of mononuclear phagocytes. A review. *Amer. J. Path.* 83:396, 1976.

43. Wagner D., Sprenger E., Merckle D., Cytophotometric studies in suspicious cervical smears. *Acta Cytol.* 20:366–371, 1976.

44. Watson M.L., The nuclear envelope. *J. Biophys. Biochem. Cytol.* 1:257, 1955.

45. Watson J.L., Further observations on the nuclear envelope of the animal cell. *J. Biophys. Biochem. Cytol.* 6:147, 1959.

46. Weissmann G., Lysosomes. *New Engl. J. Med.* 273:1084, 1965.

47. Wiener J., Spiro D., Loewenstein W.R., Ultrastructure and permeability of nuclear membranes. *J. Cell. Biol.* 27:107, 1965.

3
The Diagnosis
of Cellular Malignancy

The definition and characterization of a cancerous cell are not as easy a matter as the uninitiated might expect, and we shall attempt to explain why. It was Virchow of Berlin [12] who first confronted this problem with the publication in 1858 of his "Die Cellularpathologie," in which he elaborates on the cellular theory *(omnis cellula e cellula)* and on the nature of malignant tumor cells. His interpretations on cancer however have since proved partially false because he did not consider factors such as heredity, immunity, and environment. Today, more than a century after Virchow's work, the absolute definition of malignancy still remains a clinical one rather than a cytological one: a malignant tumor is a lesion which, if left untreated, will spread locally, and then to other parts of the body, until it finally kills the individual. Cytologically, a cancer cell is characterized by the loss of the normal mechanisms which control its harmonious multiplication.

In most cases, there is indeed a characteristic histologic image, which indicates either the malignant or the benign nature of the lesion, and a diagnosis thus may be made without hesitation. However, in a certain percentage of cases, perfectly malignant lesions do not present all the morphologic criteria of malignancy and, on the contrary, highly anaplastic cellular lesions may undergo a rather innocuous evolution. Examples of such problematic lesions are intraepithelial carcinoma of the cervix, certain sarcomas whose extension remains temporarily localized before subsequent metastasis, and basal cell carcinoma, which metastasizes only rarely. We are thus sometimes confronted with frankly equivocal cases where an immediate and prognostic diagnosis cannot be made. This confusion is translated by the variety of sometimes equivocal terms used to qualify the lesions — for example, "malignant adenomas," "local malignancy," "metastatic goiter," "severe dysplasia," and "bladder papilloma" (to designate low grade carcinoma).

An ideal tumor classification in theory would consider the origin of the lesion, its etiology, its morphologic aspect, and its biological behavior. However, such an ideal classification does not exist because certain parameters are still lacking, notably an adequate definition of the cancerous cell itself.

Progress has been made in this field by correlating the macroscopic and microscopic images of lesions with their degree of extension and their evolution. It is in this direction that cooperating research groups all over the world are concentrating their efforts to collect as much relevent data as possible and thus to better define what is malignant and what is not. If we possessed a genuine definition of the tumor cell, we could do away with the problem of borderline cases.

Since the era of van Leeuwenhoek (seventeenth century) and his primitive microscope, considerable quantities of information have been acquired on the morpology of the normal and the pathologic cell. (Let us recall that van Leeuwenhoek produced microscopes more or less as an artisan; he is reported to have produced a total of 247 during his life.[11]) Cytologic progress at first was rather slow: it was not until 1830 that the first achromatic lenses were produced. In 1870, the oil immersion objective and the microtome came into use, and in 1878, the first truly modern microscope appeared. It was designed by Abbe and produced by Zeiss at Jena[4], and it had a power of resolution approaching 0.2μ.

Thus, generally speaking, it was near the end of the nineteenth century that the rate of discoveries could accelerate. In 1879 W. Flemming of Prague introduced the concept of chromatin; in 1882 Strasburger of Bonn used the terms nucleoplasm and cytoplasm; in 1883 Metschnikoff described phagocytes in his work "Leçons sur la pathologie comparée de l'inflammation"; in 1888, Van Beneden of Liège and Boveri described the centrosome; in 1898 Golgi of Pavia discovered the complex that bears his name. After the beginning of the twentieth century, no new cellular structures were to be discovered by light microscopy and no genuine definition of the cancerous cell was proposed.

During the years that followed, progress in fixation, inclusion, and staining began to allow inquiry into the physicochemical nature of certain of the cellular constituents. Histochemistry was thus born,[7] marking a turning point in the study of the cell and of the functional significance of its constituents. However, all the data obtained thanks to these techniques still do not resolve the basic problem: cancerous cellular modifications remain too subtle to be completely unmasked as the process of cancerization begins well before the morphologic modifications are observed.

The advent of the electron microscope during the 1940s gave us a tool whose power of resolutions goes below 10 Å; this indeed opened up a whole new field for investigation. New cellular structures were described, and the hopes of finding specific morphologic characteristics in the neoplastic cell were reborn. However, in terms of cancer diagnosis, this new field of investigation has been rather disappointing: no specific ultrastructural criterion of malignancy has been observed in the very numerous investigations conducted over the last 20 years [9].

Finally, the recent introduction of automated systems of structural evaluation constitutes a new attempt to arrive at a valid definition of the cancer cell. This topic will be discussed further in Chapter 12.

Although there is no definite morphologic criteria of malignancy, we are nonetheless able to recognize human cancer cells by light microscopy and with a very good approximation (Figure 3.1). There are certain structural modifications that may be present in all types of human cancers and that allow cytologic recognition (Figure 3.2). In the nucleus, these criteria are:

- modifications of size and shape (anisonucleosis)
- abnormal membrane configuration
- abnormal chromatin distribution
- changes in the number, size, and shape of nucleoli

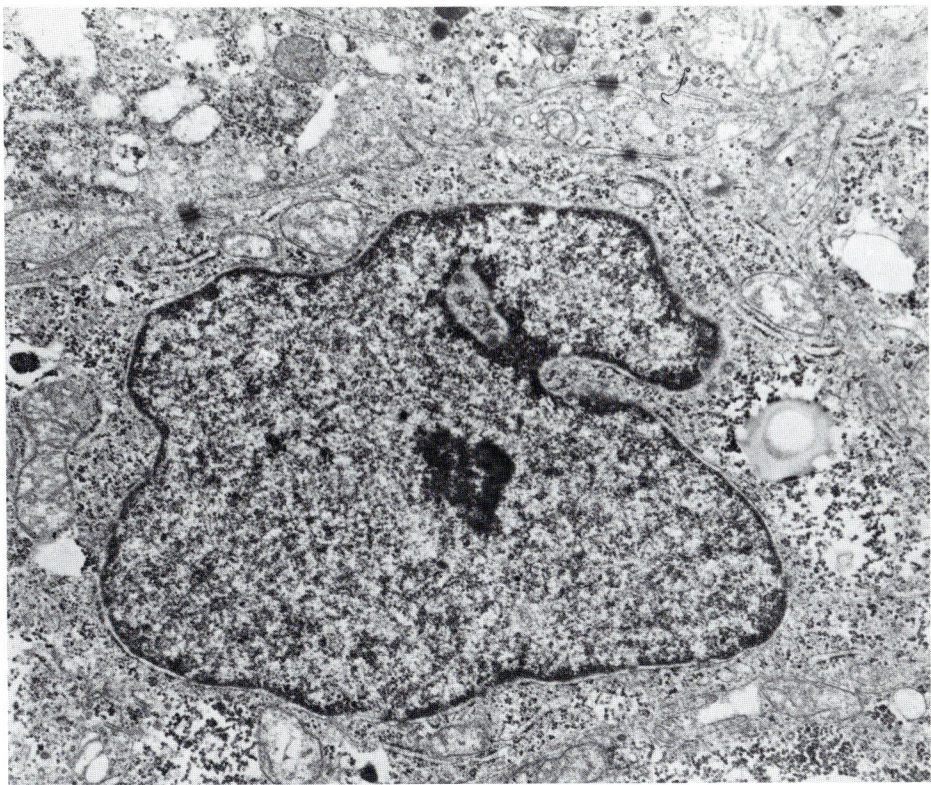

Figure 3.1. Malignant epithelial cell showing an irregular nucleus with marked indentations of the nuclear membrane. Cytoplasmic organelles do not show any structural anomalies (× 13,500).

- atypical mitoses
- multinucleation

In the cytoplasm, the modifications include:

- variations of size and shape with alteration of the nucleocytoplasmic volume ratio
- alterations of staining characteristics and abnormal presence of inclusions and/or vacuoles

Also, the mode of cellular desquamation is modified by the loss of contact inhibition. Some more specific cellular anomalies appear in particular types of tumors.

In each of the following chapters, we shall expand upon these particular morphologic anomalies. A few words of caution are necessary, however, before we begin. The cellular modifications we have listed are not always flagrant ones. The inexperienced cytopathologist may wonder, for example, just what constitutes an "abnormal chromatin distribution" and what does not. And while diagnosis certainly

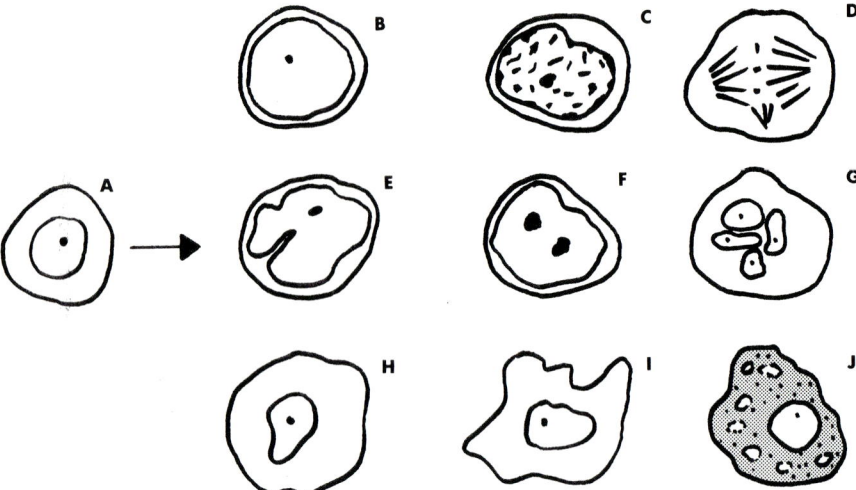

Figure 3.2. Schematic representation of most frequent structural modifications in neoplastic cells. *A*, normal looking cell; *B*, increase of nuclear size; *C*, nuclear hyperchromatism; *D*, atypical mitosis; *E*, modification of nuclear shape; *F*, increase of nucleolar volume; *G*, multinucleation; *H*, increase of cellular size; *I*, modification of cellular shape; *J*, modification of cytoplasmic staining affinities and vacuolation.

becomes easier and more exact with experience, there are some cases where a certain degree of subjectivity is inevitable. Such subjectivity constitutes one of the limitations of cytology. Moreover, the histologic criteria of malignancy, such as loss of cellular orientation, invasiveness, and destruction of adjacent structures cannot be appreciated in cytologic evaluations. Finally, it must be pointed out that the criteria we have listed are by no means restricted solely to cancer cells. Different cellular metabolic disorders may produce nuclear and cytoplasmic changes that are very similar to those of cancer cells. We shall see, for example, that certain perfectly benign lesions have the obnoxious habit of providing the cytopathologist with highly atypical cellular elements.

BIBLIOGRAPHY

1. Abercrombie M., Ambrose E.J., The surface properties of cancer cells: a review. *Cancer Res.* 22:525, 1962.

2. Busch H., Byvoet P., Smetana K., The nucleolus of the cancer cell: a review. *Cancer Res.* 23:313, 1963.

3. Caspersson T., *Cell Growth and Cell Function.* Norton, New York, 1950.

4. Castiglioni A., *Histoire de la médecine.* Payot, Paris, 1931.

5. Cowdry E.V., Properties of cancer cells. *Arch. Path.* 30:1245, 1940.

6. Lajtha L.G., The nature of cancer. In Harris R.J.C. *What we Know About Cancer.* Allen & Unwin, London, 1970, p. 34.

7. Lison, L., *Histochimie et Cytochimie Animales.* Gauthier-Villars, Paris, 1953.

8. McGee-Russell, Ross K.F.A., *Cell Structure and Its Interpretation.* Arnold, London, 1968.

9. Oberling C., Bernard W., The morphology of the cancer cells. In Brachet J., Mirsky A. (Ed.) *The Cell.* Vol. 5. Academic Press, New York, 1961, p. 405.

10. Pito H.C., Some aspects of the development biology of neoplasia. *Cancer Res.* 28:1880, 1968.

11. Singer C., *A History of Biology.* Abelard-Schuman London, 1959.

12. Virchow R., *Die Cellularpathologie.* Berlin, 1858.

13. Warburg O., On the origin of cancer cells. *Science* 123:309, 1956.

4

Cytogenetics: Terminology and Diagnostic Aid in Malignancy

A discussion of cytogenetics within the framework of a cytology atlas is justified by the fact that chromosomal analyses have become an essential part of the morphologic diagnostic process. In this chapter we shall give a brief description of chromosomal lesions and their relationship to cancer diagnosis.

GENERAL METHODOLOGY

Chromosomal analyses are performed by the observation of mitoses at metaphase: at this point, the chromosomes are condensed, individualized, and positioned on the equatorial plane of the spindle. Samples collected for such analysis are submitted to direct examination or are placed in culture media for short-term incubation. The chromosomes are harvested by colchicine incubation which inhibits spindle formation and thus accumulates mitoses in metaphase. The cells are swollen in hypotonic solution to make their membranes more fragile and then are fixed with Carnoy's solution and spread on a slide. It is during spreading that cellular membranes burst. The composition of the media involved and the details of the technical procedure are extensively described in specialized textbooks.[82-97]

Although mitoses theoretically may be observed in any type of tissue, cell cohesiveness often prevents proper spreading on the slide. Therefore, the lymphocytes of peripheral blood samples are universally used for this purpose. Phytohemagglutinin (PHA), a mitogenic protein extract of the *Phaseolus vulgaris* seed, provokes a return of these mature lymphocytes to their immature mitotic state (blastic transformation). Cultured fibroblasts from muscular aponeuroses or desquamated fetal cells obtained via amniocentesis also may prove suitable for analysis.

By this means, a chromosome count and a structural analysis provide a rapid determination of the patient's genotype, i.e., his chromosomal identity. Any congenital anomalies of chromosomal structure or number may or may not lead to clinically observable malformations. If the patient's phenotype (the observable characteristics of an individual) remains normal in spite of an abnormal chromosomal complement or karyotype, the anomaly is said to be "balanced."

If a chromosomal anomaly results not from gametic factors but rather from mitotic malfunctions occurring during the zygote's first divisions, the patient's cellular make-up is said to be "mosaic" or mixed, i.e., formed of variable proportions of normal and pathologic cells depending on the tissue examined. The phenotypic expression of such mosaicism will be more or less pronounced, and the resulting clinical syndromes may be more or less complete.

Cancerous tissues exhibit chromosomal modifications of both structure and number. Such aneuploidy, which occurs in all the cells of a tumoral clone (cell population originating from a single cell), is usually distributed around a modal number of chromosomes particular to the tumor in question. Tumors also may be characterized by the presence of a particular chromosomal rearrangement leading to a "marker" chromosome common to each of the tumor cells.

CHROMOSOME STRUCTURE AND ORGANIZATION

Chromosomes are visible in the nucleus only during mitosis. It is agreed that chromatin, like the chromosome, is an amalgam of DNA, RNA, basic proteins (histones), and acid proteins (nonhistones). However, the precise arrangements of and relationships among these constituents are far from being completely understood.[34, 88]

Basophilic nuclear stains will color the chromatin, which appears to be packed in densely staining clumps of variable size, resting on a faintly colored or clear background. The clumps represent the heteropycnotic portion of the chromatin. We thus distinguish euchromatin and heterochromatin. Euchromatin is the functioning fraction of the nuclear complement; it is in an extended state and stains only weakly. Heterochromatin is the coiled, highly condensed, nonfunctioning portion; it stains more deeply than the extended chromatin. Heterochromatin, tightly bound by nonhistone proteins, replicates later than the euchromatin.[74]

Chromosomal DNA, the chemical equivalent of genes, constitutes the cell's library of genetic information. Such information, however, would be ineffectual without the corroboration of the cytoplasmic ribosomes, which constitute the various sites of cytoplasmic protein synthesis. The genetic information is transcribed in the messenger RNA nucleotide-strand. mRNA molecules are reversibly fixed to cytoplasmic ribosomes and determine the amino acid sequence of the protein molecules to be elaborated.

Schematically, we may summarize this process as follows:

$$\text{DNA} \xrightarrow{\text{transcription}} \text{RNA} \xrightarrow{\text{translation}} \text{proteins}$$

The biochemical nature of the various constituents involved in this procedure, as well as their complex interrelationships, are described by Watson.[95]

The mitotic chromosome is a doubled structure whose two halves, the chromatids, are united by a primary constriction called the centromere. The chromatin fibers that make up the chromosome are maximally coiled and condensed to form ribbons whose average width is 250 A. Other constrictions—called secondary constrictions—may occasionally appear along the chromatids: these translate a localized lack of coiling. The length of the maximally coiled human chromosome varies between 2 and 8 μ.

CELL DIVISION

Mitosis is the process of nuclear division, followed by cytoplasmic constriction or cytokinesis, which provides each daughter cell with a genetic complement identical to that of the parent cell.

Because the chromosome is the vector of the cell's genetic information, any anomaly of chromatin distribution or structure can have rather serious consequences. In a balanced chromosomal set of genetic information, each gene occupies its own, constant position or "locus" on the homologous chromosomes. These loci may bear similar (homozygous) forms of the gene or different (heterozygous) forms.[74, 88] The heterozygous forms or alleles are either dominant or recessive with respect to each other. Any imbalance of this genetic distribution leads either to the birth of a potentially abnormal cell line or to completely prohibitive genetic combinations, in which case the cell simply dies. We can understand, then, that a harmonious and consistent mitotic mechanism is essential to the cell's survival.

Despite its importance, mitosis constitutes only a brief portion of the cellular cycle. The interphase cell—often called the resting cell—is in fact in its active metabolic period: the chromosomes are no longer maximally condensed (except for the heterochromatic portion) and molecular interactions between DNA, RNA, and proteins may occur.

Interphase

Interphase begins with a preduplication period called the first gap or G_1 (Figure 4.1). The duration of the G_1 stage depends on the tissue considered. During this stage the chromosome is a single, more or less unraveled structure bearing no resemblance to the doubled, condensed mitotic chromosome described above. G_1 is followed by the S (for synthesis) phase. It is during this phase that the DNA double helix replicates itself, resulting in two identical sister chromatids. This duplication occurs with a certain degree of asynchrony, as the heterochromatin in general and the inactivated X-chromosome in particular replicate later than the euchromatin. The G_2 stage (postduplication) occurs after DNA duplication and before the beginning of the immi-

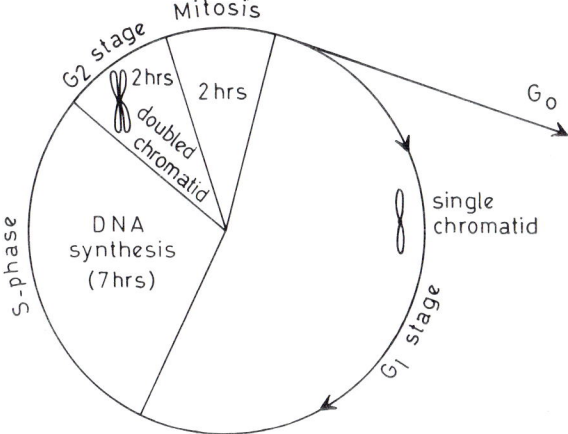

Figure 4.1. Diagram of the cell cycle.

nent mitosis. During this stage, therefore, the quantity of DNA is twice the normal value (Figure 4.2).

The concept of a cellular cycle implies a constant duration, or generation time. However, a cell may stop synthesizing new DNA, for example, after having reached its ultimate level of differentiation. It is by convention said to be in the G_0 stage.

Mitosis itself is divided into four distinct phases phases that nonetheless continuously and gradually succeed one another [74,88] (Figure 4.3).

Prophase

Prophase lasts from the beginning of mitosis until the nuclear envelope has disappeared. During this phase, the condensing chromatin gradually appears as double ribbons, and the chromosome centromeres begin to take shape. The centrioles divide and begin migrating to opposite poles of the cell. It is during this separation that microtubules will organize themselves into the mitotic spindle. The nucleolus as well as the nuclear membrane disappear.

Metaphase

At metaphase, the centrioles have reached their respective poles, and the chromosomes, which are fully condensed and positioned on the equatorial plane, are attached to the spindle by their centromeres. The next stage begins with the division of the chromosomal centromere, now more properly called a kinetochore.

Anaphase

Anaphase marks the separation of the two sister chromatids and their migration toward the opposite poles. The chromatids are attached to the spindle fibers by their kinetochore. The migration is occasioned by an apparent stretching of the spindle. As this nears completion, the cell membrane begins to constrict and cytokinesis begins.

At this point each chromatid may be properly termed a chromosome; the double DNA complement of G_2 has been equally shared, and the genetic content of both new cells is a normal one.

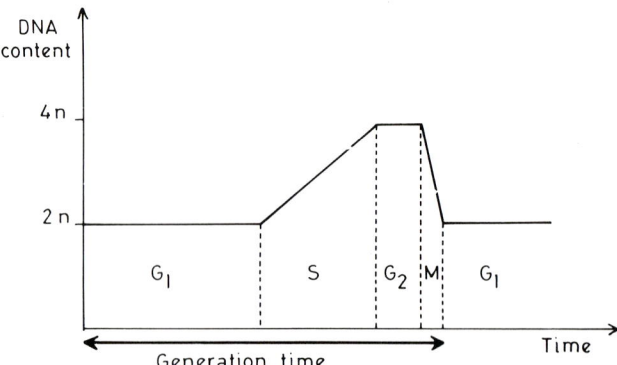

Figure 4.2. During the generation time, the DNA is synthesized during S phase, to reach twice the diploid (2n) value in G_2.

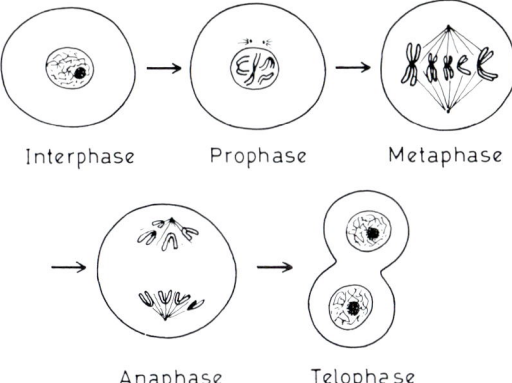

Interphase   Prophase   Metaphase

Anaphase   Telophase

Figure 4.3. Schematic representation of the successive phases of mitosis.

Telophase

Telophase begins with the neoformation of the respective nuclear membranes; the chromosomes unwind and become less apparent, the fusorial fibers disappear, and the nucleolus reappears. The cytoplasmic constriction or cytokinesis separates the two cells completely.

THE NORMAL KARYOTYPE

In 1956 the number of human chromosomes—46—was correctly determined by Tjio and Levan.[90] This number, the diploid value, applies to somatic cells. At about the same time, Ford and Hamerton [35] determined that germ cells contain 23 bivalents (associations of homologous chromosomes). Twenty-three is the haploid number.

Twenty-two pairs of homologous chromosomes—termed autosomes—are common to both sexes. The twenty-third pair are the gonosomes: XX in females, XY in males, and these determine the individual's genotypic sex.

At the Denver Conference of 1960,[23] it was agreed to classify chromosomes systematically according to a diagram called the karyotype. Until 1971, chromosomes were arranged by order of decreasing size and were grouped by the position of their centromere.

Chromosome Number	Chromosome Group	Centromere Position	Chromosome Morphology
1–3	A	median	metacentric
2	A	submedian	submetacentric
4–5	B	subterminal	telomeric
6–12 X	C	submedian	submetacentric
13–15	D	terminal	acrocentric
16	E	median	metacentric
17–18	E	submedian	submetacentric
19–20	F	median	metacentric
21–22 Y	G	terminal	acrocentric

In practice, however, this system did not eliminate major difficulties of classification, in spite of efforts at metric and autoradiographic evaluations.

In 1970, Caspersson [19] treated human chromosomes with various DNA-binding fluorochromes resulting in sequences of brilliant and faint striations along their arms (Plate 4.1). This "banding pattern" allowed him to identify the individual chromosome types and to classify the homologues with confidence. [17]

Following Caspersson's discoveries, numerous researchers working along the same lines produced an abundance of highly varied denaturation techniques—procedures that in spite of their technical differences happily provided comparable results. [18, 27, 82, 97] Essentially, the goal of such techniques is to produce bands of different tinctorial affinity along the length of the chromatids, giving them a characteristic "footprint" that is statistically unique and adaptable to computer analysis. [16]

Banding techniques made it possible to ratify in Paris [65] a universal nomenclature applicable to and descriptive of each chromosome, its regions, and its bands. Thus, we successively define four parameters: chromosome number, arm, region, and band. The regions and bands of the short arms (p) and long arms (q) are numbered, counting from the centromere (Figure 4.4). For example, the fluorescence labeling characteristic of the Y-chromosome concerns band Yq12.

Depending on the techniques used, one can distinguish the following types of banding.

Q Banding

These bands were first observed by Caspersson, [17, 19] who stained his preparations with quinacrine dihydrochloride and examined them under fluorescent light. The configuration of Q bands serves as a point of reference. Their distribution parallels the dry mass distribution along the chromatides. [39] Routinely, this technique remains useful for analysis of the Y-chromosome and its anomalies. Its terminal brillance is such that it may even be observed in the interphase nucleus.

R Banding

R banding, for "reverse banding," may be described as the negative of Q banding. The chromosomes are stained with acridine orange for fluorescence [15] (Plate 4.2) or with Giemsa for ordinary light microscopy [31] (Figure 4.5).

T Banding

A longer thermal pretreatment allows elective fluorescence enhancement of the terminal structure of the chromosome arms (telomeric banding) [26] (Plate 4.3).

G Banding

The chromosomes are stained with Giemsa after denaturation in saline solution [87] or in a trypsin solution. [83] Because of its simplicity, stability, and sensitivity, [98] this type of banding is the most widely used (Figure 4.6). G bands are with a few exceptions the morphologic equivalent of Q bands. They are probably the reflection of the concentration and arrangement of nucleoproteins along the chromatid. [7, 98]

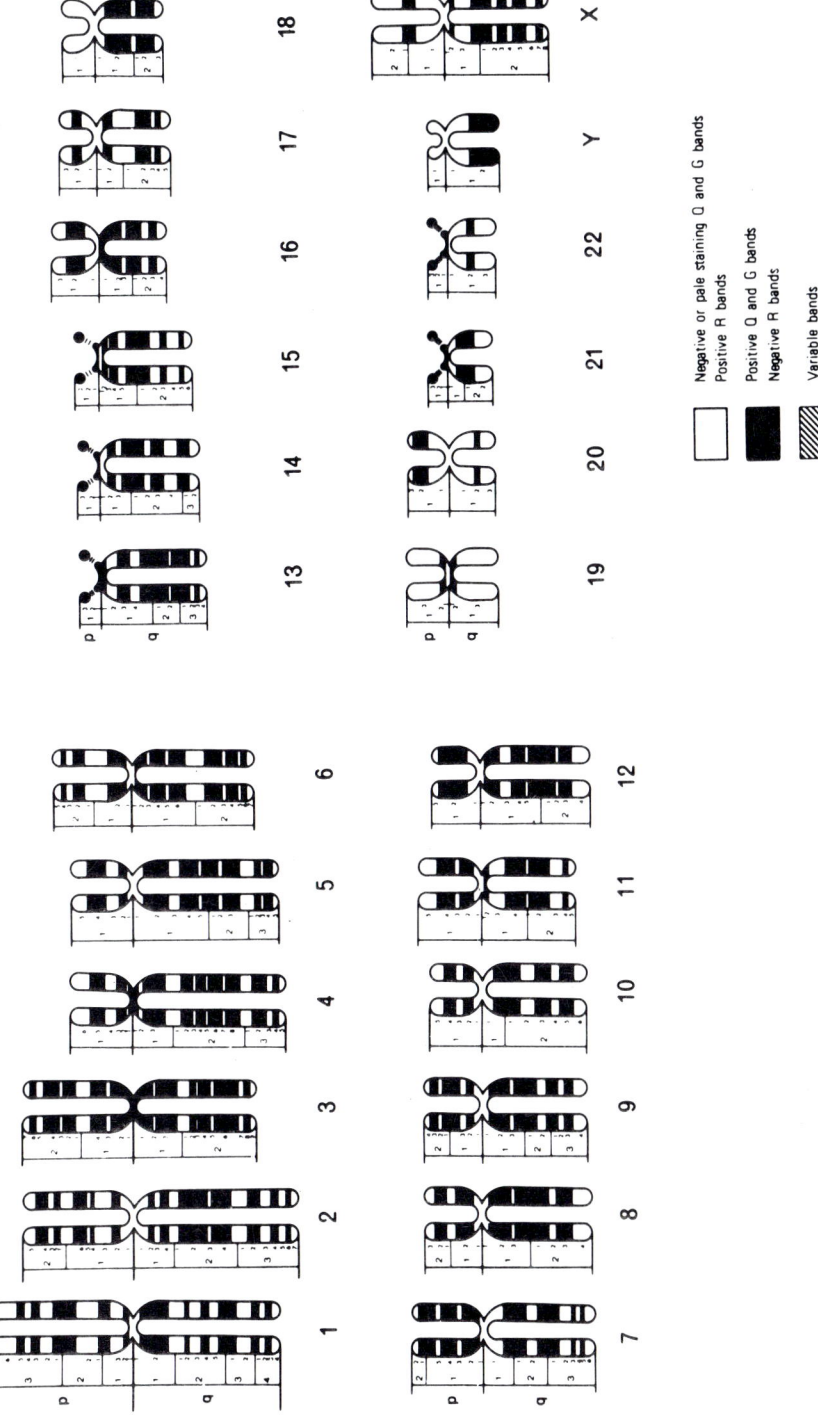

Figure 4.4. Karyotype nomenclature reproduced from the Paris Conference 1971.[65]

35

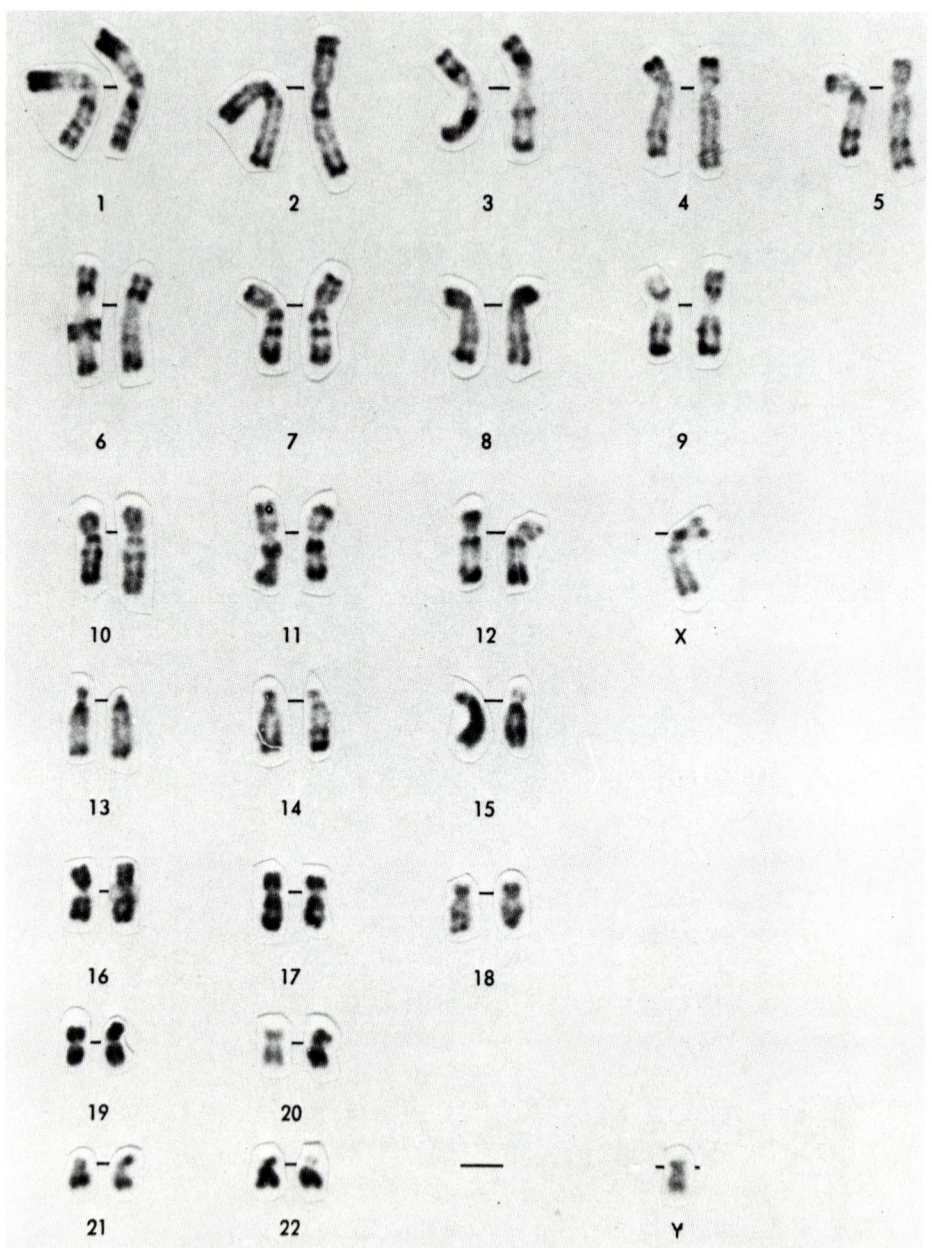

Figure 4.5. Male karyotype stained with Giemsa; R banding.

36

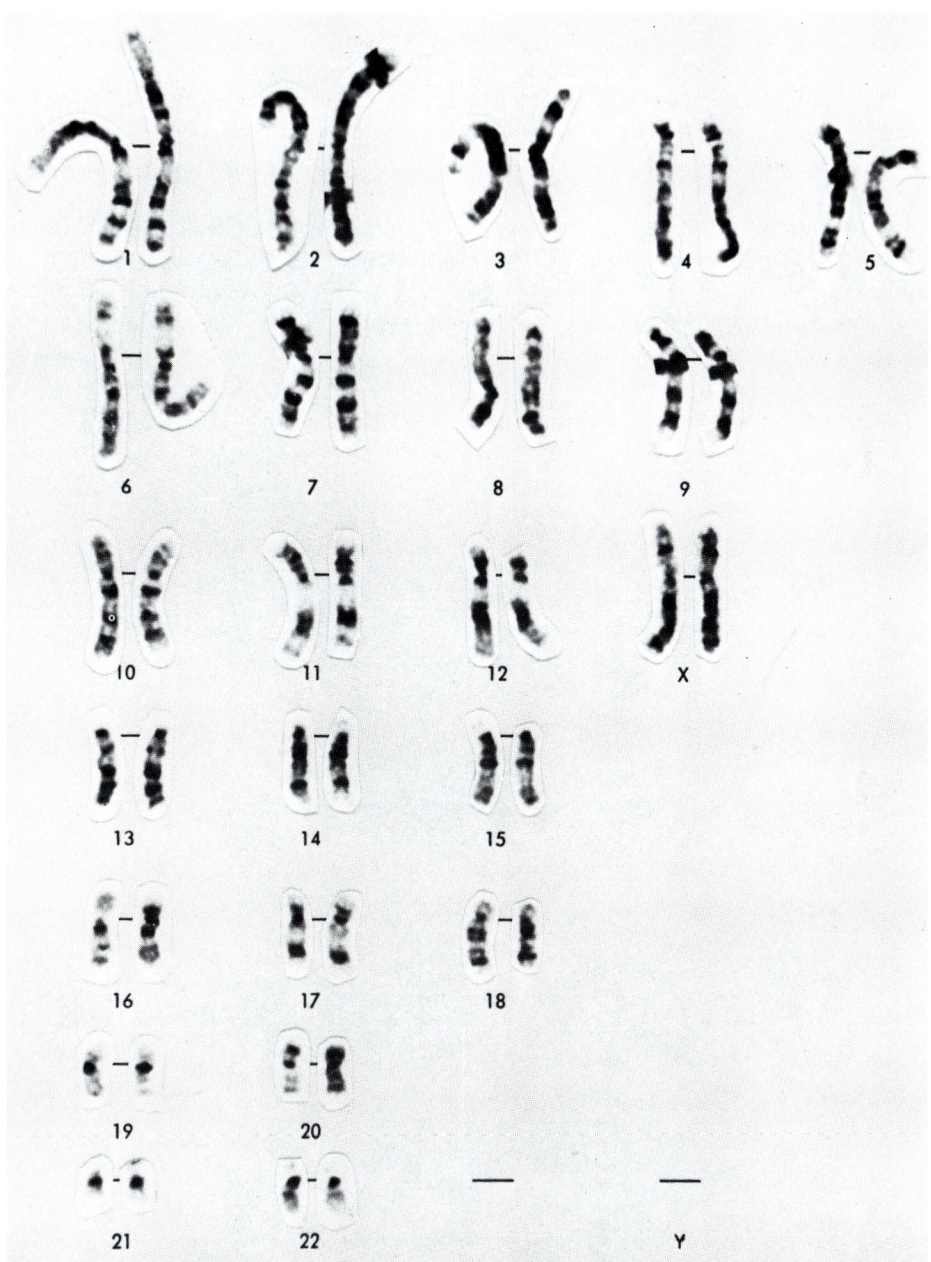

Figure 4.6. Female karyotype; G banding.

C Banding

C bands are characteristic only of juxtacentromeric regions and of the heterochromatic regions of the Y-chromosome [1] (Figure 4.7).

BrdU Treatment

5-Bromodeoxyuridine (BrdU), when introduced to cultures near the end of the S phase, will selectively arrest condensation of several chromatid segments. The resulting banding is the R type, which is better stained by acridine orange than by Giemsa[28] (Figure 4.8).

It is known that homologous chromosomes undergo condensation in a synchronous manner, with the exception of the X-chromosomes.[8] These exhibit an asynchrony of replication, the tardily unraveled, inactived X-chromosome completing its duplication later than the actived X. For this reason, the inactived X-chromosome can be easily recognized (Plate 4.4). This technique is therefore ideal for the determination of an individual's chromosomal sex (Figure 4.9), as well as for study of structural anomalies of the X-chromosome[30] and their relation to phenotypic deviations.

When BrdU is added to chromosome cultures for periods longer than 36 hours, it

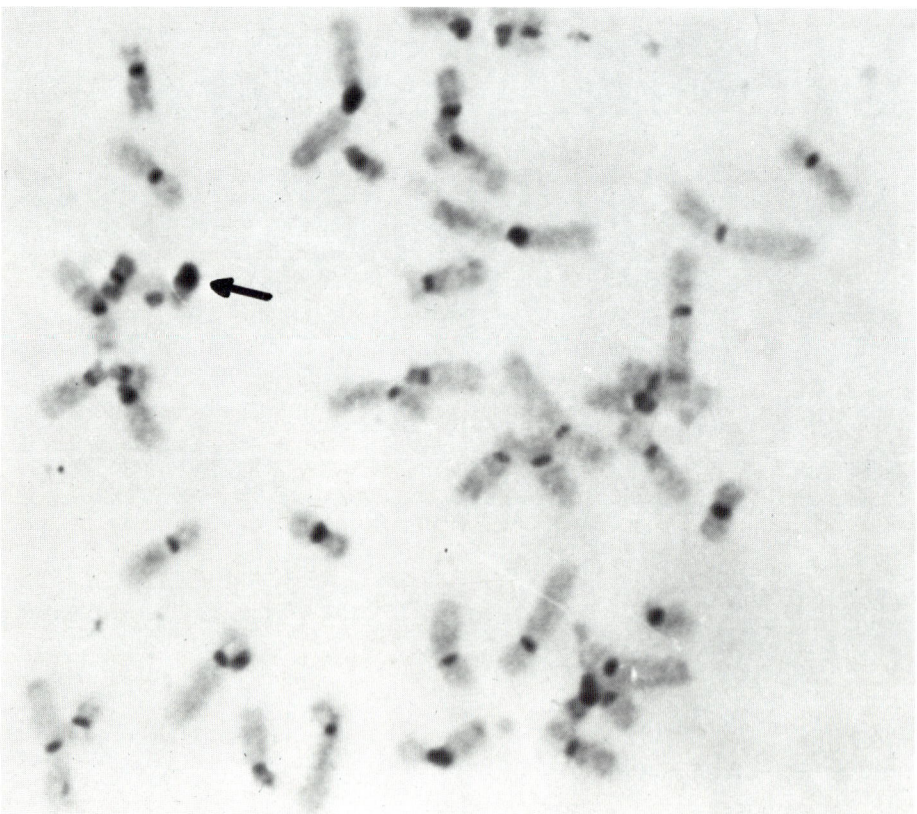

Figure 4.7. C banding. The heterochromatic portion of the Y is arrowed.

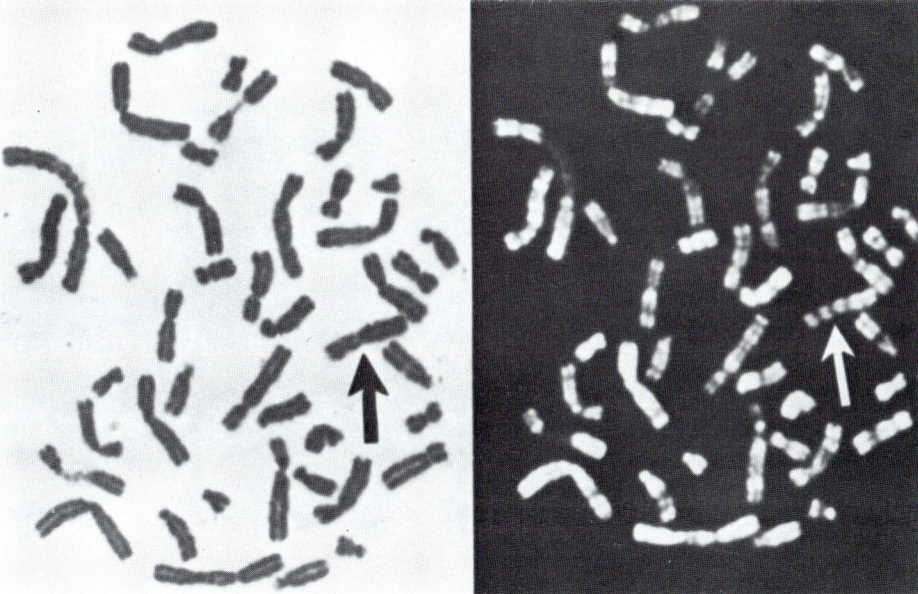

Figure 4.8. BrdU treatment; comparison of successive Giemsa staining and acridine orange on the same mitosis in a case of Turner's syndrome: 45x0 female with a single X (*arrow*).

will be incorporated gradually during successive S phases[99] (Figure 4.10). The chromosomes will appear asymmetric, one chromatid being highly fluorescent because of the persistence of one original DNA strand and the other, fluorescent pale red because both DNA strands are substituted by BrdU incorporation (Plate 4.5).

The frequency and the site of exchanges between sister chromatids also can be studied with this technique.[29, 50] Should such regions prove to be genetically active loci, their subdivision and inadequate repair via such exchanges may provide possible explanations for certain pathologic behavior.[66]

SEX CHROMATIN

The sex chromatin is that portion of the condensed heterochromatin that appears as a convex granule adhering to the inner membrane of the nuclear envelope (Plate 4.6). Barr[9] first observed that the sex chromatin is found only in the nuclei of females and represents a single, positively X-chromosome.[10] This chromosome is inactivated early in embryonic life, and it is the one that replicates more slowly during the S phase.

Sex Determination

All women exhibit a sex chromatin or Barr body in their interphasic nuclei, and men do not. The male Y-chromosome does however appear in the interphase nucleus as a brightly fluorescent point[68,69] (Plate 4.7), the Y-body, and any subject possessing several Y-chromosomes will accordingly exhibit as many Y-bodies. Similarly, a subject possessing several X-chromosomes will exhibit as many Barr bodies as there are inac-

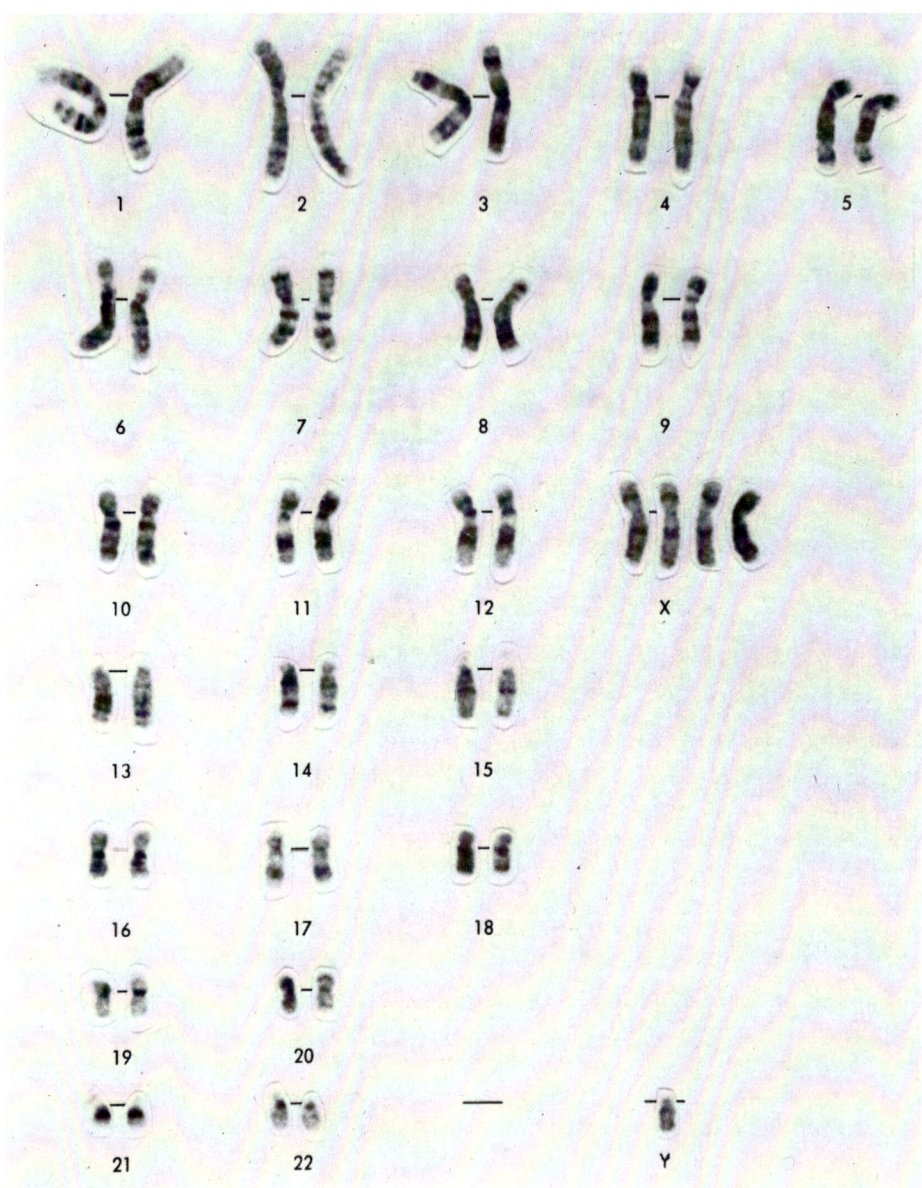

Figure 4.9a. Karyotype of 49XXXXY male (Klinefelter's syndrome is commonly 47XXY). Full karyotype of G banding.

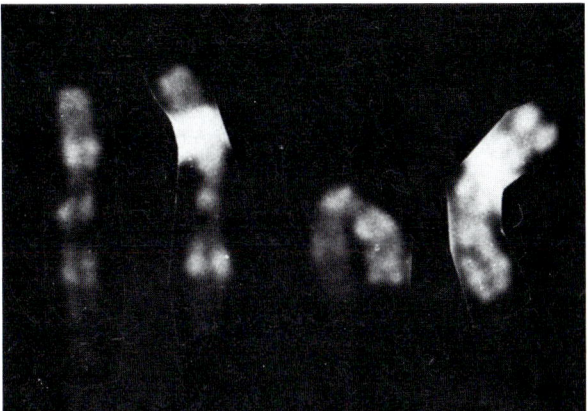

Figure 4.9b. After BrdU treatment the three inactivated Xs appear longer than the fourth one. These constitute the sex chromatins in interphase nuclei.

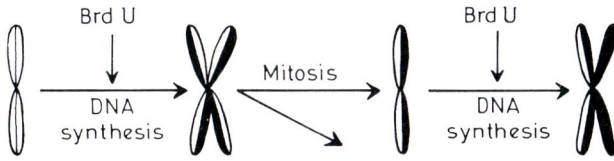

Figure 4.10. Schematic incorporation of BrdU during successive cell cycles.

tivated X-chromosomes. In other words, the number (C) of such sex chromatins equals the total number (N) of X-chromosomes minus 1:

$$C = n \times - 1$$

The detection of a sex chromatin or of a Y-body is a rapid and simple means of sex determination[37, 46] applicable to population studies[67] or prenatal diagnoses.[21]

These sex chromatin techniques are of limited interpretative value[93] in the study of mosaics and structural anomalies, and they contribute little when compared with banding and BrdU techniques.

Sex Chromatin and Cancer

Few attempts have been made to assess the persistence of a Barr body in cancer cells.[38, 73] However, those breast cancers in which a Barr body is lacking show a worse survival rate independent of tumor grade or invasion of lymph nodes.[81] Y-chromatin detection in male or female cancer cells is not yet sufficiently documented to provide prognostic information.[3,54]

THE CHROMOSOMAL ANOMALIES OF CANCER

A cancer is by definition a population of cells that proliferates by uncontrolled mitoses and that therefore no longer respects the integrity of the tissue in which it occurs.

Numerical and Structural Anomalies

Such run-away mitotic phenomena are accompanied by gross chromosomal alterations in both number and structure.[84] Unlike congenital anomalies, cancerous acquired chromosomal aberrations are limited to the cells of the tissue involved.. What is
more, these modifications are by no means stable ones: they are dynamic and may be
aggravated with the successive steps of the tumor progression (clonal evolution).[12, 42]

Structural anomalies lead to complex chromosome rearrangements resulting in
marker chromosomes to which the conventional nomenclature does not apply. Their
original homologues of these chromosomes can be ascertained only by banding
analyses.[48]

Specificity and Consistency of the Anomalies

The systematic discovery of a deleted G-chromosome (Ph')[63] in more than 90 percent
of all cases of chronic myelogenous leukemia, seemed a decisive argument for the concept of specific chromosomal imbalance as the primary event in tumorigenesis. On-
cogenesis could be produced by the loss of growth controlling genes or by position effect due to translocations.[44] Indeed, banding techniques did reveal the Philadelphia
chromosome to be a translocation between chromosome 22 and, most often,
chromosome 9[75] (Figure 4.11).

Cytogeneticists hoped to find other such associations between a specific marker
chromosome and a specific neoplastic state. This hope was not realized. Similar
markers are discovered only sporadically in certain tumors,[52, 55] usually metastatic
ones.[14, 47] Generally speaking, such markers can only characterize an individual
neoplastic growth, and tumors classified under the same heading are histologically
comparable but karyotypically unique. Thus, for the present, only the Philadelphia
chromosome has a definitive diagnostic value, and chromosomal analysis does not improve histologic tumor classification.

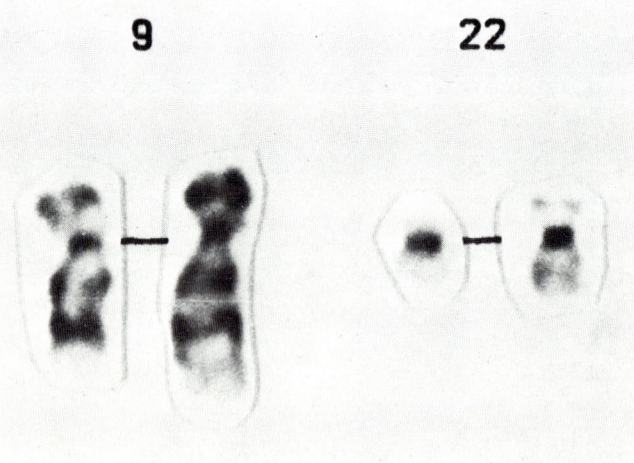

Figure 4.11. The Philadelphia chromosome; specific anomaly of chronic myelogenous
leukemia is commonly the result of a translocation of the two-thirds of the long arms of
chromosome 22 to the q end of chromosome 9.

Chromosomal Alterations: Cause or Epiphenomenon?

Let us now investigate whether cytogenetics can determine the crucial moment of cancer transformation before the appearance of cytologic criteria.

More than 50 percent of all acute leukemias exhibit diploid cells with no specific marker chromosomes; thus chromosomal changes may or may not occur during the progression of the disease. Such considerations seem to cast some doubt on the primary role of chromosomes in cancerization.[80] We should point out, however, that an apparently normal mitotic chromosome set, in sometimes highly malignant tumors,[57] could theoretically conceal the genetic aberrations of cancer initiation. A functional anomaly on the gene level leads to a genetic instability and a different biologic behavior. This in turn may or may not lead to a more flagrant chromosomal imbalance which can be detected by current techniques.

The study of the transformation of normal cells exposed in vitro to chemical or viral carcinogens is another means of investigating the causal mechanism. Breakages that are induced by such treatment generally occur between G bands and never in heterochromatic regions. Sometimes such breakages are agent-specific and most often result in cell death.

In general, gross chromosome aberrations are less informative compared with sister chromatic exchanges (SCE), which prove to be much more sensitive indices of chromosome damage.[25, 70,44, 66]

Whether specific chromosome changes are required for malignant transformation is not yet known. Moreover, since malignant transformation can occur without visible chromosome damage, systematic testing programs of SCE frequencies and localizations would provide more information on the possible mutagenicity of incomplete or abnormal repair in both parental strands.

Fanconi's anemia is characterized by a high number of chromosome breaks; they are located in negative G bands and show a definite nonrandom distribution.[49] Partial homozygosity during malignant hemopathies could prove more frequent than expected.

A remarkable rearrangement has been emphasized recently using banding techniques.[13] Some homologues appear identical (i.e., homozygous) in bone marrow cells of Fanconi's anemia patients; it is also probably that such anomalies still escape analysis by current techniques. The important fact is that recessive genes with this acquired somatic segregation can reach a homozygous state.

Precancerous States

The practical value of chromosomal analysis lies in the recognition of aneuploidies in precancerous conditions and of clonal evolution in malignant transformation. These topics require more research, and since technical difficulties have limited the number of observations, generalizations cannot be drawn yet.

Most of the cases reported were published prior to the availability of banding techniques, and therefore the rare karyotypes obtained are not much more informative than quantitative DNA measurements of interphase nuclei, which provide a reliable means of estimating the modal chromosome number.

Dysplasias of the breast[32,91] and cervix[40,85] degenerating intestinal polyps,[33] adenomatous endometrial hyperplasias[41] ovarian borderline lesions,[5,36] have been investigated by several authors. Aneuploidies are the rule. Dysplasias carry minor chromosomal anomalies as compared with carcinoma in situ or invasive cancer. Since dysplastic lesions are as likely to regress as to progress, observed aneuploidies may similarly disappear or progress. They are thus a reflection of genetic instability in precancerous conditions that have not yet totally gone out of growth control. They may remain stable for years, as does their histologic counterpart.[86]

Certain dysplasias exhibiting extremely wide ranges of chromosome number could be polyclonal in origin. If the disease progresses, various new combinations can arise in the altered epithelium and can be tested for malignant potential and possible clonal evolution: a favorable clone, i.e., one that favors more malignancy, will arise from a single cell with properly altered chromosomal complement and antigenic status. Chromosomal analysis could prove to be of diagnostic value in this field if it could detect the emergence of such a new clone prior to the histologic evidence of invasion.

Statistical study of cervical neoplasia[20] reveals a change in the distribution of the chromosome counts in the different stages of the disease; karyotypic rearrangements become more complex even with a similar chromosome number.[2, 22, 40] In invasive carcinomas, the ploidy level will be of different prognostic significance, depending on the side involved.[4,61, 2]

From experimental work with virus-induced tumors there is evidence that loss of growth control is reversible and is accompanied by reversible shifts in chromosome number;[72] specific chromosomes are involved in the expression of invasive properties,[45] and the suppression of the malignant potential occurs if such chromosomes are lost or others are gained in excess.

Preleukemia and Leukemia

The study of preleukemic states is of a more practical interest. These ill-defined myeloproliferative disorders include myelosclerosis, all subtypes of anemias, thrombocytosis, essential thrombocytopenias, granulocytopenias, and dysmyelopoietic conditions. The detection of particular chromosomal markers could allow us to classify seemingly unrelated morphologic or clinical observations into more specific etiologic tableaux.

Among the more frequent chromosomal anomalies are trisomies 8 (Figure 4.12) and 9, monosomy 7, and partial deletions of long arms 5q-, 7q-, 20q-.[51, 78, 96]

Whether these anomalies are indeed characteristic of any specific myeloproliferative syndromes is still a matter of controversy, but if data from the literature are compiled and compared, it becomes evident that chromosome aberrations are nonrandom,[77] and that only a few chromosomes are systematically involved and others are not.[56]

The prognostic significance of specific anomalies in preleukemias is not yet clear.[64] An elevated risk of developing leukemia seems to exist only if an acute transformation occurs within 3 months to a year after the appearance of the abnormal clone.[62,78]

Leukemia arising de novo, or after a preleukemic phase, is characterized in more than 50 percent of the cases by karyotypic changes; they occur in addition to the ones detectable during the chronic phase of the disease and are more complex. Such changes illustrate the clonal evolution during blastic transformation[58, 94] (Figures 4.13 and 4.14), and they also appear to be nonrandomly distributed.[58, 60, 77]

Similar results are obtained in experimental tumors: only a few specific

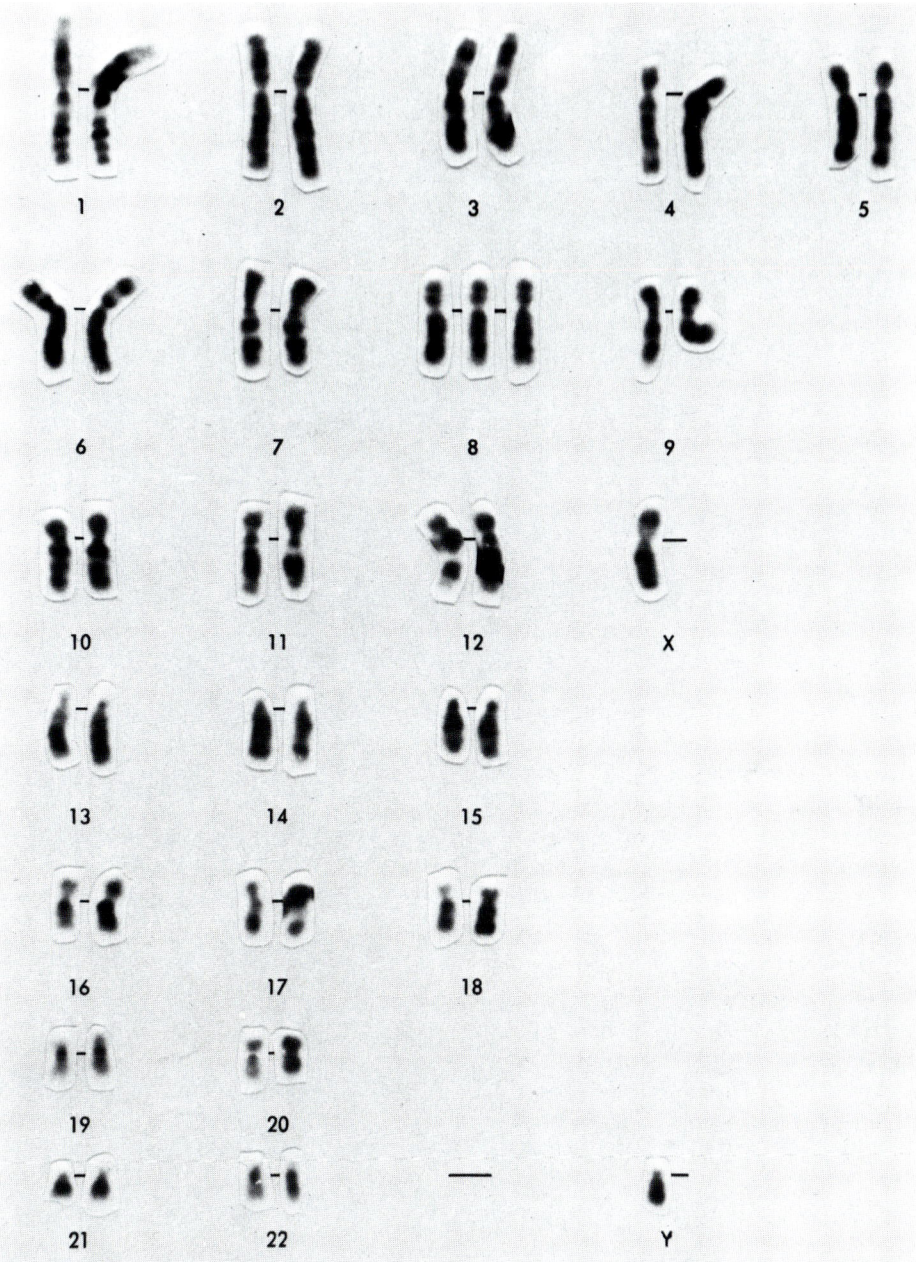

Figure 4.12. 46XY, + 8; G banding. Trisomy 8 is nonrandomly encountered in nonmalignant and malignant hemopathies.

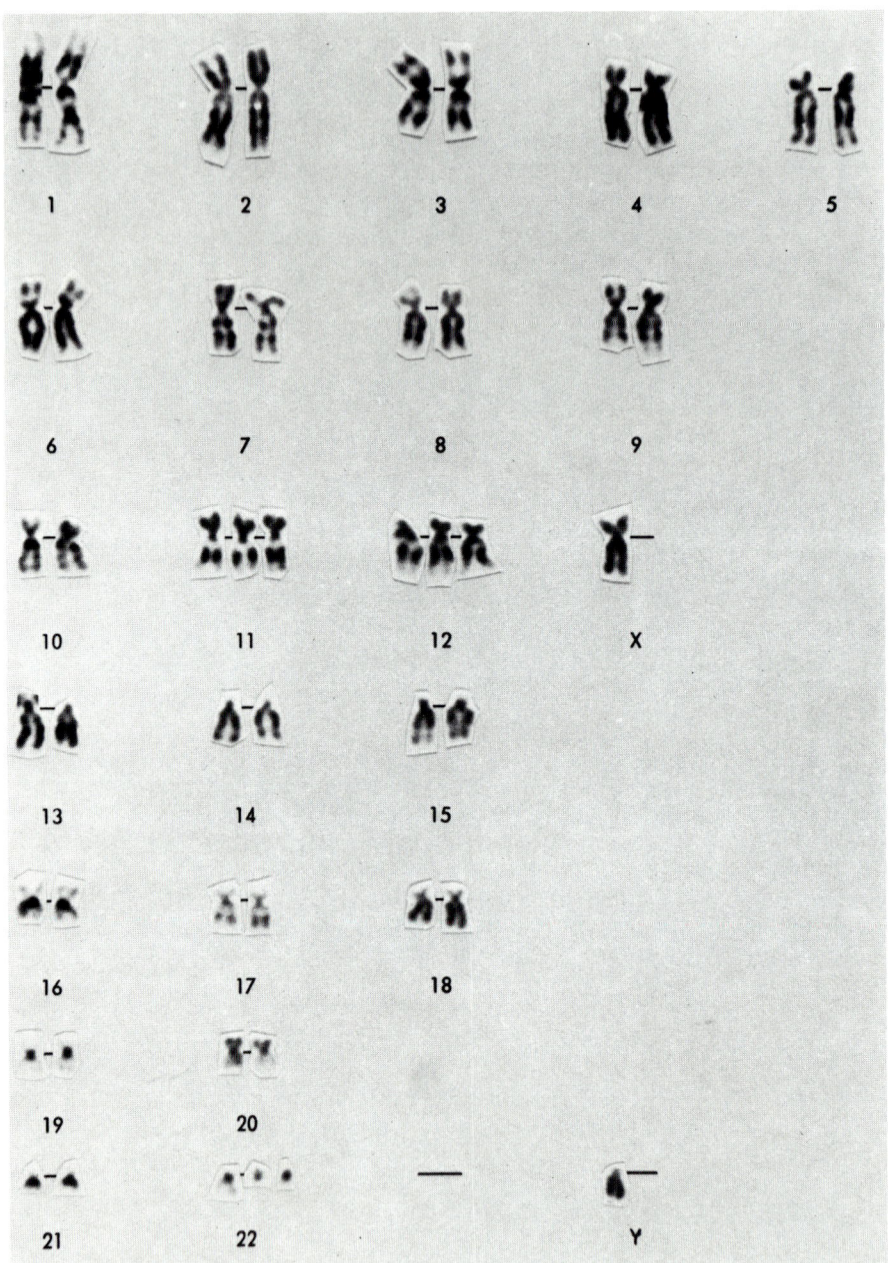

Figure 4.13. Example of clonal evolution in the blastic crisis of chronic myeloid leukemia. The modal value of this abnormal cell line is 49. The Ph'chromosome is duplicated late. It explains why the translocation to chromosome 9 is single. Trisomies 11 and 12 appeared only in the acute phase.

46

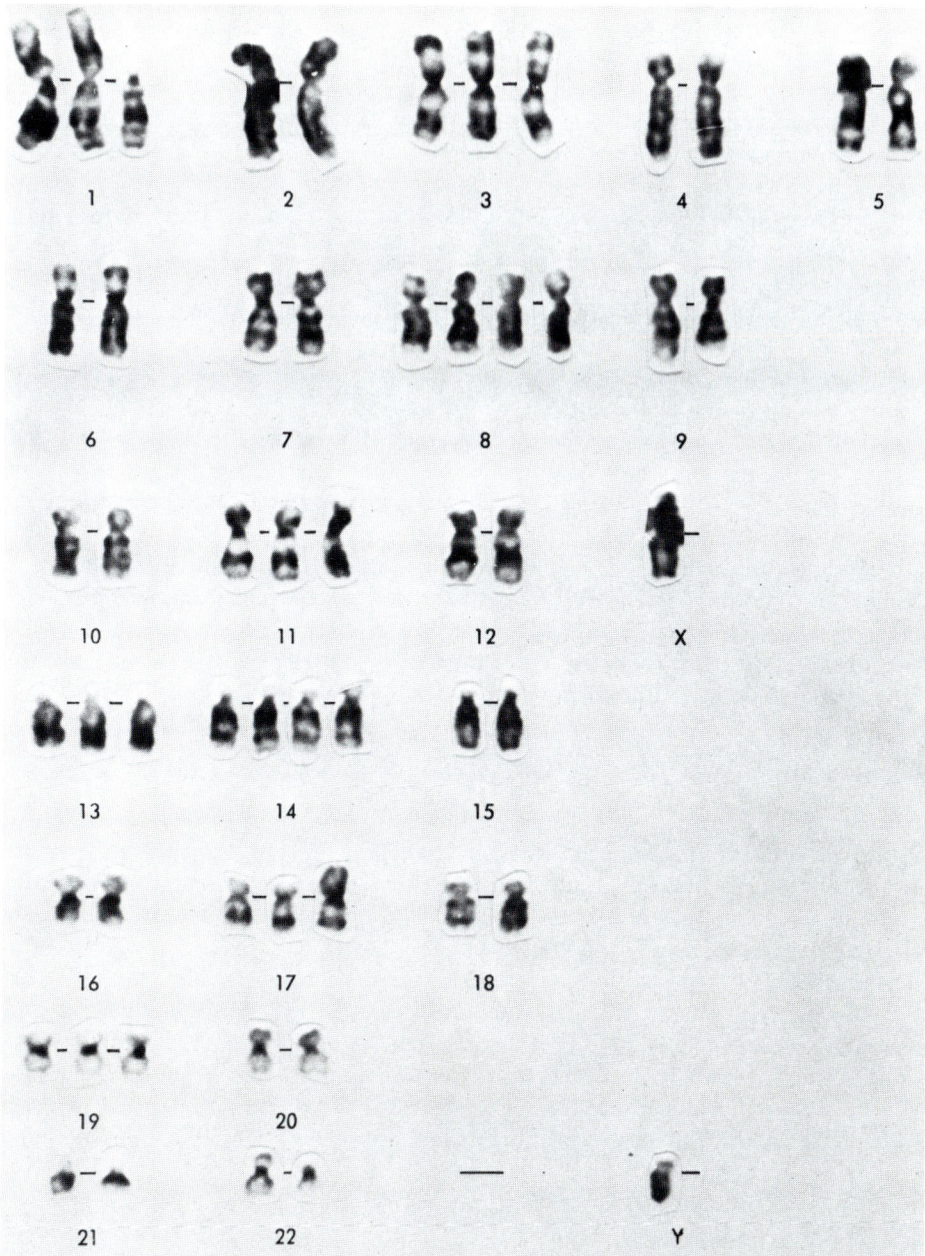

Figure 4.14. More complex karyotypic rearrangement than in Figure 4.13. This blastic crisis of chronic myeloid leukemia is characterized by a clone at 56 chromosomes. Besides the persistent classical 9–22 translocation, the blastic cell line has acquired many supernumerary chromosomes, some are reduplicated. A marker 1 q$^+$ is constant in the clone. Note the isochromosome 17 (both the long arms are present and are separated by the centromere, short arms are lost). This marker is frequent in acute leukemia and of bad prognosis.

chromosomes participate in structural rearrangements and formation of markers.[53] The induced tumors, although histologically indistinguishable, are by their karyotypic pattern specific to the causal agent.[59]

What is observed in animals may be reasonably postulated for the human, and one can imagine that different mutagenic agents can produce identical clinical or histologic diseases. In other words, the importance of similar chromosome changes (for example trisomy 8 encountered in both nonmalignant hematologic disorders and leukemia) could be better appreciated in malignant transformation if understood as agent-specific and consistent for a single etiologic factor.[76] The Ph' chromosome remains the ideal example.[63, 75] It is specifically associated to an unknown mutagenic factor and is consistently associated with clinical manifestations of chronic myeloid leukemia.

As stated before, in acute leukemia, half the patients at the time of diagnosis show no abnormalities in all the mitoses obtained from marrow aspiration. This is of favorable prognosis for the expected survival rate.[79] Patients exhibiting only abnormal metaphases have a shorter survival than those who keep a few normal mitoses among the abnormal ones, probably due to their capacity to repopulate the marrow with normal cells after response to chemotherapy.

If the type, and not only the presence, of karyotypic abnormalities is taken in account, some chromosomal changes may have definitive prognostic implications.[92] It is thus understandable that bone marrow chromosomal analysis has become one of the routine diagnostic and follow-up procedures of hematologic disorders. It may serve both as an indicator of prognosis and of response to treatment.

Cancer Effusions

Neoplastic pleural and peritoneal effusions constitute spontaneous suspensions of malignant cells and thus serve as a convenient tool for the study of cancer mitoses. It is surprising to note, however, that the use of this tool is not particularly widespread, even in those laboratories whose facilities allow for simultaneous cytologic and chromosomal analyses.

It is well established that metastatic human tumors consist of clones of cells of a sufficiently altered genetic make-up to constitute carriers of complex karyotypic changes.[14,43,47,71,89] If a sample proves suspicious upon cytological examination, the discovery of aneuploidies by chromosomal examination may contribute to diagnosis. On the other hand, the absence of abnormal mitoses can rule out false positives in irritated effusional mesothelial elements that resemble cancer cells. The combination of both techniques significantly improves the results.[11, 24]

Moreover, chromosomal analyses of effusions can, by fostering systematic use of banding techniques, improve our knowledge of the relations between karyotypic modifications and human tumors.[6,47] These systematic analyses will allow us to study, for example, the distribution of modal values of the different tumoral clones encountered in the same histologic types of cancer. We may thus establish karyotypic generalizations by a comparative analysis of the most frequently modified chromosomal pairs and those homologues that conversely are never modified (prohibitive combination). A more precise marker analysis [48] contributes to establishing their possible specificity.

BIBLIOGRAPHY

1. Arrighi F.E., Hsu T.C., Localization of heterochromatin in human chromosomes. *Cytogenetics* 10:81, 1970.

2. Atkin N.B., Cytogenetic studies on human tumors and premalignant lesions: the emergence of aneuploid cell lines and their relationship to the process of malignant transformation in man. *In Genetic Concepts and Neoplasia.* Williams & Wilkins, Baltimore, 1970, p.36.

3. Atkin N.B., Y bodies and similar fluorescent chromocentres in human tumors including teratomata. *Brit. J. Cancer* 27:183, 1973.

4. Atkin N.B., Prognostic significance of ploidy level in human tumors. I. Carcinoma of the uterus. *J. Natl. Cancer Inst.* 56:909, 1976.

5. Atkin N.B., Baker M.C., Chromosome and DNA abnormalities in ovarian cystadenomas. *Lancet* 1:470, 1970.

6. Ayraud N., Identification par dénaturation thermique ménagée des anomalies chromosomiques observées dans six tumeurs métastatiques humaines. *Biomedicine* 23:423, 1975.

7. Bahr G. F., Correlates of chromosomal banding at the level of ultrastructure *Tutorial Proceedings of the Intern. Acad. of Cytology* 2:58, 1973.

8. Baranovskaya L.T., Zakharov A.F., Dutrillaux B., Carpentier S., Prieur M., Lejeune J. Différenciation des chromosomes X par les methodes de despiralisation au 5-bromodeoxyuridine (BUDR) et de dénaturation thermique ménagée. *Ann. Génét.* 15:271, 1972.

9. Barr M.L., Bertram E.G., A morphological distinction between neurones of the male and female and the behaviour of the nucleolar satellites during accelerated nucleoprotein synthesis. *Nature* 163:676, 1949.

10. Barr M.L., Carr D.H., Correlations between sex chromatin and sex chromosomes. *Acta Cytol.* 6:34, 1962.

11. Benedict W.F., Porter I.H., The cytogenetic diagnosis of malignancy in effusions. *Acta Cytol.* 16:304, 1972.

12. Berger R., Chromosomes et leucémies humaines. La notion d'évolution clonale. *Ann. Génét.* 8:70, 1965.

13. Berger R., Briere J., Clauvel J.P., Homozygotie et syndrome myéloprolifératif. *Nv. Rev. Fr. Hémat.* 15:667, 1975.

14. Berger R., Lacour J., Anomalies chromosomiques dans un cancer de l'ovaire. *Path. Biol.* 22:603, 1974.

15. Bobrow M., Madan K., The effects of various banding procedures on human chromosomes studied with acridine orange. *Cytogenet. Cell Genet.* 12:145, 1973.

16. Caspersson T., Lomakka G., Moller A., Computerized chromosome identification by aid of the quinacrine mustard fluorescence technique. *Hereditas* 67:103, 1971.

17. Caspersson T., Lomakka G., Zech L., The 24 fluorescence patterns of the human metaphase chromosomes distinguishing characters and variability. *Hereditas* 67:82, 1971.

18. Caspersson T., Zech L., Chromosome identification technique and application in biology and medicine. *Nobel Symposium 23.* Nobel Foundation, Stockholm 2nd Academic Press, New York, 1973.

19. Caspersson T., Zech L., Johansson C., Modest E.J., Identification of human chromosomes by DNA-binding fluorescent agents. *Chromosoma* 30:215, 1970.

20. Cellier K.M., Kirkland J.A., Stanley M.A., Statistical analysis of cytogenetic data in cervical neoplasia. *J. Natl. Cancer Inst.* 44:1221, 1970.

21. Cervenka J., Gorlin R.J., Bendel R.P., Prenatal sex determination. *Obstet. Gynec.* 37:912,1971.

22. Dehnhard F., Breinl H., Schüssler J., Wehler V., Cytogenetische Untersuchungen on Praecancerosen und Carcinomen der Cervix Uteri. *Arch. Gynäk.* 220:123, 1975.

23. Denver Conference: A proposed standard system of nomenclature of human mitotic chromosomes. *Lancet* 1:1063, 1960.

24. Dewald G., Dines D.E., Weiland L.H., Gordon H., Usefulness of chromosome examination in the diagnosis of malignant pleural effusions. *N. Engl. J. Med.* 295:1494, 1976.

25. DiPaolo J.A., Popescu N.C., Relationship of chromosome changes to neoplastic cell transformation. *Amer. J. Pathol.* 85:709, 1976.

26. Dutrillaux B., Nouveau système de marquage chromosomique: les bandes T. *Chromosoma* 41:395, 1973.

27. Dutrillaux B., Sur la nature et l'origine des chromosomes humains. *Monographie des annales de génétique.* Expansion Scientifique Française, Paris, 1975.

28. Dutrillaux B., Fosse A.M., Sur le mecanisme de la segmentation chromosomique induite par le BUDR (5-bromodeoxyuridine). *Ann. Génét.* 17:207, 1974.

29. Dutrillaux B., Fosse A.M., Prieur M., Lejeune J., Analyse des échanges de chromatides dans les cellules somatiques humaines. Traitement au BUDR et fluorescence bicolore par l'acridine orange. *Chromosoma* 48:327, 1974.

30. Dutrillaux B., Laurent C., Gilgenkrantz S., Couturier J., Lejeune J., Carpentier S., Les translocations du chromosome X. Etude après traitement par le BDUR et coloration par l'acridine orange. *Helv. Paediatr. Acta Suppl* 19, 1974, p. 31.

31. Dutrillaux B., Lejeune J., Sur une nouvelle technique d'analyse du caryotype humain. *C. R. Acad. Sci.* (Paris) 272:2638, 1971.

32. Emson H.E., Kirk H., Value of desoxyribonucleic acid (DNA) in evolution of carcinomas of the human breast. *Cancer* 20:1248, 1967.

33. Enterline H.T., Arvan D.A., Chromosome constitution of adenoma and adenocarcinoma of the colon. *Cancer* 20:1746, 1967.

34. Evans H.J., Molecular architecture of human chromosomes. *Brit. Med. Bull.* 29:196, 1973.

35. Ford C.E., Hamerton J.L., The chromosomes of man. *Nature*178:1020, 1956.

36. Fraccaro M.A., Mannini A., Tiepolo L., Gerli M., Zara C., Karyotypic clonal evolution in a cystic adenoma of the ovary. *Lancet* 1:613, 1968.

37. Francois J., Matton-Van Leuven M.T., Acosta J., Male and female sex determination in hair roots. *Clin. Genet.* 2:73, 1971.

38. Ghosh S.N., Shah P.N., Prognosis and incidence of sex chromatin in breast cancer. A preliminary report. *Acta Cytol.* 19:58, 1975.

39. Golomb H.M., Bahr G.F., Correlation of the fluorescent banding pattern and ultrastructure of a human chromosome. *Exptl. Cell. Res.* 84:121, 1974.

40. Granberg I., Chromosomes in preinvasive, microinvasive and invasive cervical carcinoma. *Hereditas* 68:165, 1971.

41. Granberg I., Traneus A., Silfversward C., Chromosome pattern in a patient with cervical carcinoma in situ and atypical hyperplasia of the endometrium. *Acta Obstet. Gynec. Scand.* 51:47, 1972.

42. Grouchy J.de, Nava C. de, Cantu J.M., Bilski-Pasquier G., Bousser J., Models for clonal evolutions. A study of chronic myelogenous leukemia. *Amer. J. Hum. Genet.* 18:485, 1966.

43. Hansson A., Korsgaard R., Cytogenetical diagnosis of malignant pleural effusions. *Scand. J. Resp. Dis.* 55:301, 1974.

44. Hirschorn K., Chromosomes and cancer. In Bergsma D. (Ed.),*Cancer and Genetics,* Birth Defects: Original Article Series. Vol. 12. No. 1. 1976, p. 113.

45. Hitotsumachi S., Rabinowitz Z., Sachs L., Chromosomal control of reversion in transformed cells. *Nature* 231:511, 1971.

46. Hollander D.H., Borgaonkar D.S., The quinacrine fluorescence method of Y-chromosome identification. *Acta Cytol.*15:452, 1971.

47. Kakati S., Hayata I., Oshimura M., Sandberg A.A., Chromosomes and causation of human cancer and leukemia. X Banding patterns in cancerous effusions. *Cancer* 36:1729, 1975.

48. Kakati S., Hayata I., Sandberg A.A., Chromosomes and causation of human cancer and leukemia. XIV. Origin of a large number of markers in cancer. *Cancer* 37:776, 1976.

49. Koskull H. von, Aula P., Nonrandom distribution of chromosome breaks in Fanconi's anemia. *Cytogenet. Cell. Genet.*12:423, 1973.

50. Latt S.A., Localization of sister chromatid exchanges in human chromosomes. *Science* 185, 74, 1974.

51. Lawler S.D., Reeves B.R., Chromosome studies in man: past achievements and recent advances. *J. Clin. Path.* 29:569, 1976.

52. Lejeune J., Berger R. Sur une méthode de recherche d'un variant commun des tumeurs de l'ovaire. *C.R. Acad. Sci.*(Paris) 262: 1885, 1966.

53. Levan G., Levan A., Specific chromosome changes in malignancy: studies in rat sarcomas induced by two polycyclic hydrocarbons.*Hereditas* 79:161, 1975.

54. Litton L.E., Hollander D.H., Borgaonkar D.S., Y-Chromatin of interphase cancer cells. A preliminary study. *Acta Cytol.*16:404, 1972.

55. Martineau M., A similar marker chromosome in testicular tumours. *Lancet* 1:829, 1966.

56. Miltelman F., Levan G., Clustering of aberrations to specific chromosomes in human neoplasms. II. A survey of 287 neoplasms.*Hereditas* 82:167, 1976.

57. Mitelman F., Levan G., Brandt L., Highly malignant cells with normal karyotype in G-banding. *Hereditas* 80:291, 1975.

58. Mitelman F., Levan G., Nilsson P.G., Brandt L., Non-random karyotypic evolution in chronic myeloid leukemia. *Int. J. Cancer* 18:24, 1976.

59. Mitelman F., Mark J., Levan G., Levan A., Tumor etiology and chromosome pattern. *Science* 176:1340, 1972.

60. Mitelman F., Nilsson P.G., Levan G., Brandt L., Non-random chromosome changes in acute myeloid leukemia, chromosome banding examination of 30 cases at diagnosis. *Int. J. Cancer* 18:31, 1976.

61. Ng A.B.P., Atkin N.B., Histological cell type and DNA value in the prognosis of squamous cell cancer of the uterine cervix. *Brit. J. Cancer* 28:322, 1973.

62. Nowell P.C., Marrow chromosome studies in "preleukemia." Further correlation with clinical course. *Cancer* 28:513, 1971.

63. Nowell P.C., Hungerford D.A., A minute chromosome in human chronic granulocytic leukemia. *Science* 132:1497, 1960.

64. Nowell P., Jensen J., Gardner F., Murphy S., Chaganti R. S. K., German J., Chromosome studies in "preleukemia." III. Myelofibrosis. *Cancer* 38:1873, 1976.

65. Paris Conference (1971). *Standardization in Human Cytogenetics.* Birth Defects: Original Article Series. Vol. 8. No. 7. 1972.

66. Passarge E., Bartram C.R., Somatic recombination as possible prelude to malignant transformation. In Bergsma D. (Ed.), *Cancer and Genetics.* Birth Defects: Original Article Series. Vol. 12. No. 1. 1976, p. 177.

67. Paulson J.D., Keller D.W., Muhlendorf K., Eatherly C., A rapid screening technic for detection of Y chromosome. *Obstet. Gynec.* 48:707, 1976.

68. Pearson P.L., A fluorescent technique for identifying human chromatin in a variety of tissues. *Bull. European Soc. Hum. Genet.* 4:35, 1970.

69. Pearson P.L., Bobrow M., Vosa C.G., Technique for identifying Y chromosomes in human interphase nuclei. *Nature* 226:78, 1970.

70. Perry P., Evans H.J., Cytological detection of mutagen-carcinogen exposure by sister chromatid exchange. *Nature* 258:121, 1975.

71. Pickthall V.J., Detailed cytogenetic study of a metastatic bronchial carcinoma. *Brit. J. Cancer* 34:272, 1976.

72. Pollack R., Wolman S., Vogel A., Reversion of virus-transformed cell lines: hyperdiploidy accompanies retention of viral genes. *Nature* 228:938, 1970.

73. Rajeswaki S., Ghosh S.N., Shah P.N., Borah V.J., Barr body frequency in the human breast cancer tissue. *Europ. J. Cancer* 13:99, 1977.

74. Rothwell N.V., *Understanding Genetics.* Williams & Wilkins, Baltimore, 1976.

75. Rowley J.D., A new consistent chromosomal abnormality in chronic myelogenous leukemia identified by quinacrine fluorescence and Giemsa staining. *Nature* 243:290, 1973.

76. Rowley J.D., Do human tumors show a chromosome pattern specific for each etiologic agent? *J. Natl. Cancer Inst.* 52:315, 1974.

77. Rowley J.D., Non-random chromosomal abnormalities in hematologic disorders of man. *Proc. Nat. Acad. Sci.*: 72:152, 1975.

78. Rowley J.D., The role of cytogenetics in hematology. *Blood* 48:1, 1976.

79. Sakurai M., Sandberg A.A., Chromosomes and causation of human cancer and leukemia. IX. Prognostic and therapeutic value of chromosomal findings in acute myeloblastic leukemia. *Cancer* 33:1548, 1974.

80. Sandberg A.A., Sakurai M., Chromosomes in the causation and progression of cancer and leukemia. In The *Molecular Biology of Cancer.* Academic Press, New York, 1974, p. 81.

81. Savino A., Koss L.G., The evaluation of sex chromatin as a prognostic factor in carcinoma of the breast. *Acta Cytol.* 15:372, 1971.

82. Schwarzacher H.G., Wolf U., Passarge E., *Methods in Human Cytogenetics.* Springer-Verlag, Berlin, 1974.

83. Seabright M., A rapid banding technique for human chromosomes. *Lancet* 2:971, 1971.

84. Sonta S., Oshimura M., Evans J.T., Sandberg A.A., Chromosomes and causation human cancer and leukemia. XX. Banding patterns of primary tumors. *J. Natl. Cancer Inst.* 58:49, 1977.

85. Spriggs A.I., Bowey B., Cowdell R.H., Chromosomes of precancerous lesions of the cervix uteri. *Cancer* 27:1239, 1971.

86. Stanley M.A., Kirkland J.A., Chromosome and histologic patterns in pre-invasive lesions of the cervix. *Acta Cytol.* 19:142, 1975.

87. Sumner A.T., Evans H.J., Buckland R.A., New technique for distinguishing between human chromosomes. *Nature New Biol.* (London) 232:31, 1971.

88. Sutton H.E. (Ed.). *An Introduction to Human Genetics.* Holt, Rinehart & Winston, New York, 1975.

89. Tiepolo L., Zuffardi O., Identification of normal and abnormal chromosomes in tumor cells. *Cytogenet. Cell. Genet.* 12:8, 1973.

90. Tjio J.H., Levan A., The chromosome number of man. *Hereditas* 42:1, 1956.

91. Toews H.A., Katayama K.P., Masukawa T., Lewison E.F., Chromosomes of benign and malignant lesions of the breast. *Cancer* 22:1296, 1968.

92. Trujillo J.M., Cork A., Hart J.S., George S.L., Freireich E.J., Clinical implications of aneuploid cytogenetic profiles in adult acute leukemia. *Cancer* 33:824, 1974.

93. Vakil D.V., Lewin P.K., Conen P.E., Value of fluorescent Y chromosome and sex chromatin tests. *Acta Cytol.* 17:220, 1973.

94. Verhest A., Lustman F., Wittek M., Van Schoubroeck F., Naets J.P., Cytogenetic evidence of clonal evolution in 5q-anemia. *Biomedicine,* 27:211, 1977.

95. Watson J.D. (Ed.). *Molecular Biology of the Gene.* Benjamin, New York, 1965.

96. Whang-Peng J., Banding techniques in leukemia: techniques and implications. *J. Natl. Cancer Inst.* 58:3, 1977.

97. Yunis J.J., *Human Chromosome Methodology.* Academic Press, New York, 1974.

98. Yunis J.J., Sanchez O., G-banding and chromosome structure. *Chromosoma* 44:15, 1973.

99. Zakharov A. F., Egolina N.A., Differential spiralization along mammalian mitotic chromosomes. I. BUDR revealed differentiation in chinese hamster chromosomes. *Chromosoma* 38:341, 1972.

Color Plates

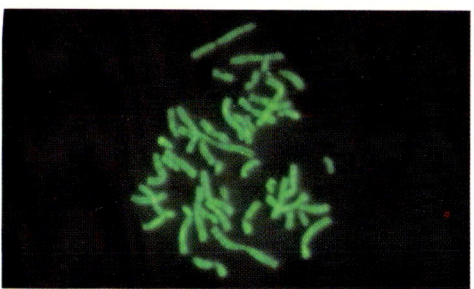

Plate 4.1. Metaphasic chromosomes stained with quinacrine mustard (Q.M). The brilliant characteristic fluorescence of the Y appears at two o'clock.

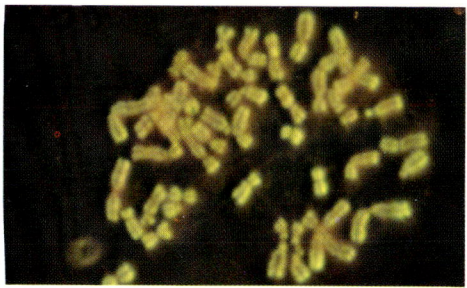

Plate 4.2. R bands; metaphase stained with acridine orange.

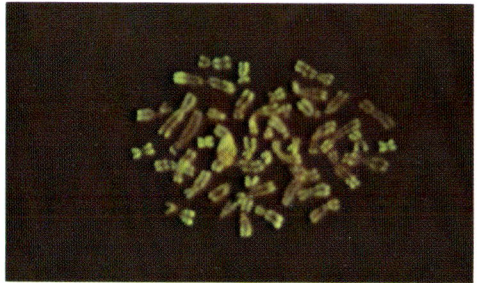

Plate 4.3. T banding in acridine orange.

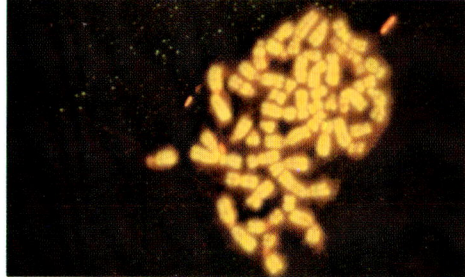

Plate 4.4. BrdU staining. The two Xs (near the red filament at one o'clock) are head to foot. The late replicating one appears elongated.

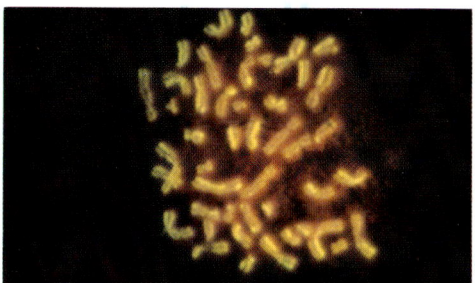

Plate 4.5. Sister chromatid stained in acridine orange after BrdU incubation.

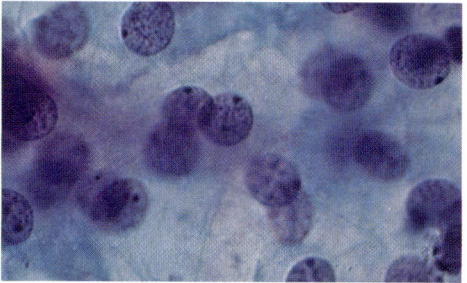

Plate 4.6. Female sex chromatin in interphase nuclei (buccal smear stained with cresyl violet) (300 ×).

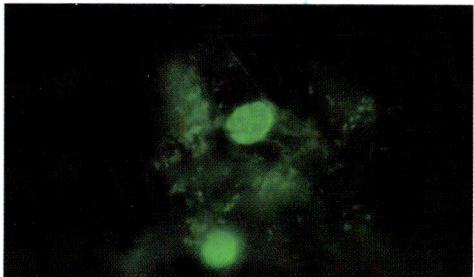

Plate 4.7. Fluorescent Y-body in male interphase nucleus.

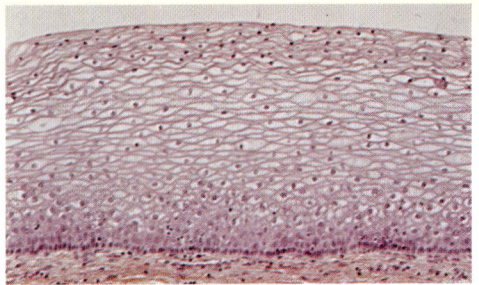

Plate 5.1. Histology of normal vaginal mucosa (40 ×).

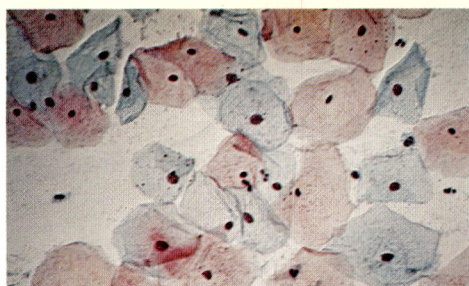

Plate 5.2. Vaginal smear: eosinophilic and cyanophilic superficial cells (100 ×).

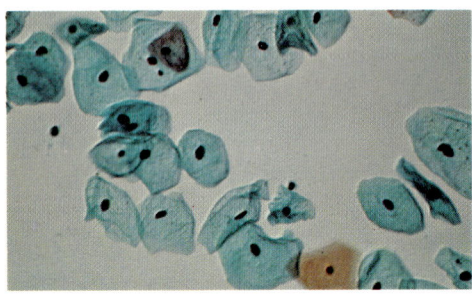

Plate 5.3. Vaginal smear: intermediate cells (100 ×).

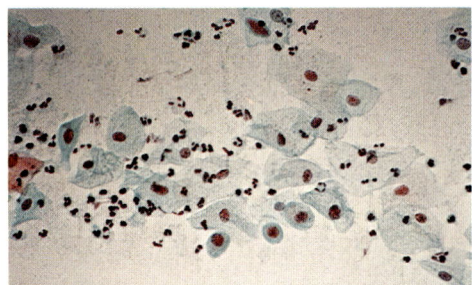

Plate 5.4. Vaginal smear: intermediate and parabasal cells atrophic type smear (100 ×).

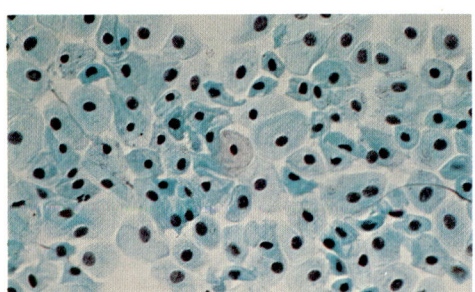

Plate 5.5. Vaginal smear: parabasal and intermediate cells; menopausal patient with atrophic smear receiving corticoid therapy.

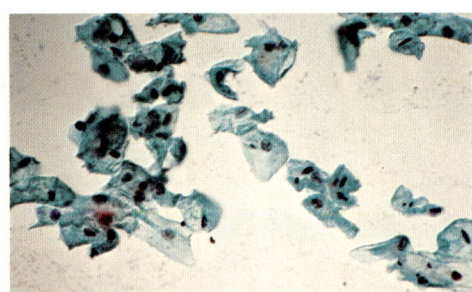

Plate 5.6. Vaginal smear: fourth month of pregnancy; clusters of cyanophilic intermediate cells with navicular (boatlike) cells (100 ×).

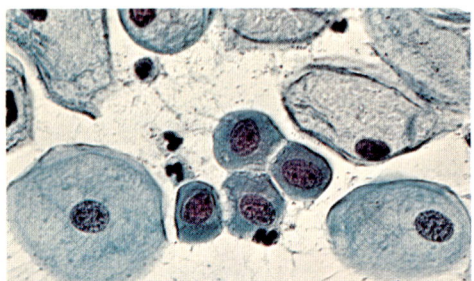

Plate 5.7. Vaginal smear: postpartum period (250 ×).

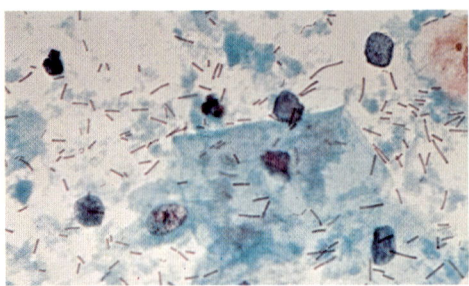

Plate 5.8. Vaginal smear: cytolytic type: twenty-fourth day of cycle (250 ×).

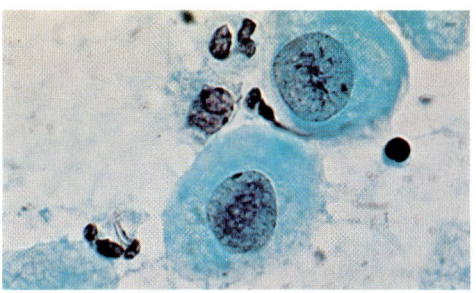

Plate 5.9. Vaginal smear: androgenic cell; large parabasal cells with vesicular nuclei having a fine chromatin pattern; the cytoplasm stains a pale blue; vacuolation may occur; there are no Doderleins bacilli present (250 ×).

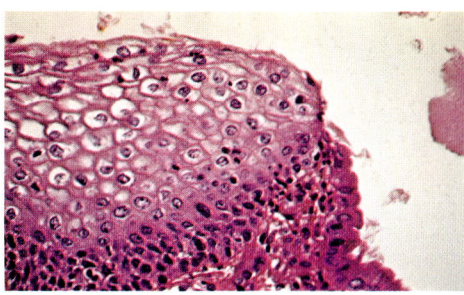

Plate 5.10. Cervical biopsy: columnar and squamous epithelial junction; in this case, the transition zone is sharp (100 ×).

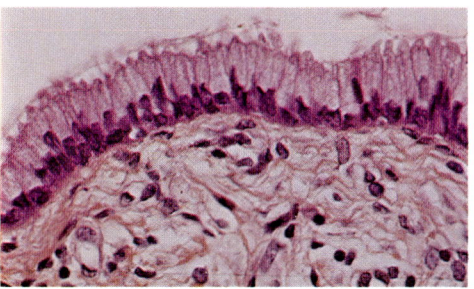

Plate 5.11. Cervical biopsy: columnar epithelium: tall columnar cells with basal nuclei and clear cytoplasm (100 ×).

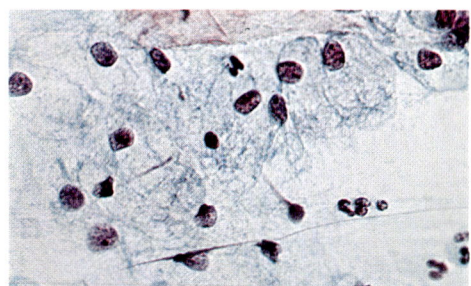

Plate 5.12. Cervical smear: columnar cells lateral view; small round nuclei and vacuolated cytoplasm (250 ×).

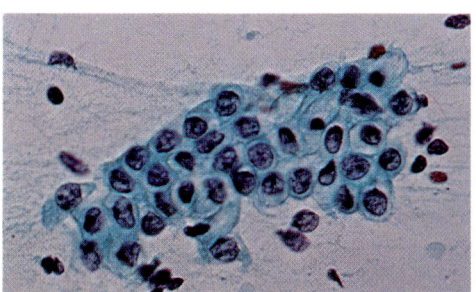

Plate 5.13. Cervical smear: columnar cells apical view: cluster of cells with central nuclei and sharp, apparent boundaries; honeycomb appearance; nucleoli are small (250 ×).

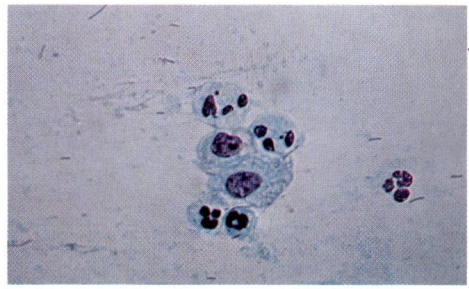

Plate 5.14. Cervical smear: histiocytes and leukocytes: in polymorphonuclear leukocytes, the small "drumstick" mass of chromatin attached to the nuclear lobes represents the sex chromatin (250 ×).

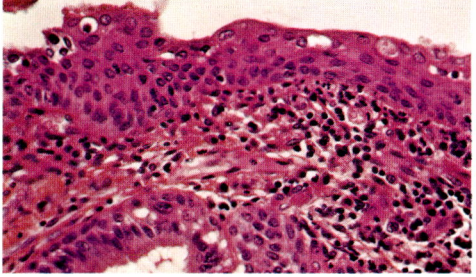

Plate 5.15. Cervical biopsy: chronic cervicitis; vacuolation of squamous cells; inflammatory reaction in the stroma (100 ×).

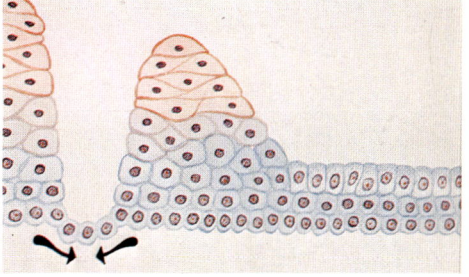

Plate 5.16. Schematic representation of cervical erosion and secondary healing with basal benign repair cells originating from the adjacent epithelium (*arrows*).

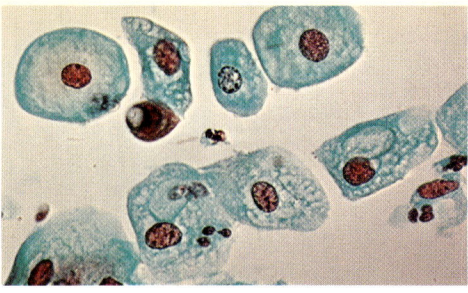

Plate 5.17. Cervical smear: chronic cervicitis; marked cytoplasmic vacuolation; increased intracytoplasmic fluid; keratin is confined to the periphery of the cells underlying the plasma membrane (250 ×).

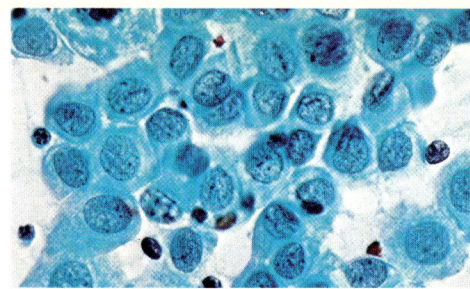

Plate 5.18. Cervical smear: menopausal woman; chronic cervicitis: clumps of hyperplastic, metaplastic cells with slight nuclear irregularity and cytoplasmic vacuolation (250 ×).

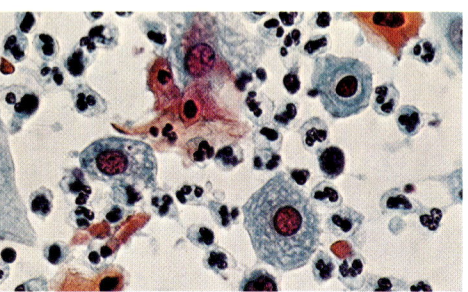

Plate 5.19. Cervical smear: inflammatory lesions: slight dyscariosis and hyperchromatism; cytoplasmic vacuolation; premature eosinophilia; leukocytes, histiocytes, and red blood cells (250 ×).

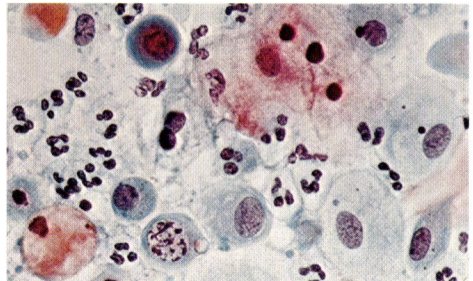

Plate 5.20. Cervical smear: inflammatory lesions: nuclear size and shape alterations; cellular shape alterations; mitosis; tinctorial cytoplasmic alterations; leukocytes (250 ×).

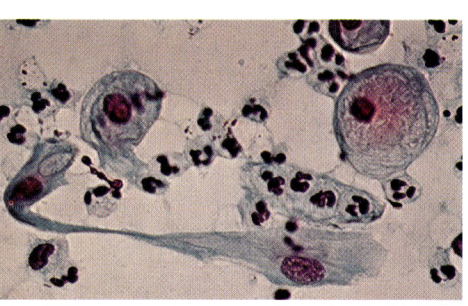

Plate 5.21. Cervical smear: chronic cervicitis: tadpole cell (250 ×).

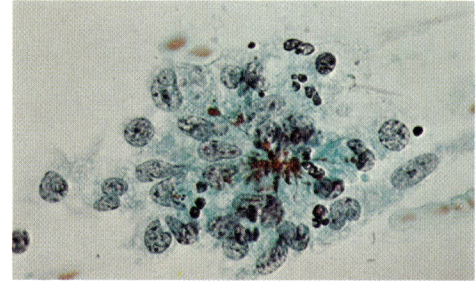

Plate 5.22. Cervical smear: chronic cervicitis: foreign body granuloma showing histiocytes and leukocytes (250 ×).

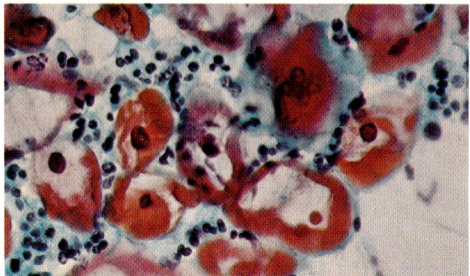

Plate 5.23. Cervical smear: chronic cervicitis: inflammatory lesions; parabasal cells showing atypical nuclei (anisokaryosis) and cytoplasmic vacuolation; keratin is displaced toward the periphery of cytoplasm (250 ×).

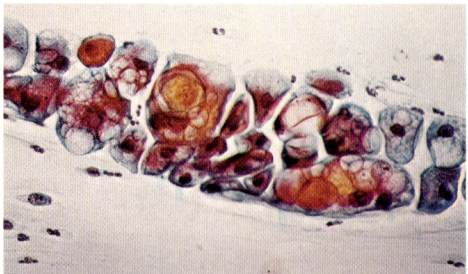

Plate 5.24. Cervical smear: chronic cervicitis: very marked vacuolation of parabasal and intermediate cells. These elements should not be confused with columnar cells (250 ×).

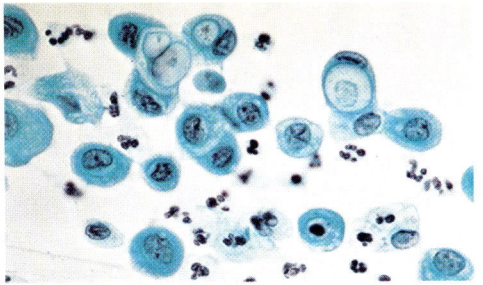

Plate 5.25. Cervical smear: chronic cervicitis: parabasal cells with large cytoplasmic vacuoles suggesting a columnar configuration (250 ×).

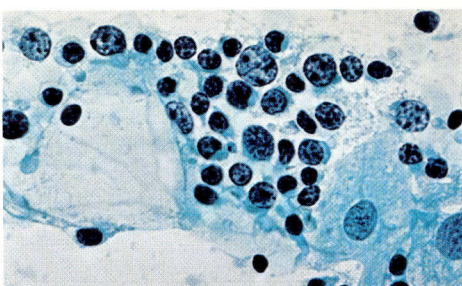

Plate 5.26. Cervical smear: chronic follicular cervicitis; presence of a large number of mature lymphocytes and lymphoblasts. The large lymphoid cells should not be confused with endometrial or leukemic cells (250 ×).

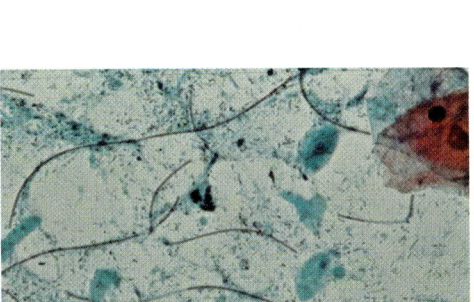

Plate 5.27. Cervical smear: Trichomonas and Leptotricia (250 ×).

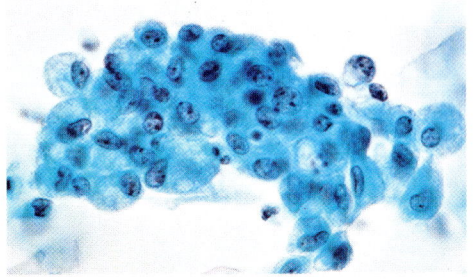

Plate 5.28. Cervical smear: presence of columnar cells and histiocytes desquamating in papillarylike structures. The biopsy showed a chronic cervicitis with papillary proliferation (250 ×).

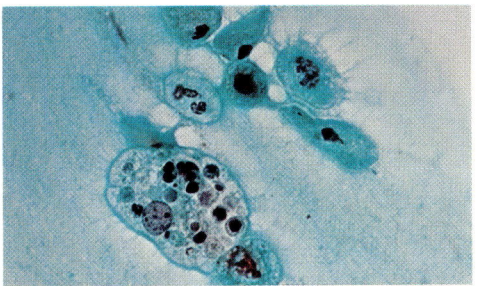

Plate 5.29. Cervical smear: chronic cervicitis: macrophages, histiocytes, and leukocytes. Note the presence of engulfed particles in one macrophage (250 ×).

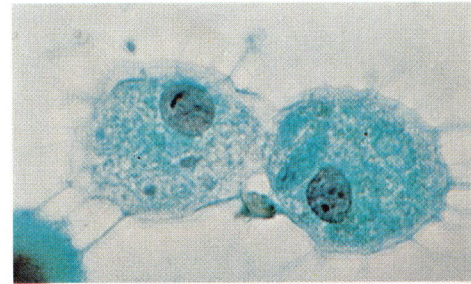

Plate 5.30. Cervical smear: chronic cervicitis with erosion: macrophages with ingested particles in the cytoplasm (400 ×).

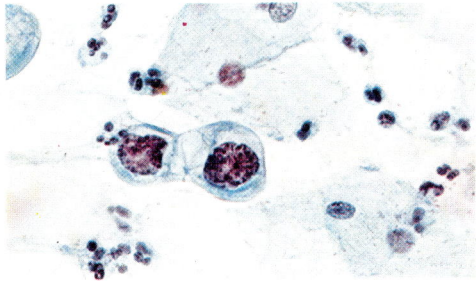

Plate 5.31. Cervical smear: chronic cervicitis: parabasal cells; hyperchromatic and irregular nuclei and cytoplasmic vacuolation (250 ×).

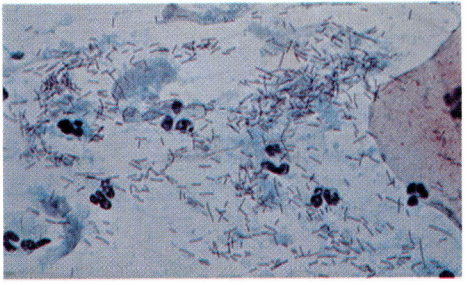

Plate 5.32. Cervical smear: Doderlein's bacillus (250 ×).

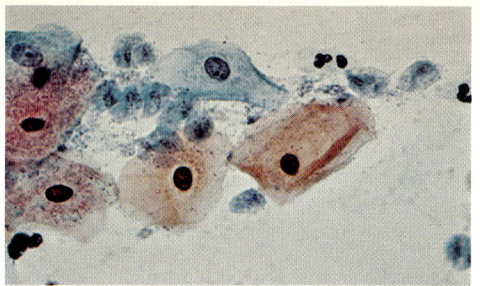

Plate 5.33. Cervical smear: *Trichomonas vaginalis* infestation: presence of perinuclear halos; cytoplasmic eosinophilia (250 ×).

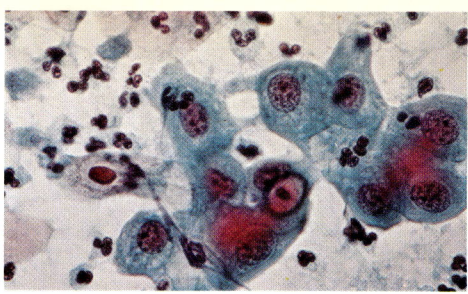

Plate 5.34. Cervical smear: *Trichomonas vaginalis* infestation; anisonucleosis; hyperchromatism; cytoplasmic vacuolation; leukocytes (250 ×).

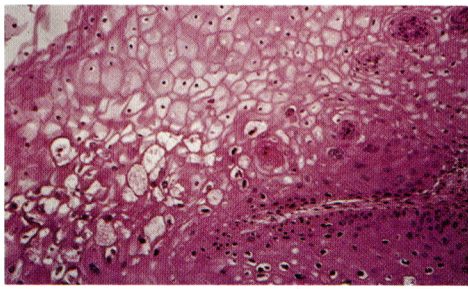

Plate 5.35. Cervical biopsy: Herpes simplex: squamous epithelium showing marked cytoplasmic vacuolation of superficial layers and multinucleation (100 ×).

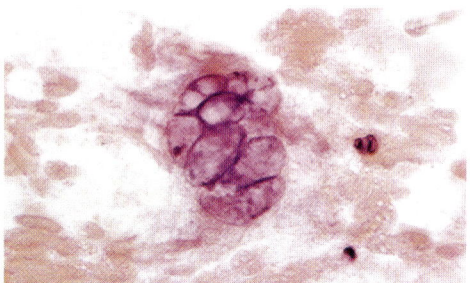

Plate 5.36. Cervical smear: Herpes simplex: multinucleated cell and "ground-glass" appearance of nuclei; molding of adjacent nuclei (250 ×).

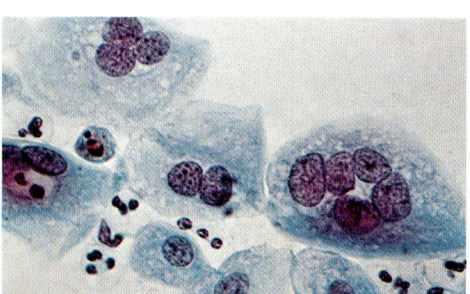

Plate 5.37. Cervical smear: Herpes simplex: multinucleated cells and cytoplasmic vacuolation (250 ×).

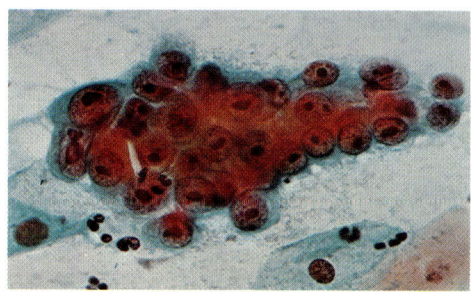

Plate 5.38. Cervical smear: Herpes simplex: intranuclear eosinophilic inclusions (250 ×).

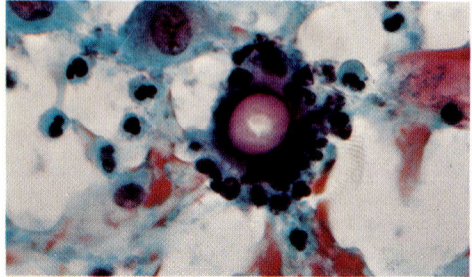

Plate 5.39. Cervical smear: chronic cervicitis: foreign body reaction with pseudorosette formation (250 ×).

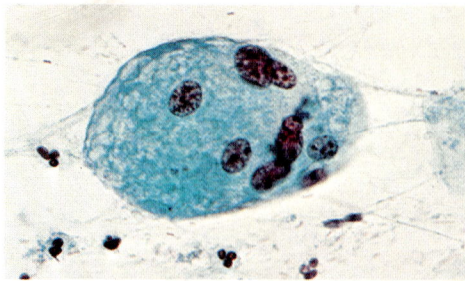

Plate 5.40. Cervical smear: chronic cervicitis: presence of giant multinucleated histiocyte; nuclei are small and regular; no molding of adjacent nuclei (250 ×).

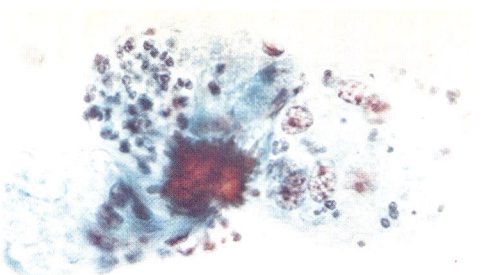

Plate 5.41. Cervical smear: chronic cervicitis: foreign body granuloma: the nature of the foreign body is impossible to identify (mycosis? altered cell?) (250 ×).

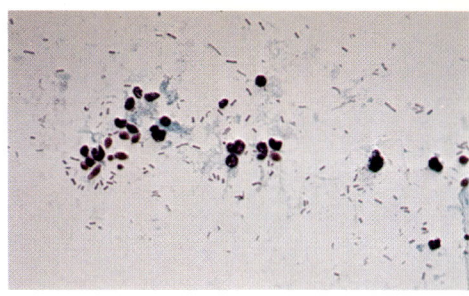

Plate 5.42. Cervical smear: *Candida albicans* and bacteria (250 ×).

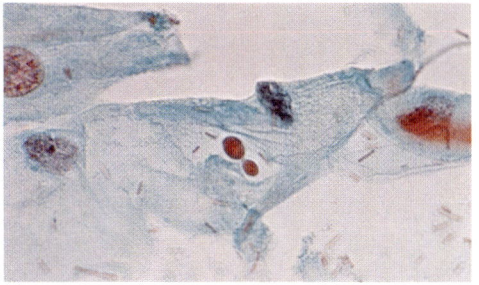

Plate 5.43. Cervical smear: fungus disease: *Candida albicans;* presence of spores (250 ×).

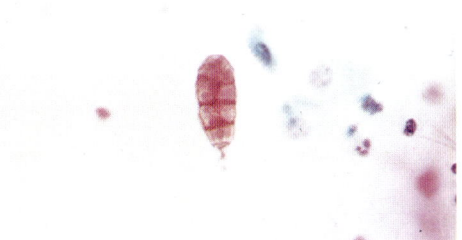

Plate 5.44. Cervical smear: fungus disease: *Candida albicans;* macronidia (250 ×).

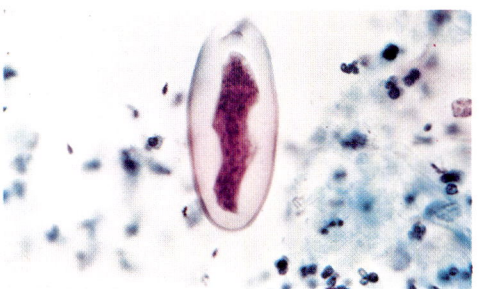

Plate 5.45. Cervical smear: egg of *Enterobius vermicularis* (pinworm), feces contamination (250 ×).

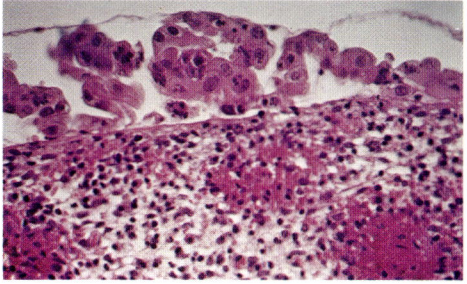

Plate 5.46. Cervical biopsy: chronic cervicitis showing spontaneous desquamation of metaplastic cells. Note the marked inflammatory and vascular congestion of the stroma (100 ×).

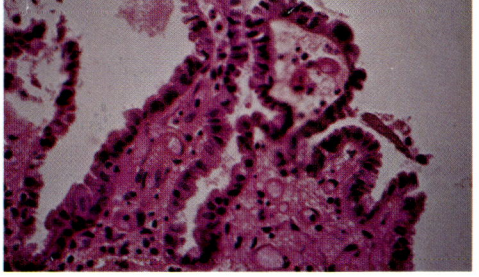

Plate 5.47. Cervical biopsy: papillomatous glandular hyperplasia: papillary proliferation of atypical columnar cells lining a stroma infiltrated by numerous leukocytes and histiocytes. These lesions are seen in ectropion and during pregnancy, or oral contraceptive therapy (100 ×).

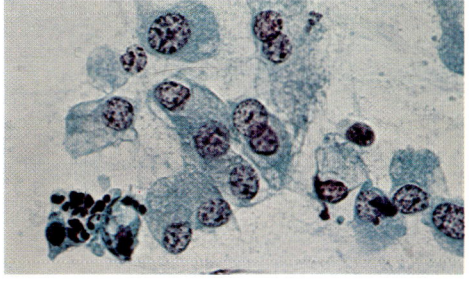

Plate 5.48. Cervical smear: cervical polyp: columnar cells from the polyp, sheets of atypical cells with vacuolated cytoplasm and hypertrophic nuclei (250 ×).

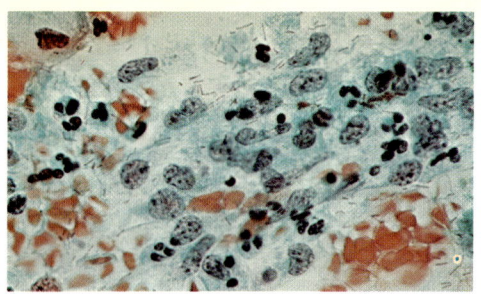

Plate 5.49. Cervical smear: chronic cervicitis, erosion: cluster of enlarged, hyperplastic columnar cells, hyperchromatic nuclei. May be easily confused with differentiated malignant columnar cells of endocervix. Absence of large nucleoli favors a benign origin (250 ×).

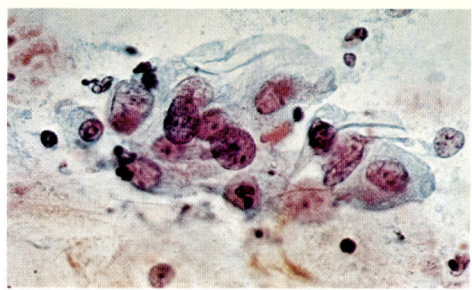

Plate 5.50. Cervical smear: (same case as Plate 5.49): the presence of large clusters of histiocytes suggests the existence of a marked inflammatory reaction with erosion and cell repair (250 ×).

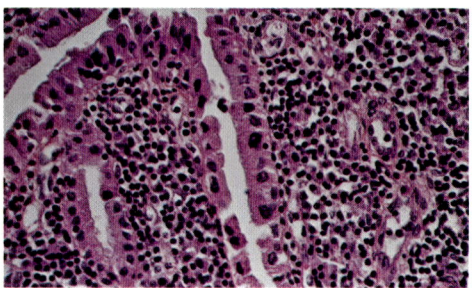

Plate 5.51. Cervical biopsy: chronic cervicitis: columnar epithelium modified by chronic cervicitis. Note the nuclear abnormalities and the chronic inflammation in cervical stroma (100 ×).

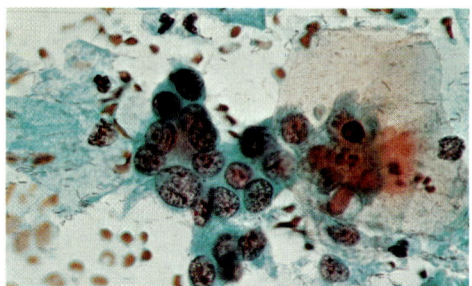

Plate 5.52. Cervical smear: chronic cervicitis (the smear corresponds to the cervical biopsy shown in Plate 5.51): columnar cells. Note the slightly enlarged and hyperchromatic nuclei and the cyanophilic cytoplasm (250 ×).

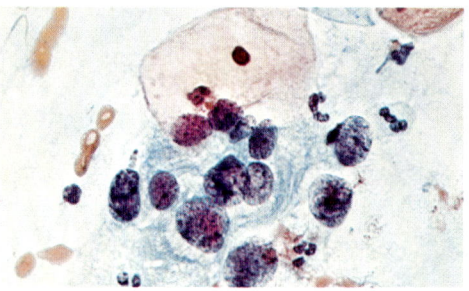

Plate 5.53. Cervical smear: fifth month of pregnancy: hyperplastic columnar cells with dense, hyperchromatic nuclei (250 ×).

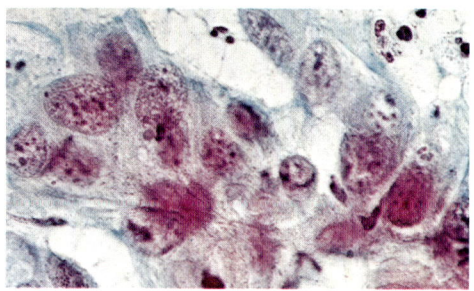

Plate 5.54. Cervical smear: 40-year-old patient; chronic cervicitis: marked hyperplasia of columnar cells; 3-year follow-up: persistence of marked inflammatory lesions (250 ×).

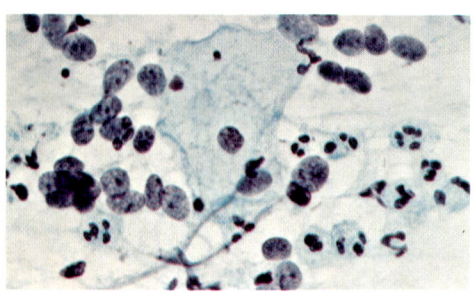

Plate 5.55. Cervical smear: naked nuclei from cervical columnar cells; cervical biopsy shows lesions of chronic cervicitis with erosion and destruction of the columnar epithelium (250 ×).

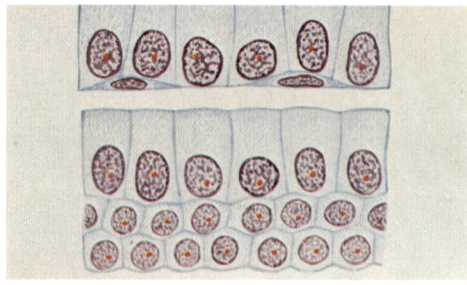

Plate 5.56. Schematic representation of reserve cells (*A*, flat cells at the base of the epithelium) and reserve cell hyperplasia (*B*, double cellular layer below the columnar cells).

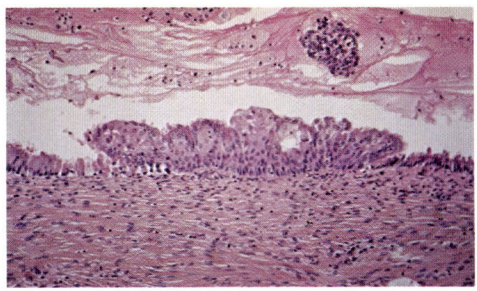

Plate 5.57. Cervical biopsy: isolated focus of squamous metaplasia (40 ×).

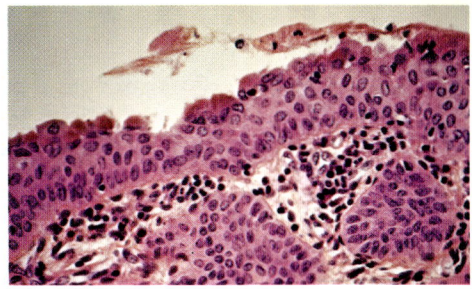

Plate 5.58. Cervical biopsy: squamous metaplasia: superficial columnar cells overlying epidermoid cells on the right; metaplastic cells originating from reserve cell hyperplasia (100 ×).

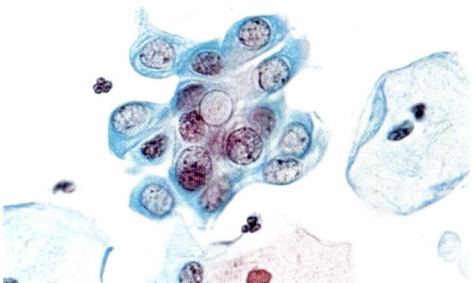

Plate 5.59. Cervical smear: immature hyperplasia: clump of columnar and squamous metaplastic cells; reserve cells have produced columnar and squamous cells. Note the marked basophilia of the cytoplasm (250 ×).

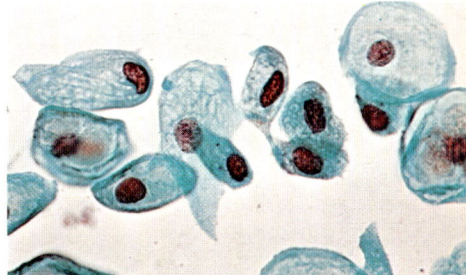

Plate 5.60. Cervical smear: squamous metaplasia; round, oval, polygonal, or elongated cells with round or oval nuclei; nucleoli are not seen; dense, microvacuolated, cyanophilic cytoplasm (250 ×).

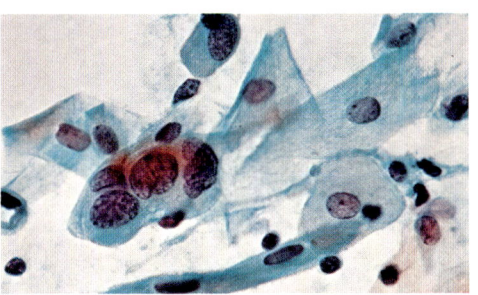

Plate 5.61. Vulvar imprint: epidermoid carcinoma of majora labia: cluster of cyanophilic cells with irregular and hyperchromatic nuclei (250 ×).

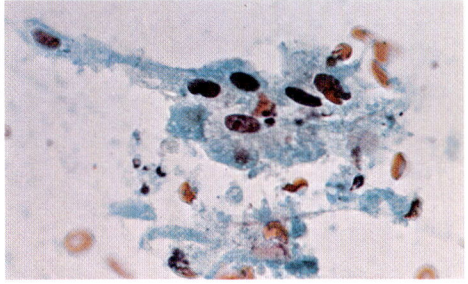

Plate 5.62. Vaginal smear: epidermoid carcinoma of the vagina: poorly preserved and undifferentiated squamous cells (250 ×).

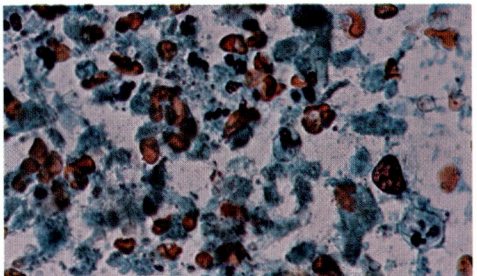

Plate 5.63. Vaginal smear: epidermoid carcinoma: necrotic smear containing rare malignant hyperchromatic nuclei (250 ×).

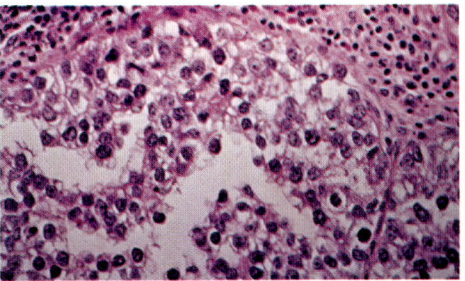

Plate 5.64. Vaginal biopsy: clear-cell carcinoma of the vagina: 16-year-old girl whose mother received stilbestrol during first third of her pregnancy (100 ×).

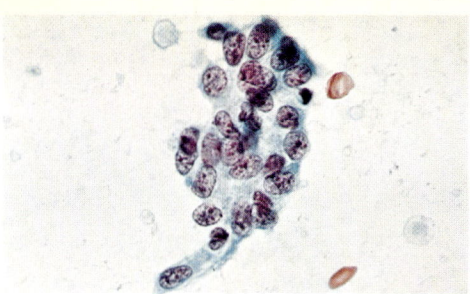

Plate 5.65. Vaginal smear (smear corresponding to the biopsy shown in Plate 5.64): clear-cell carcinoma of the vagina: cluster of columnar cells with slightly enlarged and hyperchromatic nuclei (250 ×).

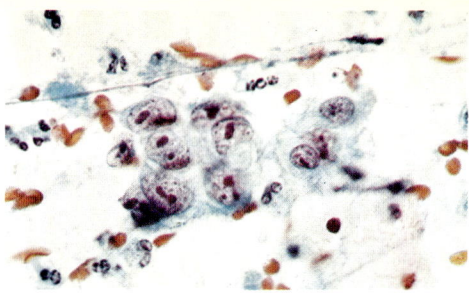

Plate 5.66. Vaginal smear: metastatic nodule in the vaginal mucosa: primary tumor is differentiated glandular carcinoma of endometrium (250 ×).

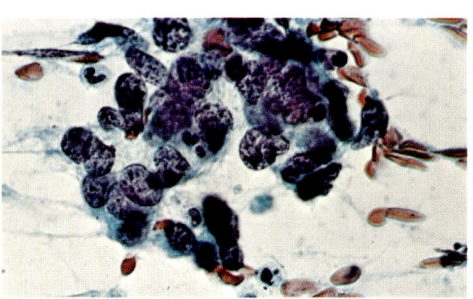

Plate 5.67. Vaginal smear: metastatic lesion of the vaginal wall: primary tumor is serous carcinoma of ovary (250 ×).

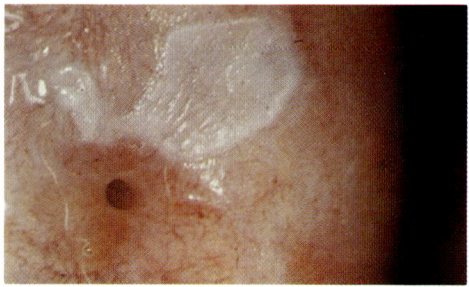

Plate 5.68. Colposcopy: leukoplakia: well-defined whitish area on the exocervix.

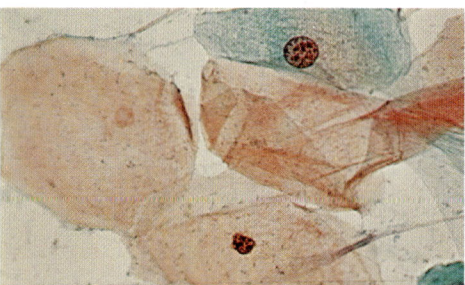

Plate 5.69. Cervical smear: clinical leukoplakia: presence of anucleated squamous cells (250 ×).

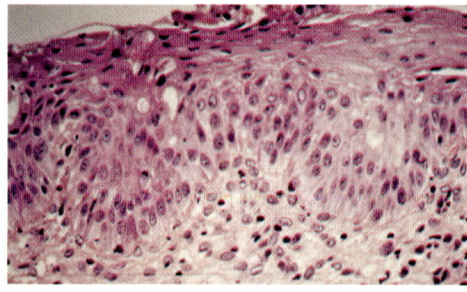

Plate 5.70. Cervical biopsy: slight dysplasia: disordered development of cellular layers; moderated nuclear abnormalities; variations in cytoplasmic maturation (premature eosinophilia) (100 ×).

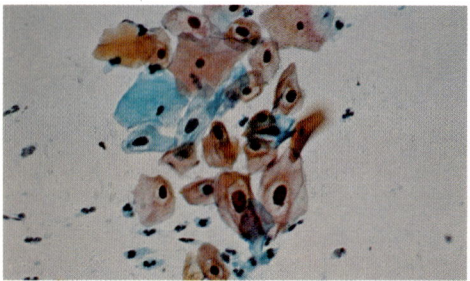

Plate 5.71. Cervical smear: slight dysplasia, keratinizing type: polygonal cells with hyperchromatic nuclei and clear perinuclear zones; the nucleocytoplasmic ratio is slightly increased (100 ×).

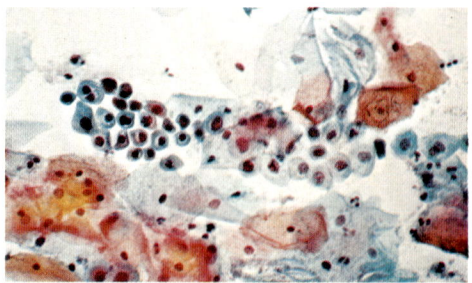

Plate 5.72. Cervical smear: slight dysplasia, nonkeratinizing type: small parabasal cells with hyperchromatic nuclei: perinuclear vacuoles; the nucleocytoplasmic ratio is slightly increased (100 ×).

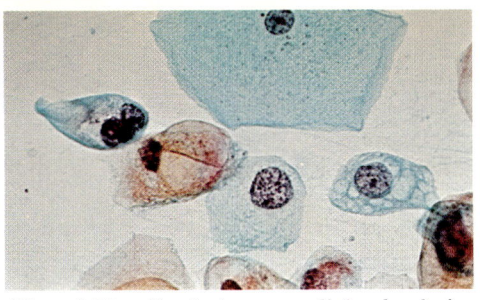

Plate 5.73. Cervical smear: slight dysplasia: large round hyperchromatic nuclei in parabasal and intermediate cells; premature keratinization and cytoplasmic vacuoles (250 ×).

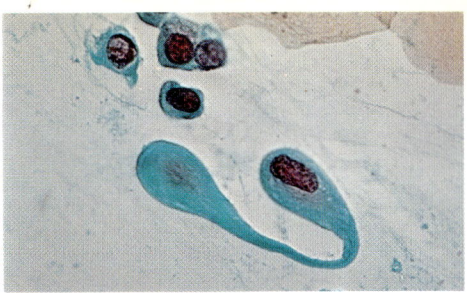

Plate 5.74. Cervical smear: slight dysplasia: presence of a tadpole cell (250 ×).

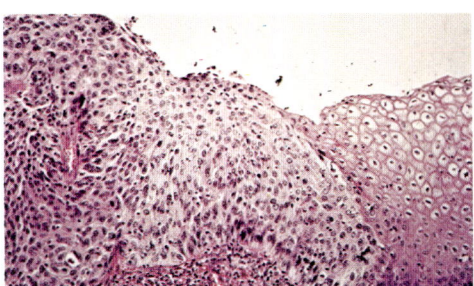

Plate 5.75. Cervical biopsy: moderate dysplasia: junction between the normal squamous epithelium (*right*) and the dysplasia (*left*) (250 ×).

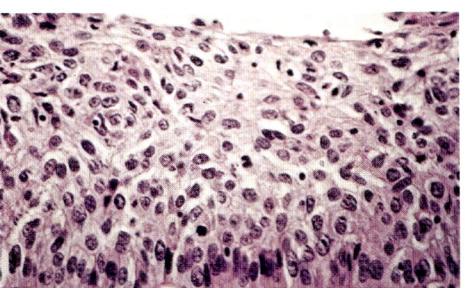

Plate 5.76. Cervical biopsy: moderate dysplasia: note abnormal differentiation of the superficial cell layer and disturbances of cellular arrangement in all layers (100 ×).

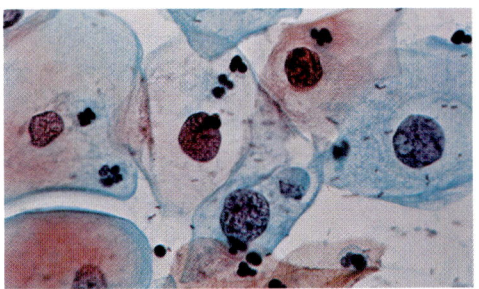

Plate 5.77. Cervical smear: moderate dysplasia: enlarged, slightly hyperchromatic nuclei; nucleoli not apparent; nucleocytoplasmic ratio is moderately increased (250 ×).

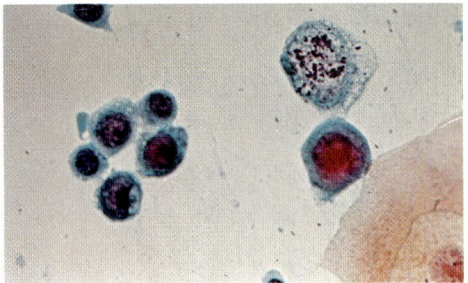

Plate 5.78. Cervical smear: moderate dysplasia: presence of a mitosis; clumps of parabasal metaplastic cells (250 ×).

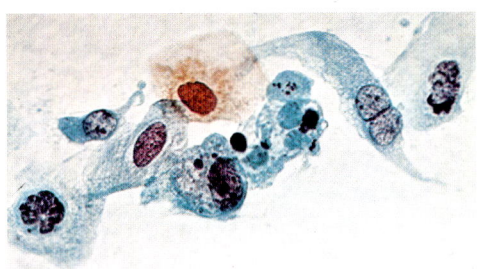

Plate 5.79. Cervical smear: moderate dysplasia: metaplastic cells (250 ×).

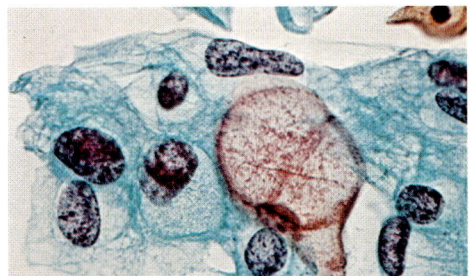

Plate 5.80. Cervical smear: moderate dysplasia: enlarged and hyperchromatic nuclei; cytoplasmic vacuolation (250 ×).

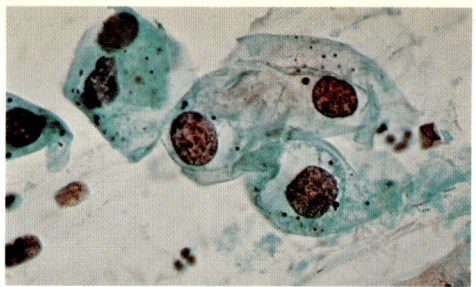

Plate 5.81. Cervical smear: moderate dysplasia: undifferentiated metaplastic cells showing perinuclear vacuoles and parakeratin granules (250 ×).

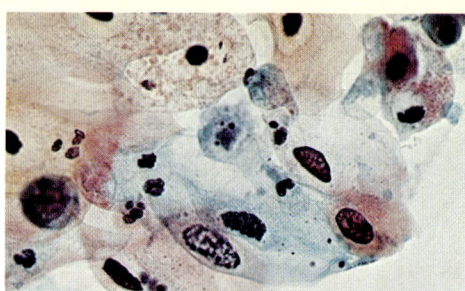

Plate 5.82. Cervical smear: moderate dysplasia: dense, irregular nuclei; premature eosinophilia; cytoplasmic vacuolation (250 ×).

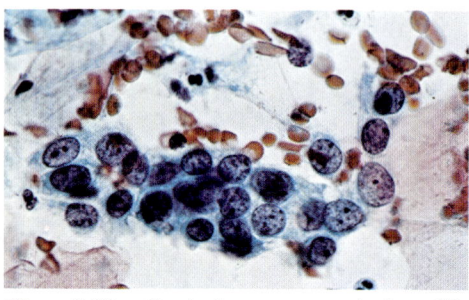

Plate 5.83. Cervical smear: atypical undifferentiated metaplasia: cluster of hyperchromatic parabasal cells (250 ×).

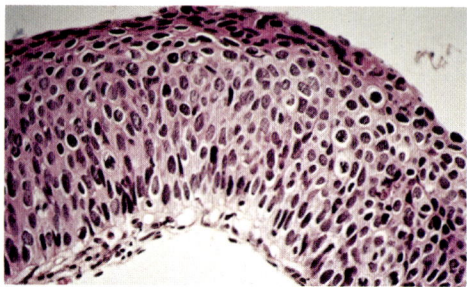

Plate 5.84. Cervical biopsy: severe dysplasia: disordered arrangement of all cellular layers; hyperchromatic nuclei; cytoplasmic vacuolation; nucleocytoplasmic ratio is moderately increased (100 ×).

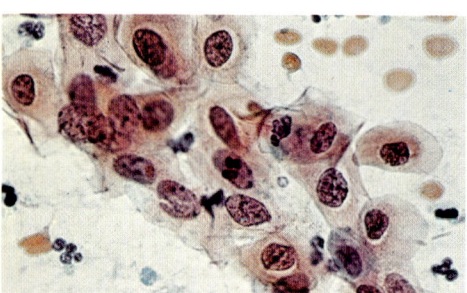

Plate 5.85. Cervical smear: severe dysplasia: squamous cells with enlarged, irregular, hyperchromatic nuclei; irregularity of cellular shape; perinuclear vacuoles, premature eosinophilia; increased nucleocytoplasmic ratio (250 ×).

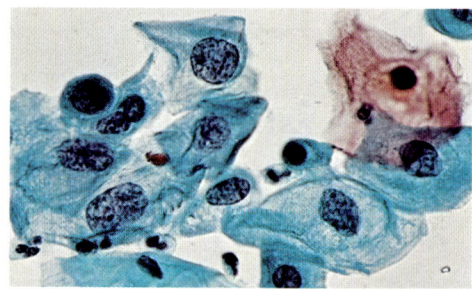

Plate 5.86. Cervical smear: severe dysplasia confirmed by biopsy; cyanophilic intermediate and superficial cells with enlarged hyperchromatic nuclei (250 ×).

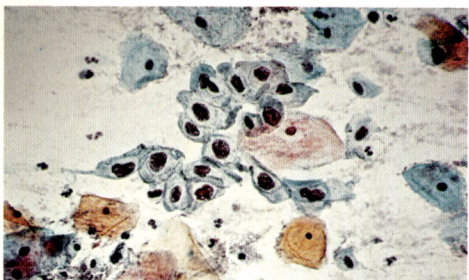

Plate 5.87. Cervical smear: severe dysplasia: low-power magnification showing clusters of atypical cells among superficial cells. Note the absence of inflammatory reaction (100 ×).

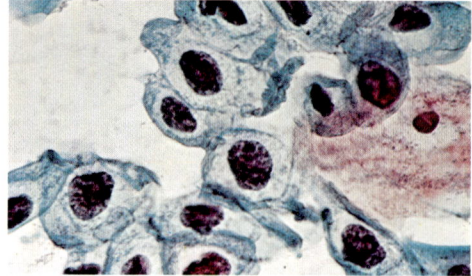

Plate 5.88. Cervical smear: severe dysplasia (detail of Plate 5.87): dense irregular nuclei; 3-year follow-up. Note the vacuolation and ballooning of the superficial cells (koilocytosis or warty atypia) (250 ×).

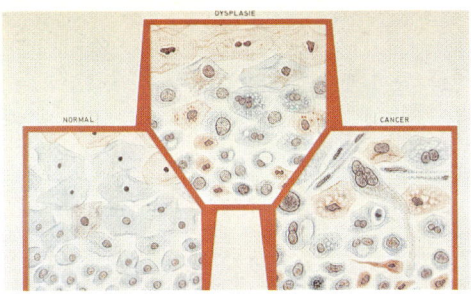

Plate 5.89. Schematic representation of cellular alteration in dysplasia and in situ carcinoma compared with normal squamous cells.

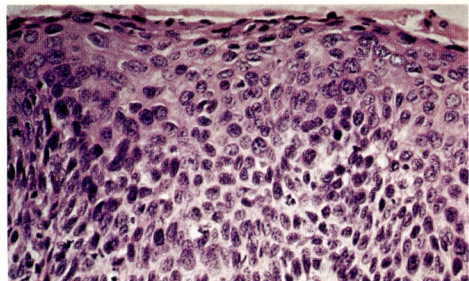

Plate 5.90. Cervical biopsy: in situ carcinoma: parabasal type cells, enlarged hyperchromatic nuclei occupying full thickness of epithelium; cellular disorientation; disturbed maturation; increased nucleocytoplasmic ratio; no glycogen in cells; mitotic figures present ($100 \times$).

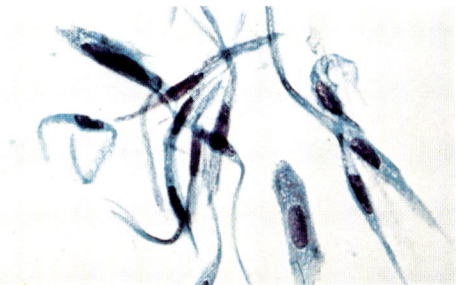

Plate 5.91. Cervical smear: in situ carcinoma: large, elongated cells with hyperchromatic, cigar-shaped nuclei corresponding to the upper layer of the epithelium (see Plate 5.90) ($250 \times$).

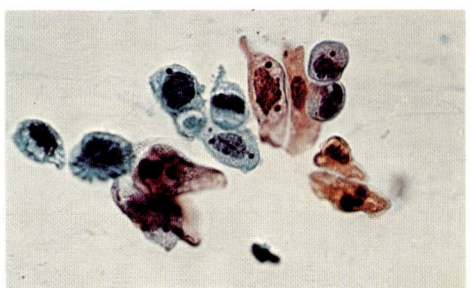

Plate 5.92. Cervical smear: in situ carcinoma, keratinizing type: absence of inflammatory reaction ($250 \times$).

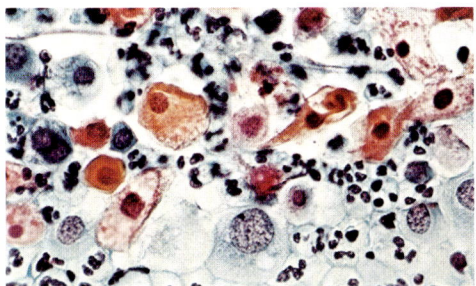

Plate 5.93. Cervical smear: in situ carcinoma, keratinizing type: presence of a marked inflammatory infiltrate and necrosis. Generally inflammatory reaction is not present; here we have the exception to the rule ($250 \times$).

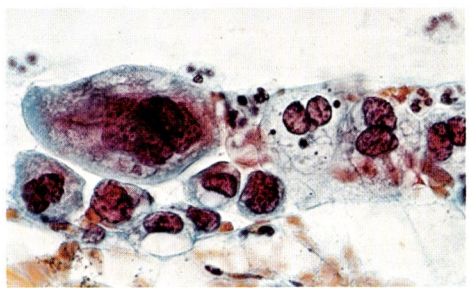

Plate 5.94. Cervical smear: in situ carcinoma: large multinucleated cells with nuclear anomalies and cytoplasmic vacuolation. (This type of malignant cell should not be confused with herpes simplex modification.) ($250 \times$).

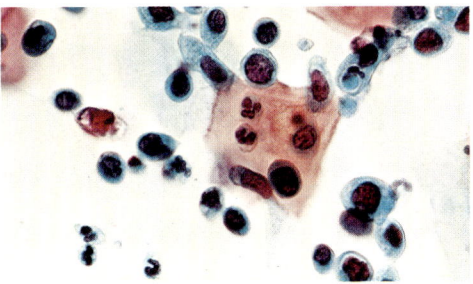

Plate 5.95. Cervical smear: in situ carcinoma; small-cell, nonkeratinizing ($250 \times$).

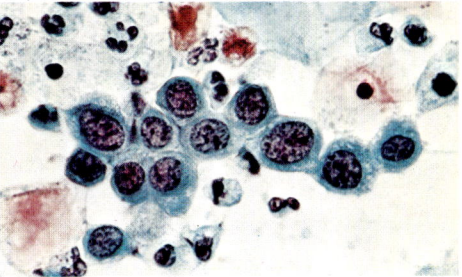

Plate 5.96. Cervical smear: in situ carcinoma, large-cell, nonkeratinizing ($250 \times$).

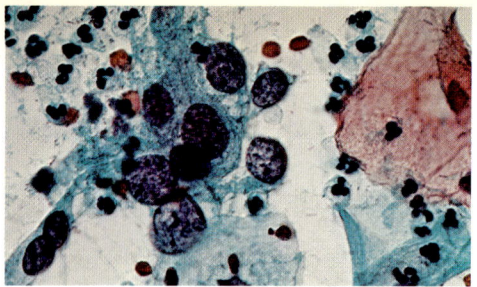

Plate 5.97. Cervical smear: in situ carcinoma, large-cell, nonkeratinizing in a pregnant woman: there was no regression of the lesion after pregnancy (250×).

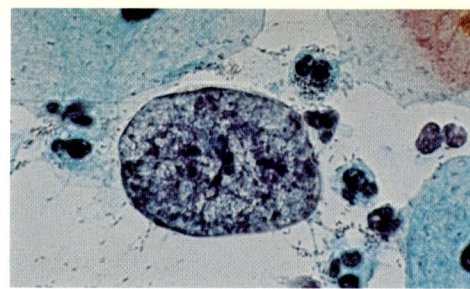

Plate 5.98. Cervical smear: in situ carcinoma: gigantic, hyperchromatic, naked, malignant nucleus (400×).

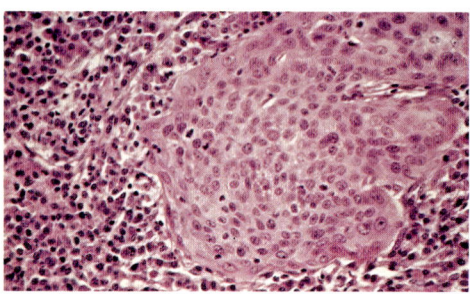

Plate 5.99. Cervical biopsy: invasive epidermoid carcinoma: epithelial cord infiltrating the connective tissue stroma; presence of a lymphoplasmacytic, leukocytic infiltration with edema and neovascularization (100×).

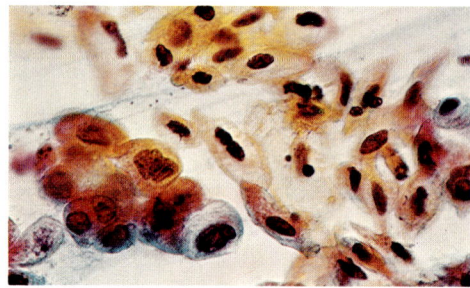

Plate 5.100. Cervical smear: invasive epidermoid carcinoma; keratinizing type: clump of spindle squamous cells; elongated, hyperchromatic nuclei with predominantly eosinophilic cytoplasm (250×).

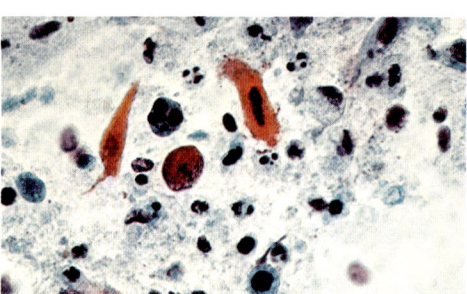

Plate 5.101. Cervical smear: invasive epidermoid carcinoma; keratinizing type; smear displaying necrosis, cellular debris, leukocytes and rare preserved eosinophilic spindle cells (250×).

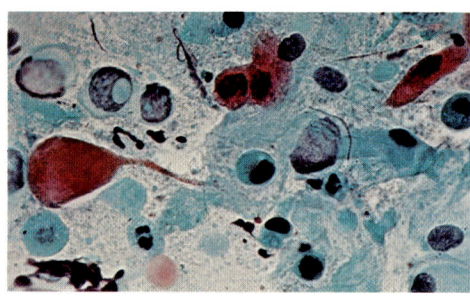

Plate 5.102. Cervical smear: invasive epidermoid carcinoma, keratinizing type: club-shaped cell with elongated eosinophilic cytoplasm (tadpole cell); presence of *Trichomonas vaginalis* (250×).

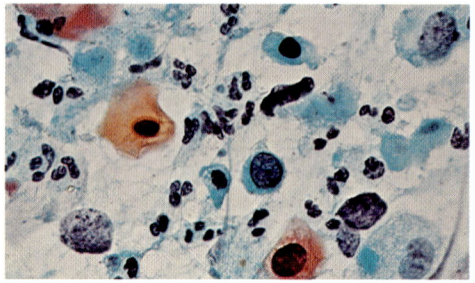

Plate 5.103. Cervical smear: invasive epidermoid carcinoma, keratinizing type: simultaneous presence of keratinized and undifferentiated cells (250×).

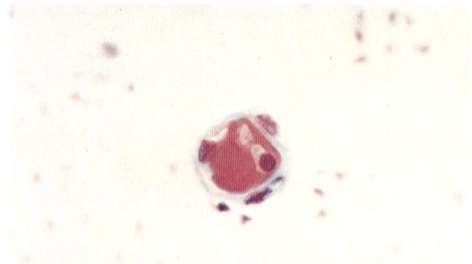

Plate 5.104. Cervical smear: invasive epidermoid carcinoma, keratinizing type: squamous pearl (250×).

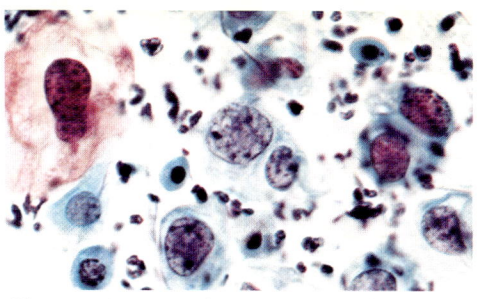

Plate 5.105. Cervical smear: invasive epidermoid carcinoma; predominantly large-cell, nonkeratinizing type: large regular, hyperchromatic nuclei with a small ring of cyanophilic cytoplasm (*upper left*) (250 ×).

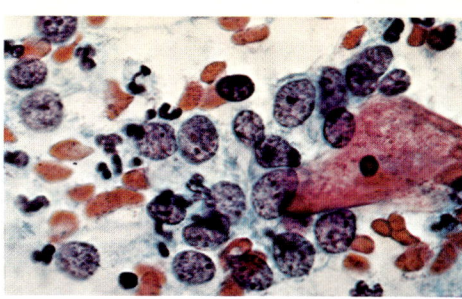

Plate 5.106. Cervical smear: invasive epidermoid carcinoma, large-cell, nonkeratinizing type: these cells may resemble those of adenocarcinoma but nuclei are larger and nucleoli are not as apparent (250 ×).

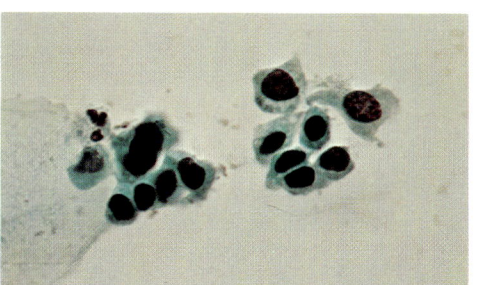

Plate 5.107. Cervical smear: invasive epidermoid carcinoma; small-cell, nonkeratinizing type: the resemblance of these cells to adenocarcinoma is striking (250 ×).

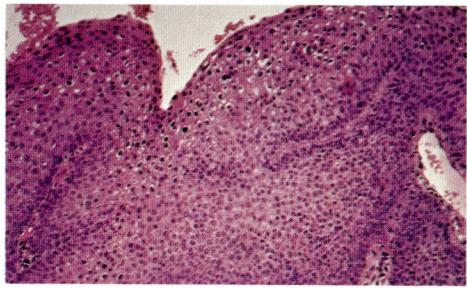

Plate 5.108. Cervical biopsy: invasive epidermoid carcinoma: note the presence of koilocytic atypia — cytoplasmic vacuolation of the superficial cells of the epithelium (100 ×).

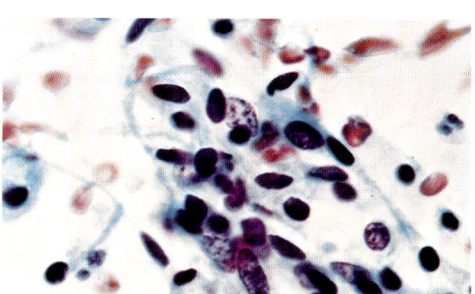

Plate 5.109. Cervical smear: invasive epidermoid carcinoma: clusters of fiber cells. These elongated cells are more frequent in karatinizing cancers, but may be observed in smears from inflammatory conditions (250 ×).

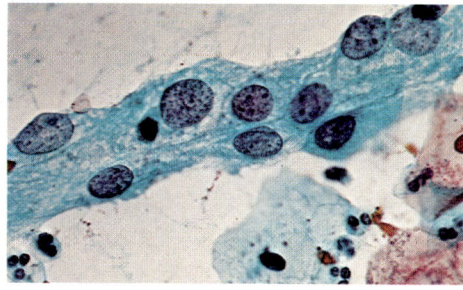

Plate 5.110. Cervical smear: severe chronic cervicitis with erosion: cluster of large columnar cells with regular hyperchromatic nuclei; prominent nucleoli; abundant cyanophilic cytoplasm. Cells originate from the squamocolumnar junction modified by marked inflammation (250 ×).

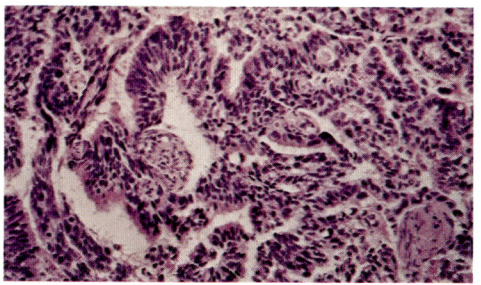

Plate 5.111. Endocervical biopsy: adenocarcinoma of the endocervix: glandular structures lined by differentiated columnar cells with hyperchromatic nuclei (100 ×).

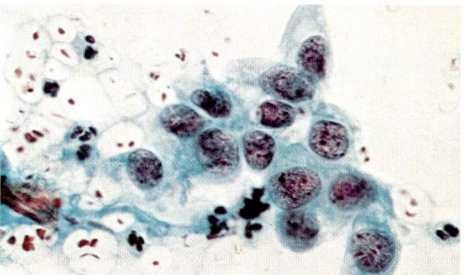

Plate 5.112. Cervical smear: adenocarcinoma of the endocervix: columnar malignant cells with regular finely granular hyperchromatic nuclei and prominent nucleoli; the cytoplasm is preserved and vacuolated (250 ×).

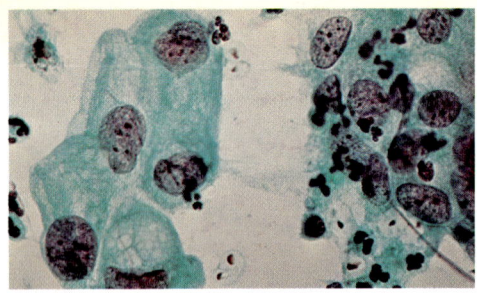

Plate 5.113. Cervical smear: adenocarcinoma of the endocervix: cellular atypias are more pronounced than in Plate 5.112; nuclei are large, hyperchromatic, and irregular. Note the vacuolated cytoplasm (250 ×).

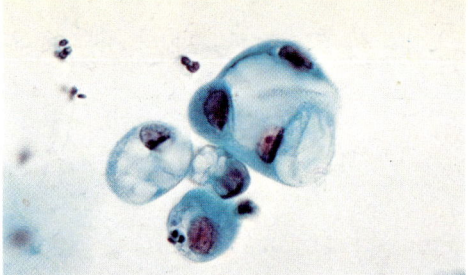

Plate 5.114. Cervical smear: well-differentiated adenocarcinoma of the endocervix: small hyperchromatic nuclei and cytoplasmic vacuolation. Diagnosis is difficult because cellular anomalies are discrete. Histology may clarify by showing invasion of stroma (250 ×).

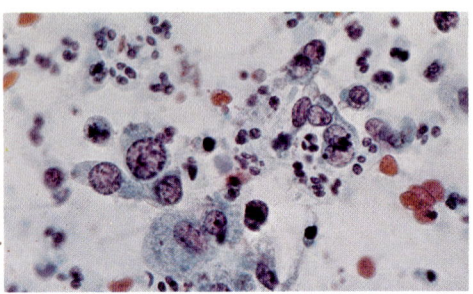

Plate 5.115. Cervical smear: mixed carcinoma of cervix (squamous cell carcinoma and adenocarcinoma): note simultaneous presence of columnar cells (*upper left*) and poorly differentiated squamous cells (*below*) (250 ×).

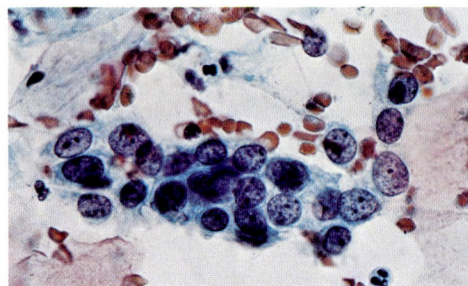

Plate 5.116. Cervical smear: epidermoid carcinoma; large-cell nonkeratinizing type: differential diagnosis with adenocarcinoma of endocervix. Note the marked grouping of cells with clumped nuclear chromatin (250 ×).

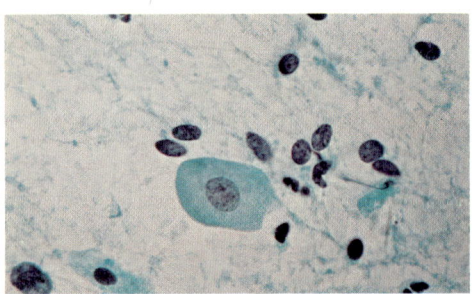

Plate 5.117. Cervical smear: atrophic epithelium: presence of a parabasal cell and naked stromal cell nuclei. Stromal cells can be seen in smears when there is erosion of the mucosa (250 ×).

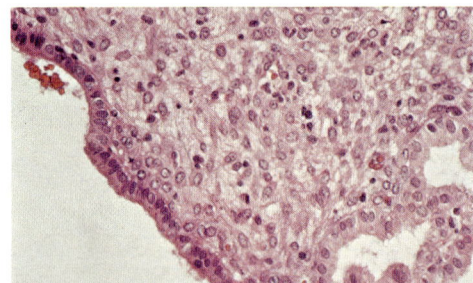

Plate 5.118. Endometrial biopsy: showing the three different cellular types: surface epithelium, glandular epithelium, and stromal cells (100 ×).

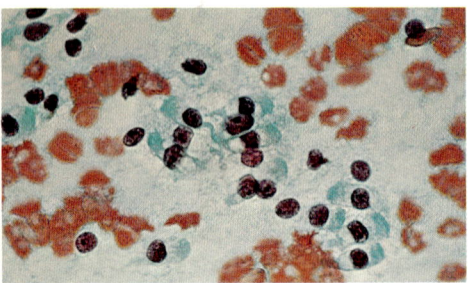

Plate 5.119. Endometrial aspiration: clusters of glandular cells (seen longitudinally) (250 ×).

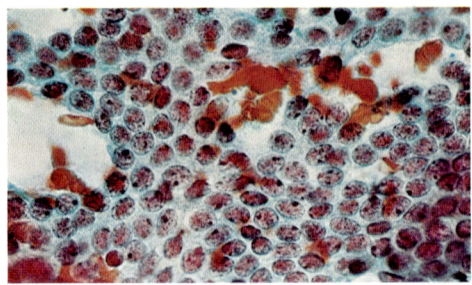

Plate 5.120. Endometrial aspiration: epithelial cell clusters: the honeycomb appearance showing apical poles of cells; finely dispersed nuclear chromatin; small visible nucleolus; cyanophilic cytoplasm forms halo around the nucleus (250 ×).

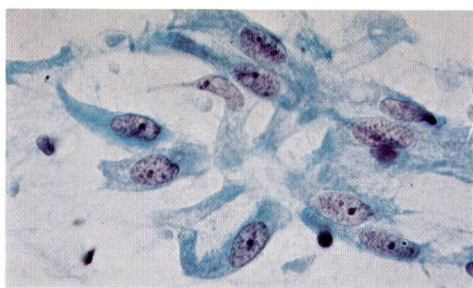

Plate 5.121. Endometrial imprint: the structure of the proliferative type is very well preserved; columnar cells with a centrally located nucleus; the nucleolus is easily seen (250 ×).

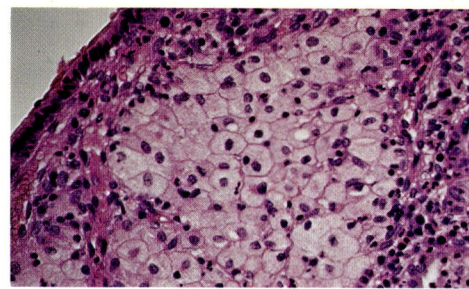

Plate 5.122. Endometrial biopsy: decidual reaction of the stromal cell: abundant, glycogen-rich cytoplasm (100 ×).

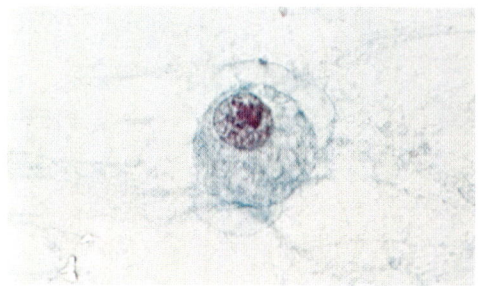

Plate 5.123. Endometrial aspiration: isolated decidual cell with large, clear cytoplasm (400 ×).

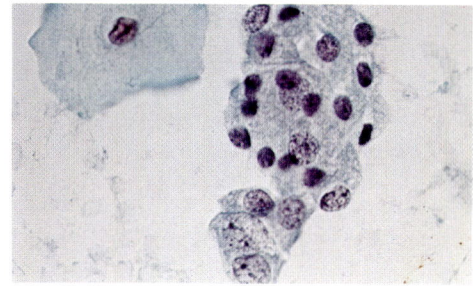

Plate 5.124. Endometrial aspiration: secretory endometrium: cluster of endometrial cells with abundant, clear cytoplasm and round nuclei. The abundant cytoplasm suggests secretory activity (250 ×).

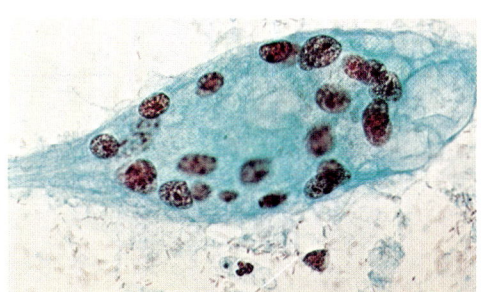

Plate 5.125. Endometrial aspiration: endometrial cells arranged in a syncytium: this disposition should not be confused with multinucleated histiocytes, cells of syncytiotrophoblast, and herpes multinucleation (250 ×).

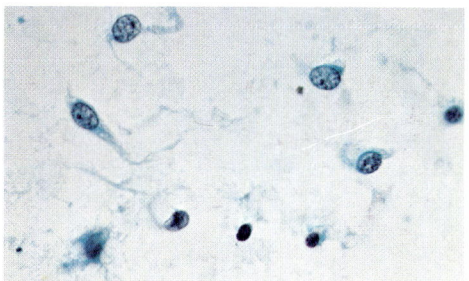

Plate 5.126. Endometrial aspiration: submucosal leiomyoma with erosion of the mucosa: presence of smooth muscle cells in the aspiration.

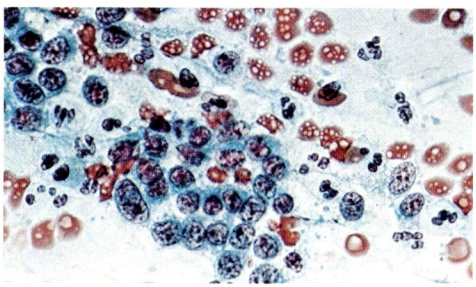

Plate 5.127. Endometrial aspiration: biopsy: adenomatous hyperplasia of endometrium: slightly enlarged, irregular endometrial cell cluster suggesting endometrial hyperplasia; must be confirmed by histology (250 ×).

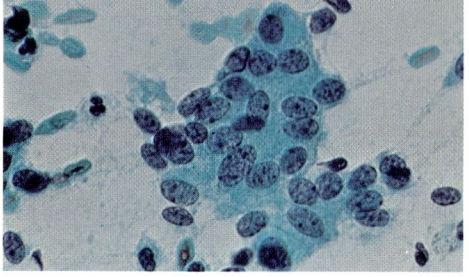

Plate 5.128. Endometrial aspiration: endometrial hyperplasia: cluster of endometrial cells showing discrete enlargement of nuclear size and cell pile-up; regular nuclei finely dispersed chromatin (250 ×).

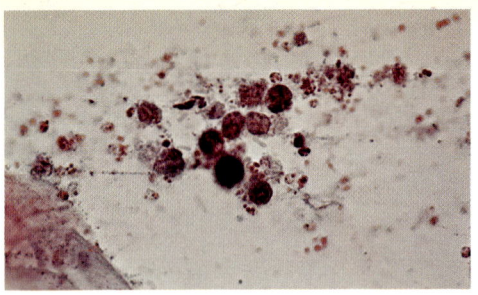

Plate 5.129. Endometrial aspiration: blood collected in the endometrial cavity: presence of endometrial cells, macrophages, and hemosiderin. Abundance of hemosiderin granules suggests chronic hemorrhage (250 ×).

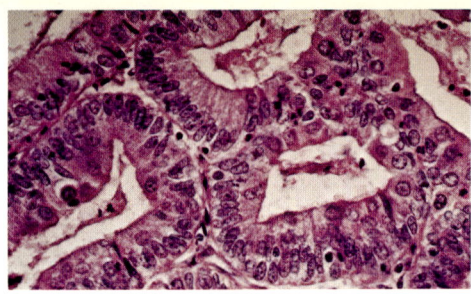

Plate 5.130. Endometrial biopsy: differentiated adenocarcinoma of the endometrium (100 ×).

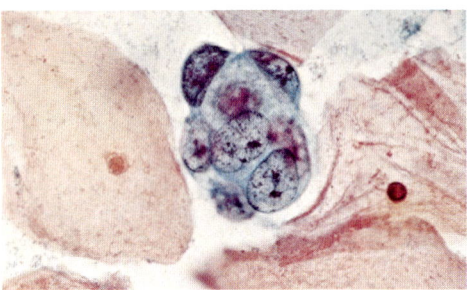

Plate 5.131. Vaginal smear: presence of endometrial cancer cells in the vaginal smear (250 ×).

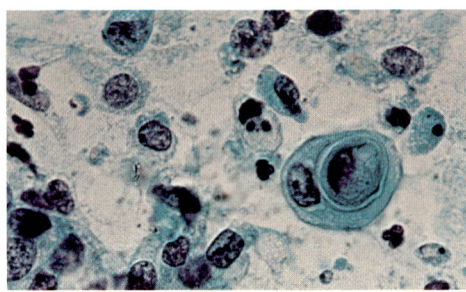

Plate 5.132. Endometrial aspiration: clusters and isolated endometrial cancer cells; cellular cannibalism (250 ×).

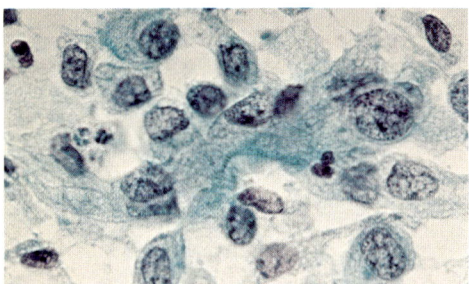

Plate 5.133. Endometrial aspiration: presence of ciliated endometrial cancer cells. Electron microscopy has confirmed that cilia may be observed in tumor cells (250 ×).

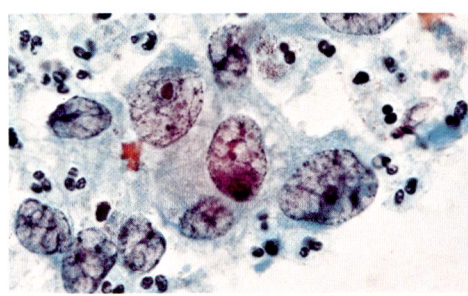

Plate 5.134. Endometrial aspiration: undifferentiated adenocarcinoma of the endometrium. Note the high degree of cellular abnormalities (250 ×).

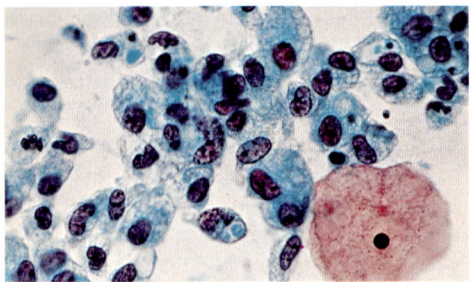

Plate 5.135. Endometrial aspiration: differentiated adenocarcinoma of the endometrium: tumor cells are small and regular and have a columnar structure (250 ×).

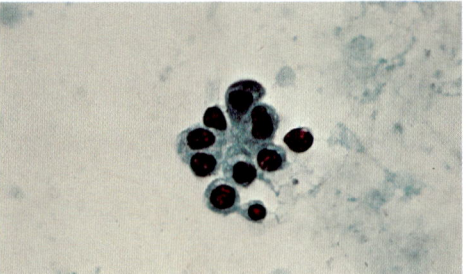

Plate 5.136. Cervical smear: contraceptive intrauterine device: presence of small clusters of irritated endometrial cells. These cells may simulate columnar malignant cells (250 ×).

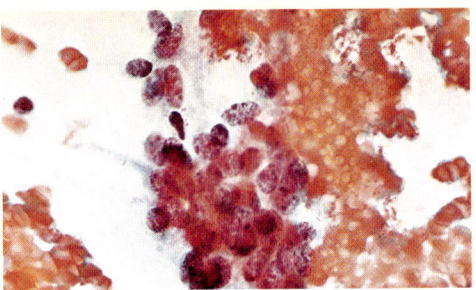

Plate 5.137. Endometrial aspiration: endometrial hyperplasia confirmed by biopsy. Enlarged endometrial cells with a hemorrhagic background. These cells simulate differentiated adenocarcinoma (250 ×).

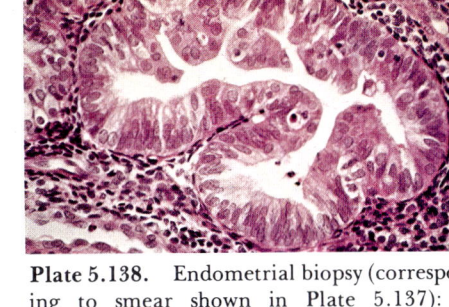

Plate 5.138. Endometrial biopsy (corresponding to smear shown in Plate 5.137): endometrial adenomatous hyperplasia (100 ×).

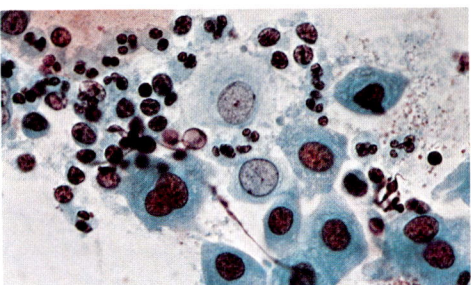

Plate 5.139. Cervical smear: coexistence of cervical squamous metaplasia and endometrial hyperplasia: note the marked difference in the size of endometrial and squamous cervical cells (250 ×).

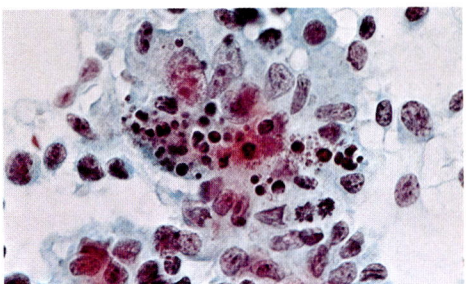

Plate 5.140. Endometrial aspiration: 68-year-old patient, marked inflammatory reaction necrosis and karyorrhexis: presence of atypical endometrial cells. The biopsy revealed an atrophic endometrium (250 ×).

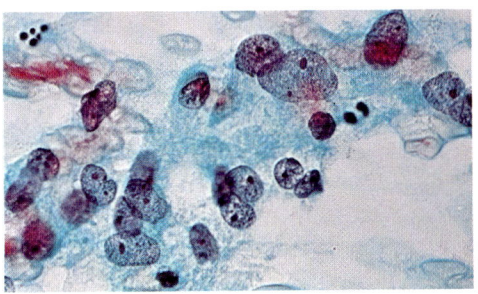

Plate 5.141. Endometrial aspiration: poorly differentiated adenocarcinoma of the endometrium. Note nuclear abnormalities and well-distinguished nucleoli. The cytoplasm is preserved in some cells (250 ×).

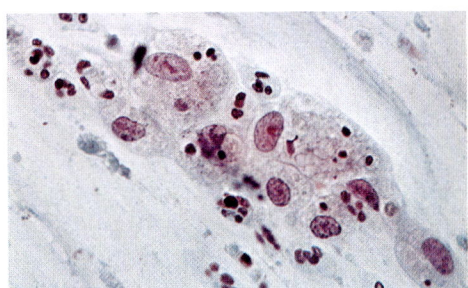

Plate 5.142. Endometrial aspiration: presence of histiocytes and macrophages in endometrial smears should raise the suspicion. These cells originating in the endometrial stroma are frequently seen in adenocarcinoma (250 ×).

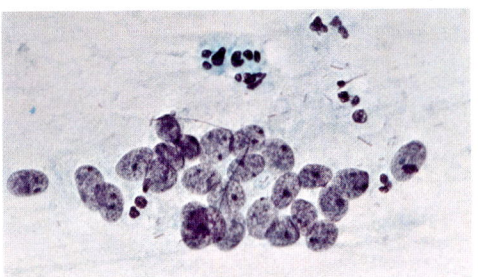

Plate 5.143. Vaginal smear: chronic cervicitis in a menopausal woman: cluster of naked nuclei with prominent nucleoli. Size and fine granular structure of nuclei favor an endocervical, not an endometrial origin (250 ×).

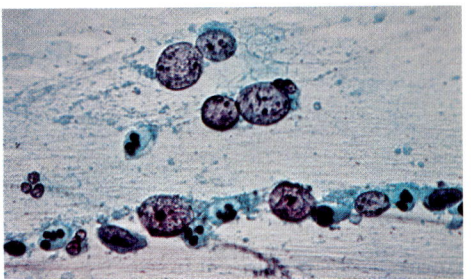

Plate 5.144. Vaginal smear: chronic cervicitis with squamous metaplasia, erosion, and tissue repair: clusters of atypical naked nuclei. Distribution of chromatin and presence of large nucleoli favor an endocervical origin (250 ×).

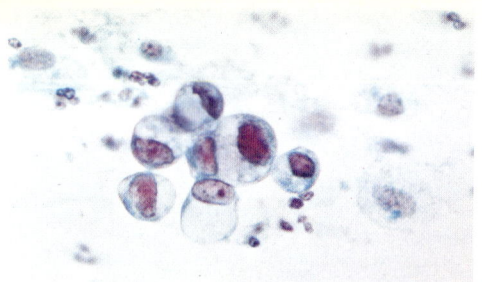

Plate 5.145. Endometrial aspiration: adenocarcinoma of the endometrium, mucinous type: note the large cytoplasmic vacuole displacing and compressing the nucleus toward the periphery (250 ×).

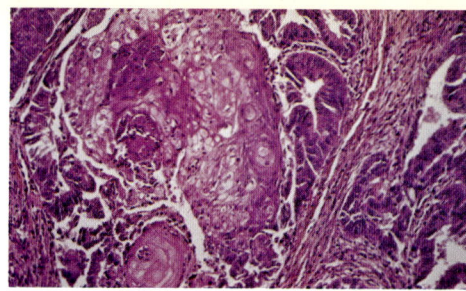

Plate 5.146. Endometrial biopsy: adenoacanthoma of the endometrium (adenocarcinoma with benign squamous elements (100 ×).

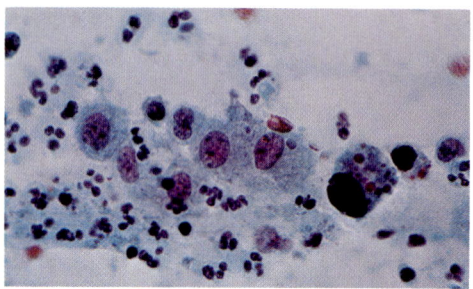

Plate 5.147. Cervical smear: adenoacanthoma of the endometrium: in this case, only undifferentiated columnar elements are present (250 ×).

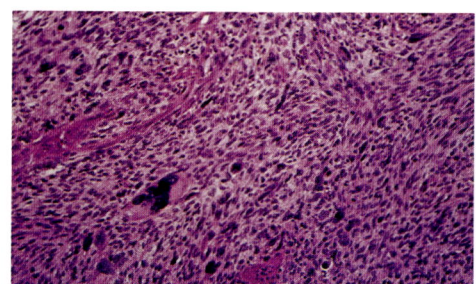

Plate 5.148. Endometrial biopsy: leiomyosarcoma of uterus (100 ×).

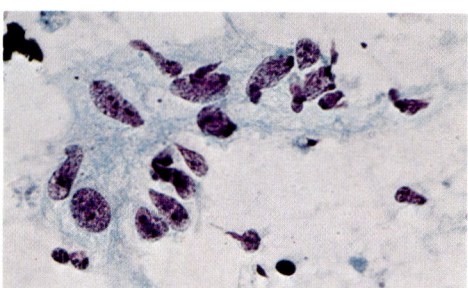

Plate 5.149. Endometrial aspiration: cluster of malignant sarcomatous cells showing irregular but mostly elongated, hyperchromatic nuclei: the cytoplasm is pale and poorly defined (250 ×).

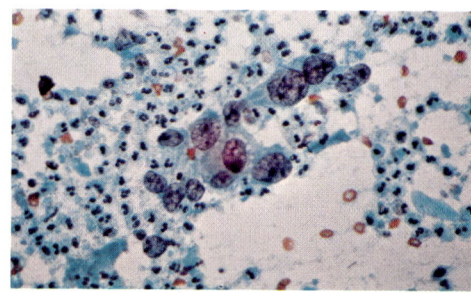

Plate 5.150. Vaginal smear: adenocarcinoma of the ovary: cells have large, regular hyperchromatic nuclei, scant, cyanophilic cytoplasm and are very similar to some observed from cervical or endometrial adenocarcinomas (250 ×).

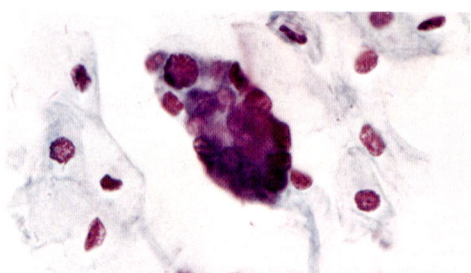

Plate 5.151. Vaginal smear: adenocarcinoma of the ovary: presence of rare clusters of cells arranged in a syncytial manner easily picked up in a clean background. Ovarian origin could not be identified by cytology (250 ×).

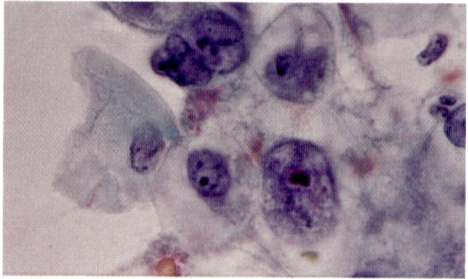

Plate 5.152. Vaginal smear: undifferentiated carcinoma of the ovary (250 ×).

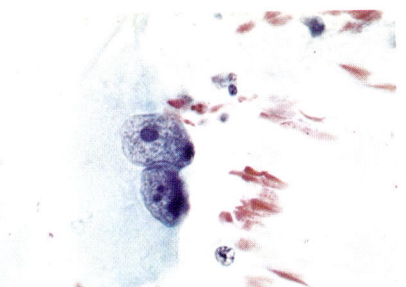

Plate 5.153. Vaginal smear: adenocarcinoma of rectum, secondary fistulization into vaginal cavity: large cells with hyperchromatic nuclei, prominent nucleoli. Cell malignancy is certain. Origin not identified by cytology alone. Large nucleoli favor glandular origin (250 ×).

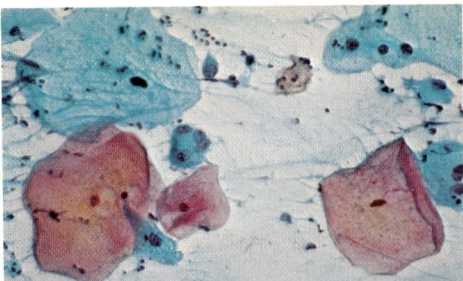

Plate 5.154. Vaginal smear: following radiotherapy, invasive squamous carcinoma of the cervix: marked cellular enlargement of benign squamous superficial cells (250 ×).

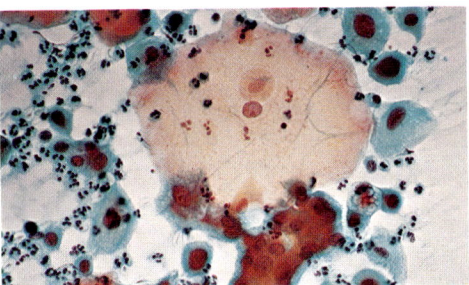

Plate 5.155. Cervical smear: following radiotherapy: markedly enlarged eosinophilic superficial cell and altered parabasal cells with irregular nuclei and vacuolated cytoplasm (250 ×).

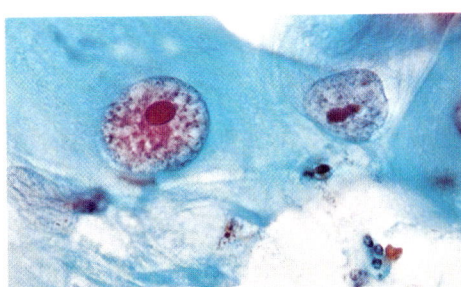

Plate 5.156. Vaginal smear: following radiotherapy: nuclear enlargement; irregular clumps of chromatin; large nucleolus. The large nucleocytoplasmic ratio is consistent with benign altered cells (400 ×).

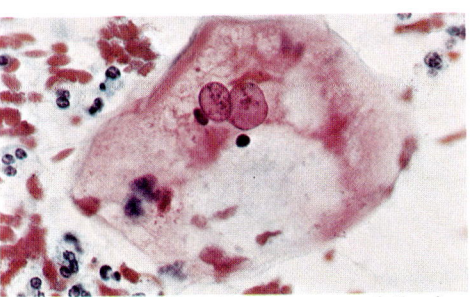

Plate 5.157. Cervical smear: following radiotherapy, invasive squamous carcinoma of cervix: enlarged binucleated benign superficial cell (400 ×).

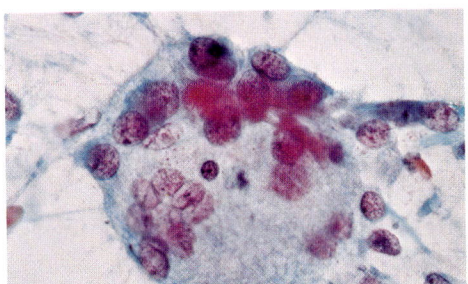

Plate 5.158. Cervical smear: following radiotherapy, giant histiocyte: note the typical cytoplasmic microvascularization (250 ×).

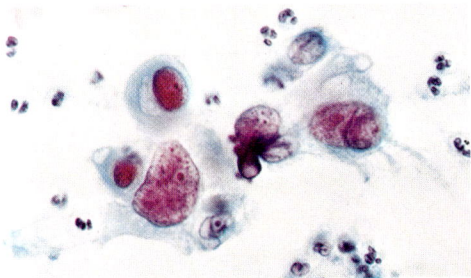

Plate 5.159. Cervical smear: acute radiation changes, invasive squamous carcinoma of the cervix: malignant altered cells; wrinkling of the nucleus with chromatin clumping, vacuolation, and necrosis of cytoplasm (250 ×).

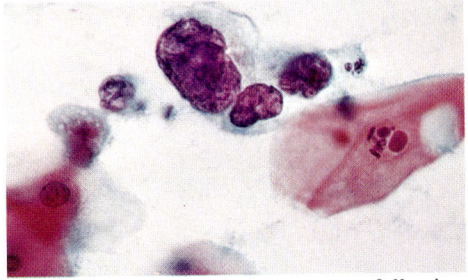

Plate 5.160. Cervical smear: following radiotherapy, invasive carcinoma of the cervix: persistent malignant disease; presence of altered malignant cells 3 months after irradiation (250 ×).

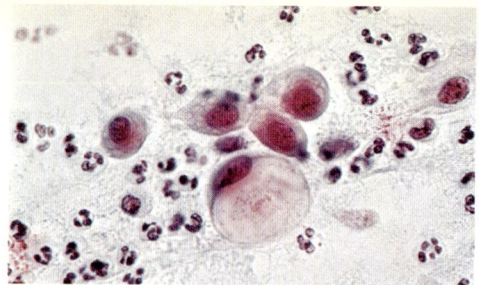

Plate 5.161. Cervical smear: following radiotherapy, invasive carcinoma of the cervix: altered benign columnar and parabasal cells (250×).

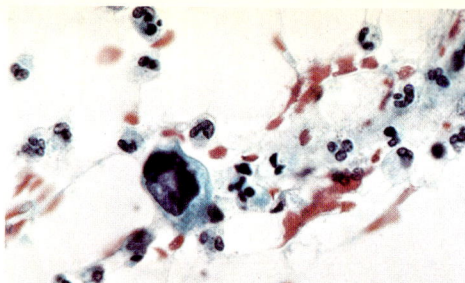

Plate 5.162. Cervical smear: following radiotherapy, adenocarcinoma of the endometrium: altered endometrial cell, probably malignant (250×).

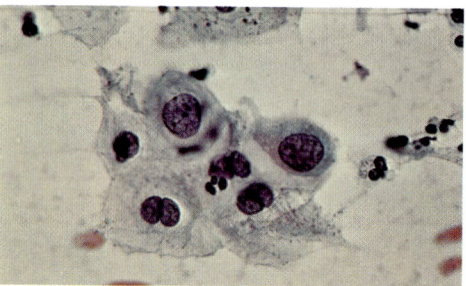

Plate 5.163. Cervical smear: carcinoma of the cervix, 5 years after irradiation: note the persistence of modified benign squamous cells (250×).

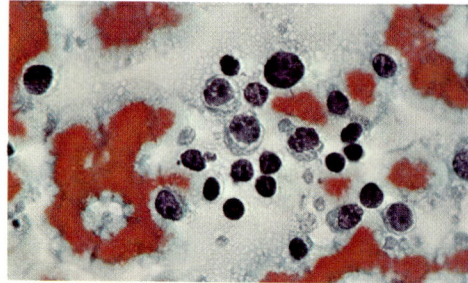

Plate 5.164. Cervical smear: poorly differentiated lymphocytic lymphoma: the cellular size, the hyperchromatic nuclei, and the abundance of these elements may lead to their confusion with endometrial cells (250×).

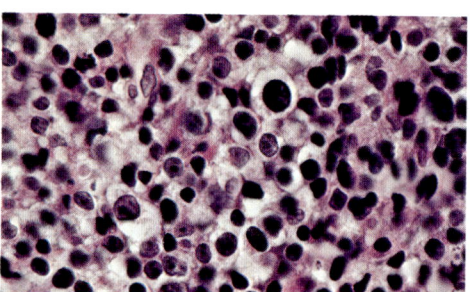

Plate 5.165. Cervical biopsy: poorly differentiated lymphocytic lymphoma (see Plate 5.164) (100×).

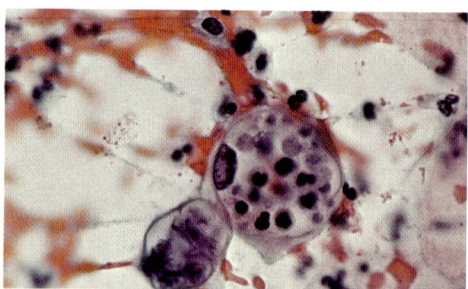

Plate 5.166. Cervical smear: mixed adenocarcinoma and epidermoid carcinoma in pregnancy: phagocytized leukocytes in malignant cell with a crescentlike nucleus; phagocytosis not specific criterion of malignancy (250×).

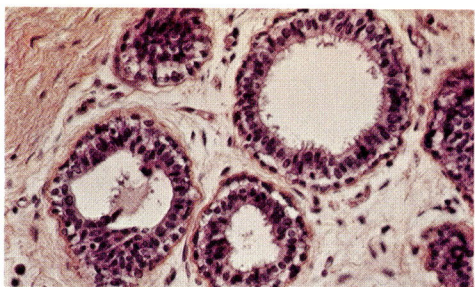

Plate 6.1. Histology of normal functional mammary gland: glandular structures disseminated in the connective tissue stroma (40 ×).

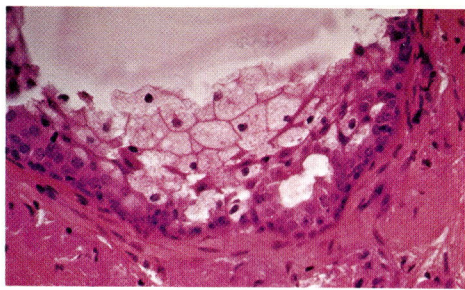

Plate 6.2. Biopsy: benign mammary dysplasia: cystic wall with ductal epithelium shows epithelial cells and foaming cells that are going to desquamate (100 ×).

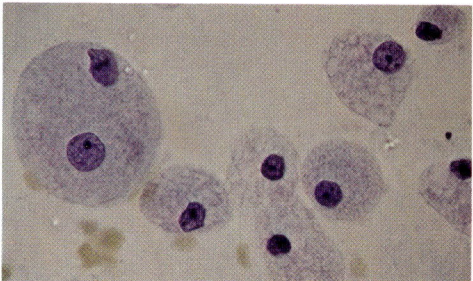

Plate 6.3. Needle aspiration: mammary dysplasia: desquamated foamy cells. These cells acquire a phagocytic capacity (250 ×).

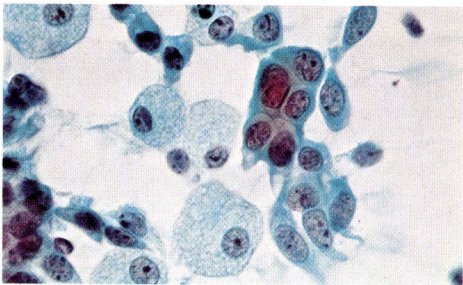

Plate 6.4. Needle aspiration: mammary dysplasia: ductal epithelial cells and desquamated foamy cells (250 ×).

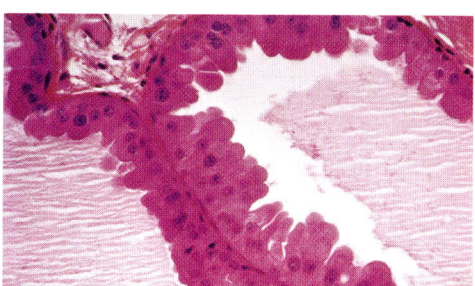

Plate 6.5. Biopsy: mammary dysplasia: apocrine metaplasia; glands lined by large, bulging cells with small regular nuclei (100 ×).

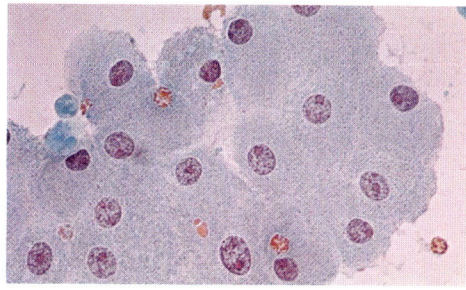

Plate 6.6. Needle aspiration: mammary dysplasia: hyperplastic ductal cells and foamy cells. Note the increased volume of some nuclei without modification of the chromatin structure (250 ×).

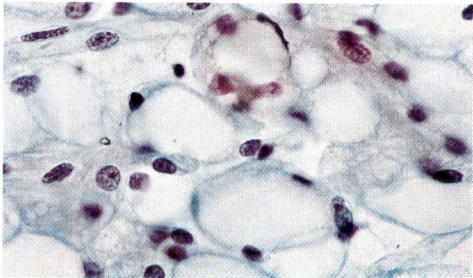

Plate 6.7. Needle aspiration: adipous tissue: adipocytes and stromal cells obtained by aspiration of a fatty mammary gland (250 ×).

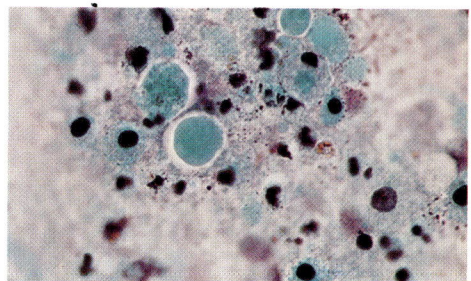

Plate 6.8. Needle aspiration: mammary cyst: cyst fluid contains leukocytes, cellular necrotic debris, histiocytes, and homogeneous nuclear-free degenerating cytoplasm. Note absence of well-preserved epithelial cells (250 ×).

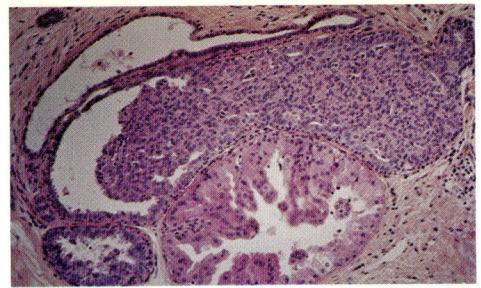

Plate 6.9. Biopsy: fibrocystic dysplasia: proliferation of duct epithelium including papillary patterns; apocrine metaplasia (100 ×).

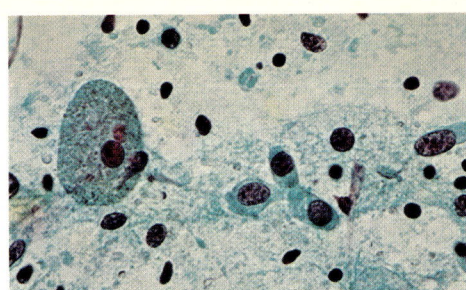

Plate 6.10. Needle aspiration: mammary dysplasia: ductal and foamy cells; some elements modified by inflammation and necrosis; nuclei slightly hyperchromatic nuclear cytoplasmic ratio not increased; small nucleoli (250 ×).

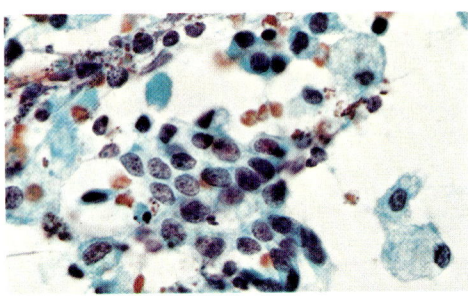

Plate 6.11. Needle aspiration: mammary dysplasia: clusters of irregularly distributed elements; round or elongated nuclei with a dense or foamy cytoplasm; presence of cellular debris, leukocytes, and red blood cells (250 ×).

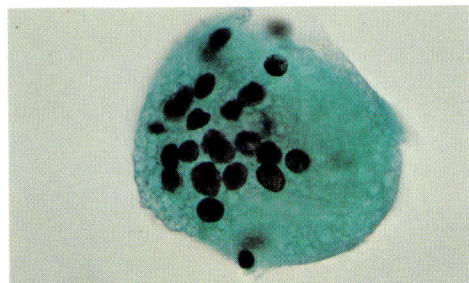

Plate 6.12. Needle aspiration: mammary dysplasia: giant multinucleated histiocyte. These cells may represent ductal cell agglomerates, foreign body, giant cell, or multinucleated histiocyte (250 ×).

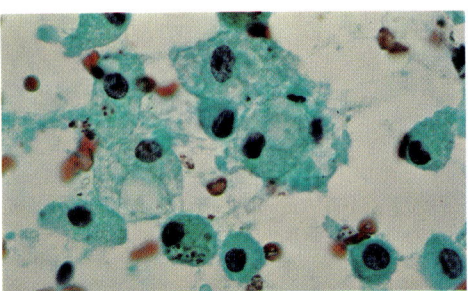

Plate 6.13. Needle aspiration: fibrocystic dysplasia: irregularly distributed vacuolated ductal elements; the nucleocytoplasmic ratio is preserved; cellular degeneration is moderate; presence of blood cells (250 ×).

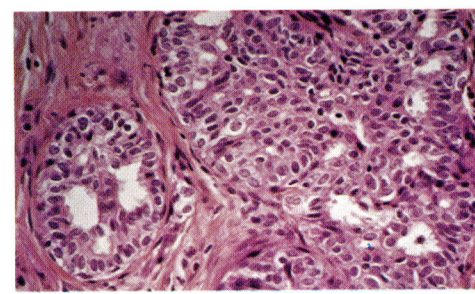

Plate 6.14. Histology of fibrocystic dysplasia: foci of hyperplastic ducts with epithelial proliferation and filling up of ductal lamina; cellular anomalies are discrete (100 ×).

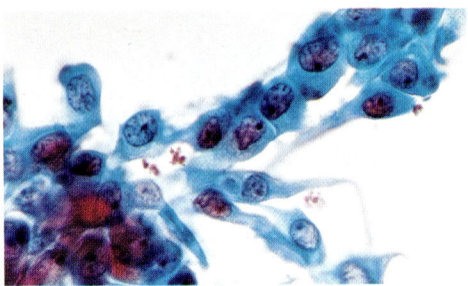

Plate 6.15. Needle aspiration: fibrocystic dysplasia: clusters of hyperplastic ductal cells; nuclei are regular and the chromatin exhibits an even distribution; nuclear membranes are apparent (250 ×).

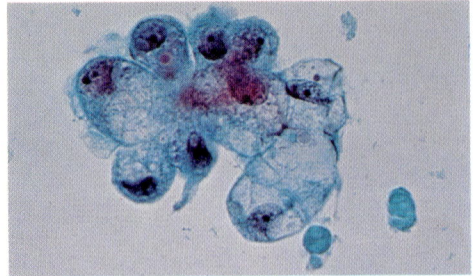

Plate 6.16. Needle aspiration: mammary cyst papillomatous proliferation of ductal epithelium: papillary cellular arrangement, abundant vacuolated, partly degenerated cytoplasm. Regular nuclei pushed aside by vacuolation (400 ×).

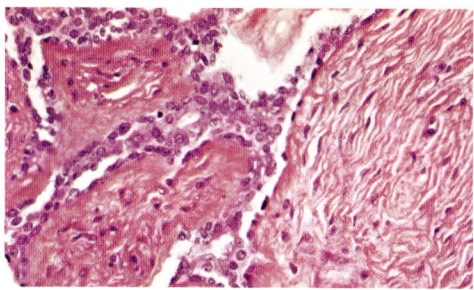

Plate 6.17. Histology of fibroadenoma: detail of epithelial cords surrounded by an abundant fibrous stroma (100 ×).

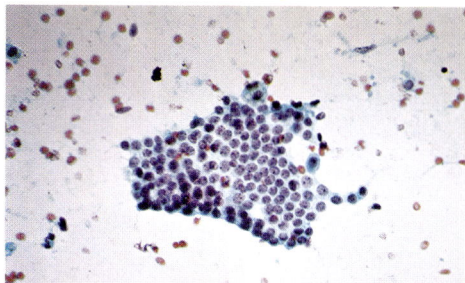

Plate 6.18. Needle aspiration: fibroadenoma: low magnification showing a dense cluster of ductal cells with regular, round nuclei. The presence of such clusters suggests the diagnosis of fibroadenoma (100 ×).

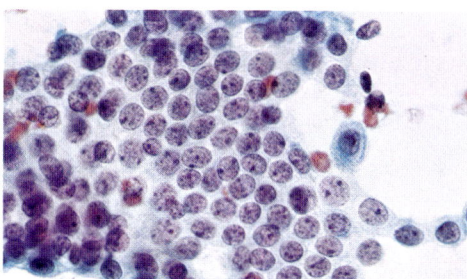

Plate 6.19. Needle aspiration: fibroadenoma: detail of cells; the nuclei are round or oval and exhibit a uniform, slightly granular chromatin; nucleoli are apparent; cytoplasm is pale and cyanophilic (250 ×).

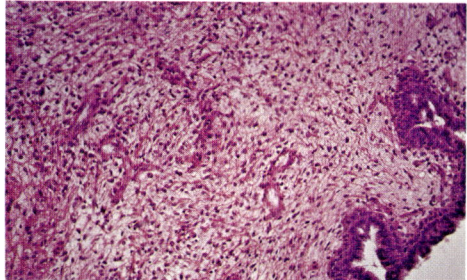

Plate 6.20. Histology: giant fibroadenoma (cystosarcoma phyllodes): bistratified ductal structures embedded in an abundant, highly cellular stroma; the stromal cells are numerous and elongated (100 ×).

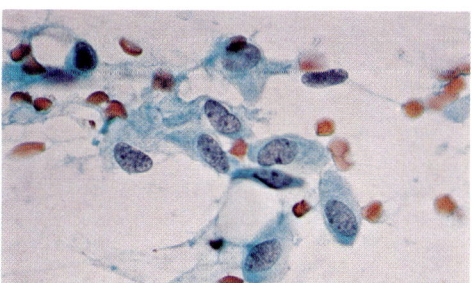

Plate 6.21. Needle aspiration: giant fibro-adenoma (cystosarcoma phyllodes): columnar ductal cells, elongated stromal cells identified; naked nuclei of stromal origin; finely dispersed chromatin; small nucleoli; normal nucleocytoplasmic ratio (250 ×).

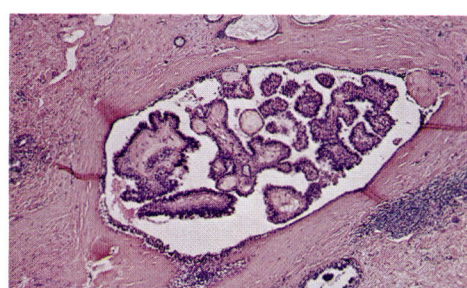

Plate 6.22. Histology of intraductal papilloma: papillary structures originating from the ductal epithelium and invading the glandular lumen; papillae have a stromal axis covered by two layers of cells (40 ×).

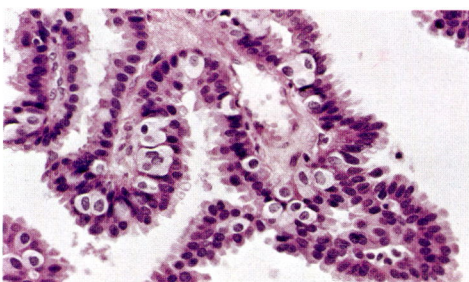

Plate 6.23. Histology of intraductal papilloma: detail of papillary structures showing the bistratified structure of the epithelium (100 ×).

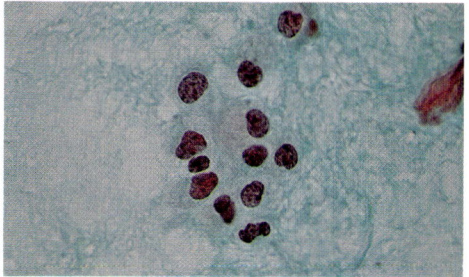

Plate 6.24. Nipple discharge: intraductal papilloma: clusters of epithelial cells disseminated in a dense exudate; slight cellular degeneration with dense nuclear structure (250 ×).

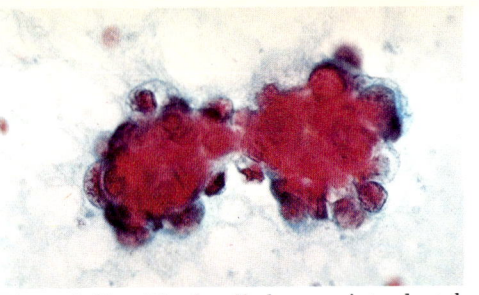

Plate 6.25. Nipple discharge: intraductal papilloma: papillary structures giving the impression of the three dimensional disposition of the cells (raspberrylike structure); absence of cellular criteria of malignancy (250 ×).

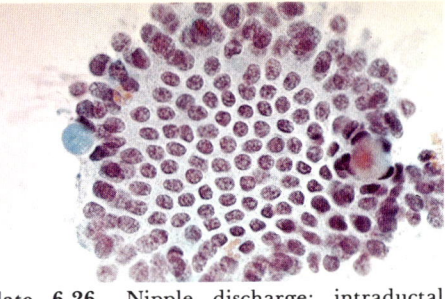

Plate 6.26. Nipple discharge: intraductal papilloma: well-preserved papillary structure; regular disposition of cells; presence of some secretion droplets (250 ×).

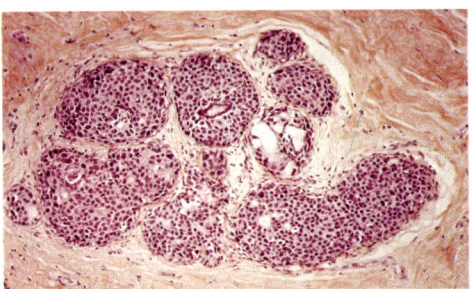

Plate 6.27. Biopsy: in situ lobular carcinoma: well-demarcated alveoli lined and filled by neoplastic cells; fibrosis of the adjacent stroma (40 ×).

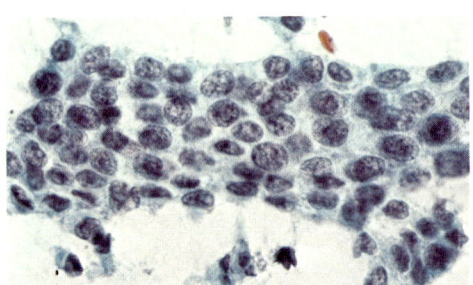

Plate 6.28. Needle aspiration: in situ lobular carcinoma (same case as Plate 6.27): cluster of regular cells with round or oval hyperchromatic nuclei; absence of marked cellular atypias (250 ×).

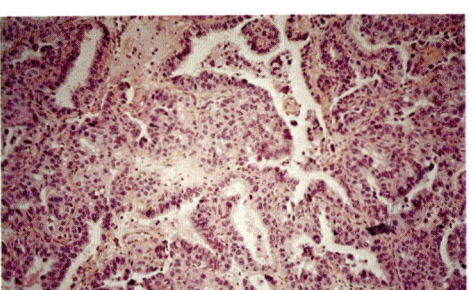

Plate 6.29. Biopsy: infiltrating duct carcinoma: cords, tubular and glandular neoplastic structures, invading the mammary parenchyma (40 ×).

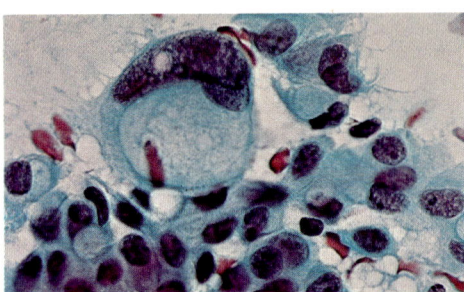

Plate 6.30. Needle aspiration: infiltrating duct carcinoma: cluster of disorderly arranged neoplastic cells with hyperchromatic, sometimes voluminous nuclei; cytoplasmic vacuolation (250 ×).

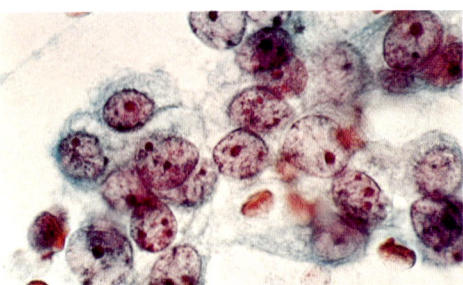

Plate 6.31. Needle aspiration: infiltrating duct carcinoma: ductal neoplastic cells showing moderate hyperchromatism and nucleolar hypertrophy; partial lysis of cytoplasm. Note the "piling up" of cells (400 ×).

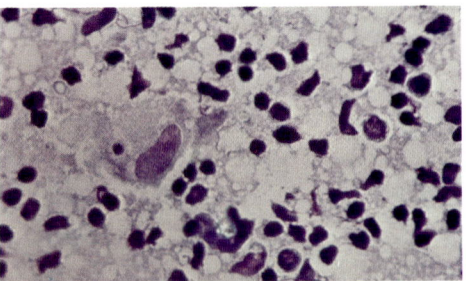

Plate 6.32. Needle aspiration (dried smear): infiltrating duct carcinoma: note the poorly fixed nuclei with the absence of structural details (May Grunwald stain, 250 ×).

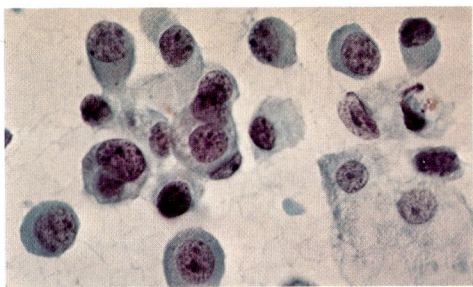

Plate 6.33. Needle aspiration: infiltrating duct carcinoma: well-preserved, isolated neoplastic cells showing round, enlarged hyperchromatic nuclei and a dense cytoplasm (250×).

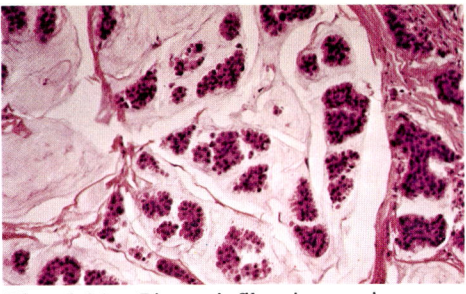

Plate 6.34. Biopsy: infiltrating mucinous carcinoma: the neoplastic cords are dispersed in an abundant mucinous secretion (40×).

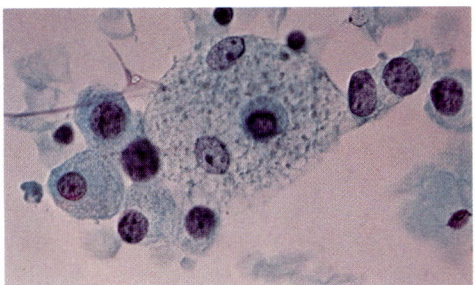

Plate 6.35. Needle aspiration: infiltrating mucinous carcinoma: differentiated cells sometimes with an abundant mucin-producing cytoplasm (250×).

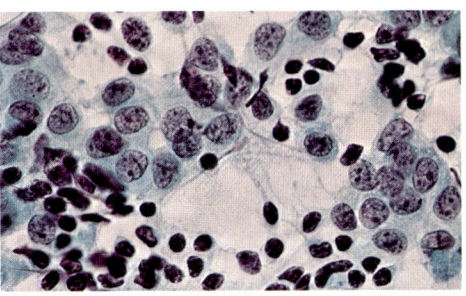

Plate 6.36. Needle aspiration: medullary carcinoma: mixture of clusters of large oval or polygonal cells with hyperchromatic, vesicular nuclei and clusters of lymphocytes (250×).

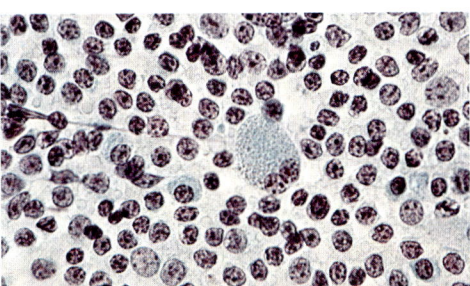

Plate 6.37. Needle aspiration: primary malignant lymphoma. Massive infiltration of the gland with lymphocytes having both cleaved and noncleaved nuclei; rare ductal cells are noted (250×).

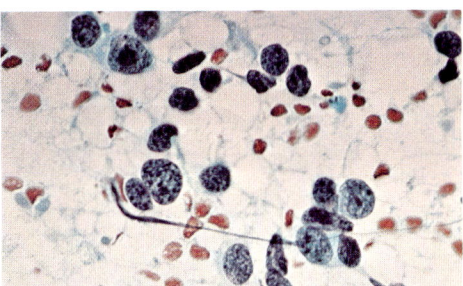

Plate 6.38. Needle aspiration: rhabdomyosarcoma: anaplastic cells with hyperchromatic large nuclei, cytoplasm poorly preserved. Cytologic diagnosis was anaplastic malignant tumor; final diagnosis of rhabdomyosarcoma rested on histologic exam (250×).

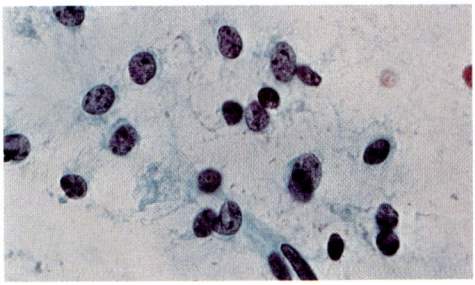

Plate 6.39. Needle aspiration: liposarcoma: presence of cells with small, regular or elongated nuclei and poorly delineated cytoplasm (250×).

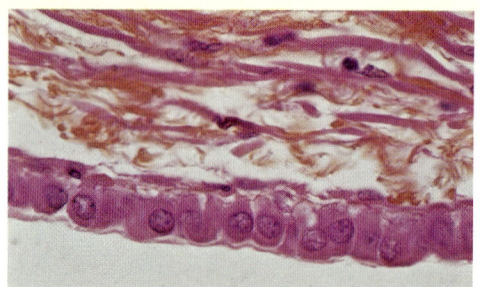

Plate 7.1. Biopsy: normal pleura: mesothelial lining and underlying connective tissue (250 ×).

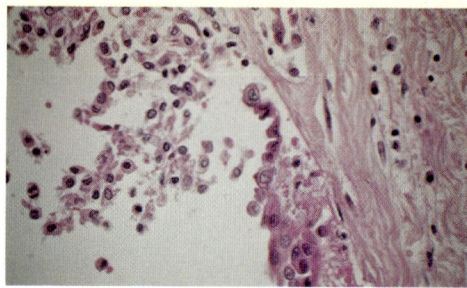

Plate 7.2. Biopsy: pleura: chronic inflammatory conditions; proliferation and exfoliation of mesothelial cells, thickening and fibrosis of the connective tissue (100 ×).

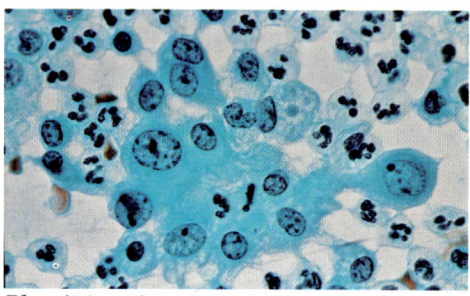

Plate 7.3. Pleural effusion smear: pulmonary embolism: "reactive" polyhedral mesothelial cells; moderate hyperchromatism and apparent nucleolus; densely stained cytoplasm; particles in cytoplasm of macrophages (250 ×).

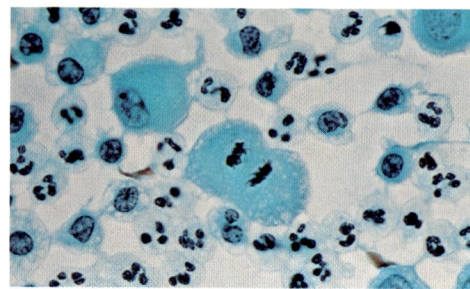

Plate 7.4. Pleural effusion smear: pulmonary embolism; presence of a mitotic figure; leukocytes and histiocytes (250 ×).

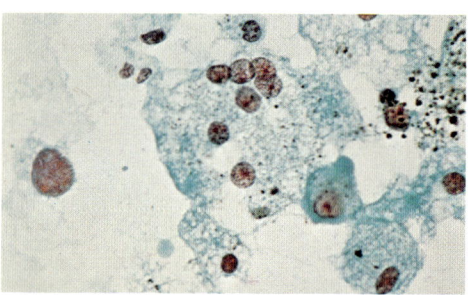

Plate 7.5. Pleural effusion smear: degenerating irritated mesothelial cells; cytoplasmic vacuolation and cytolysis; presence of blood pigment (hemosiderin) granules in the cytoplasm (250 ×).

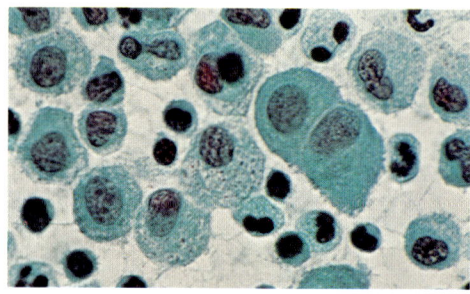

Plate 7.6. Pleural effusion: lobar pneumonia: highly cellular fluid; numerous mesothelial cells and macrophages; normal nucleo-cytoplasmic ratio; presence of cytoplasmic granules in some cells (250 ×).

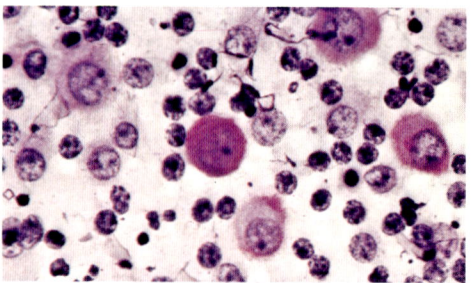

Plate 7.7. Pleural effusion smear; chronic cardiac failure: some mesothelial cells exhibit a dense PAS-positive cytoplasm; lymphocytes (PAS, 250 ×).

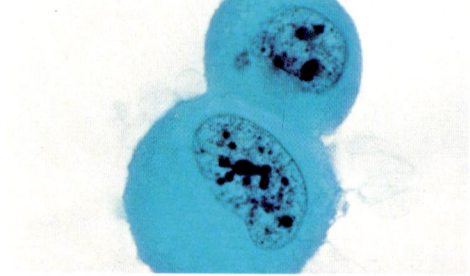

Plate 7.8. Pleural effusion smear: detail of mesothelial cells showing the nuclear structure and the dense cytoplasm with small pseudopods (400 ×).

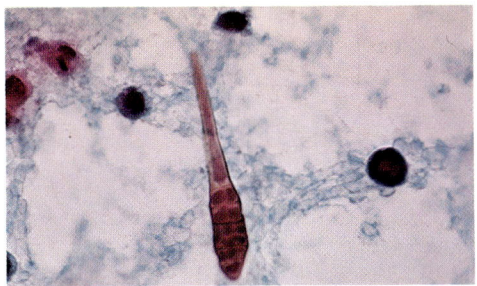

Plate 7.9. Pleural effusion smear: mycotic infection: presence of a conidium (spores on hypha buds) (250×).

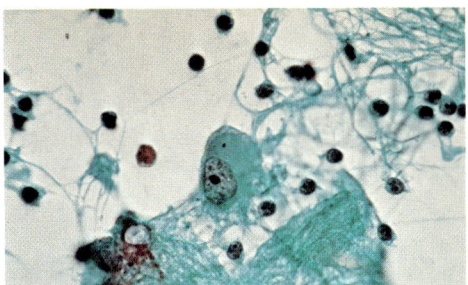

Plate 7.10. Pleural effusion smear: serofibrinous pleurisy: mesothelial cells and fibrin strands (250×).

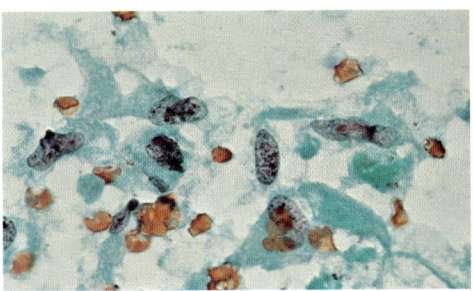

Plate 7.11. Pleural effusion smear: malignant mesothelioma: presence of bizarre, elongated cells with hyperchromatic fusiform nuclei; red blood cells (250×).

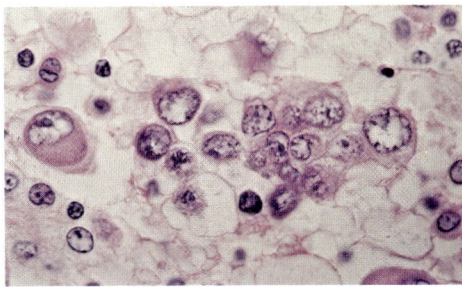

Plate 7.12. Pleural effusion, paraffin embedding, H & E stain: malignant mesothelioma: clusters of large irregular cells with irregularly distributed chromatin; the cytoplasm is rather well preserved (250×).

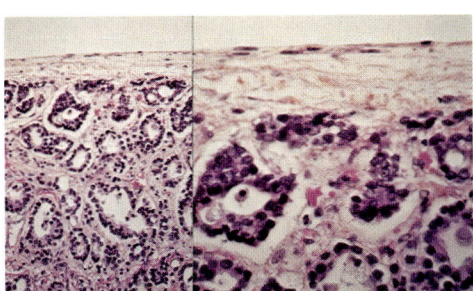

Plate 7.13. Pleural biopsy: metastatic adenocarcinoma of the breast: presence of neoplastic tubular structures infiltrating the connective tissue with conservation of the mesothelial lining (40 and 100×).

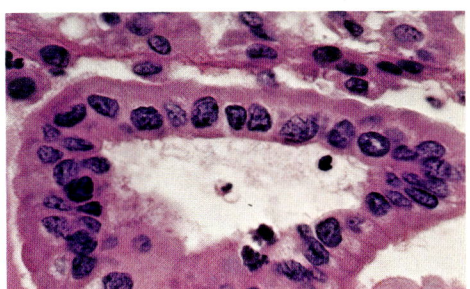

Plate 7.14. Pleural effusion, paraffin embedding: infiltrating mammary adenocarcinoma: typical glandular metastatic structures suggesting the diagnosis of adenocarcinoma (250×).

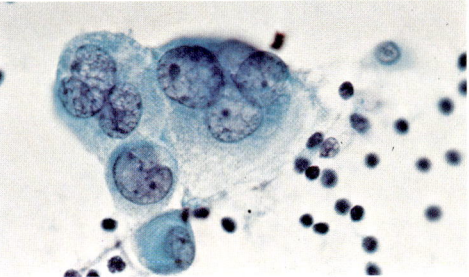

Plate 7.15. Pleural effusion smear: adenocarcinoma of the ovary: clusters of malignant cells. Compare with Plate 7.16. Note the delicate chromatin structure obtained with alcohol fixation and Shorr stain (250×).

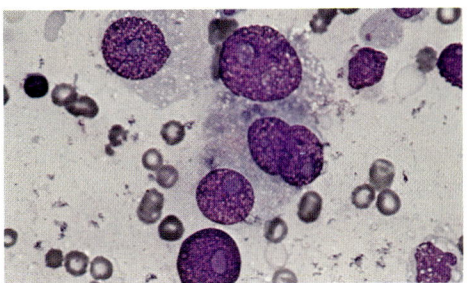

Plate 7.16. Pleural effusion smear: adenocarcinoma of the ovary: presence of a cluster of cancer cells; large hyperchromatic nuclei with enlarged nucleoli (May Grunwald-Giemsa stain, 250×).

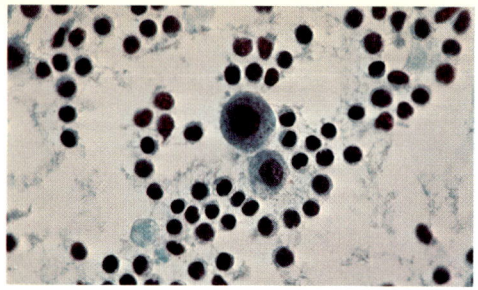

Plate 7.17. Pleural effusion: metastatic adenocarcinoma: mesothelial cells and lymphocytes; no cancer cells identified; example of mesothelial irritation by underlying tumor infiltration with preservation of mesothelial lining (250 ×).

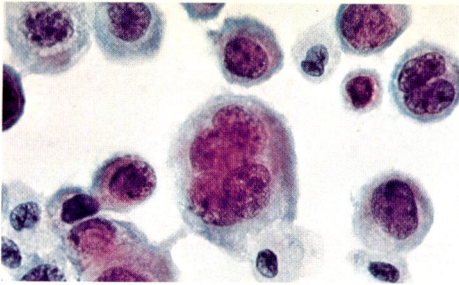

Plate 7.18. Pleural effusion smear: infiltrating duct carcinoma of the breast: clusters of malignant cells with multinucleated elements (250 ×).

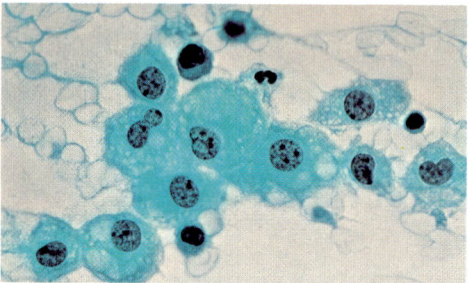

Plate 7.19. Ascites smear: adenocarcinoma of the ovary: presence of mesothelial cells showing regular nuclei and a finely vacuolated cytoplasm; normal nucleocytoplasmic ratio; absence of neoplastic cells in this field (250 ×).

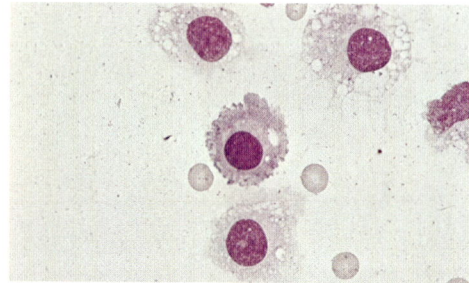

Plate 7.20. Ascites smear: mesothelial cells showing small cytoplasmic pseudopods and vacuolation (May Grunwald-Giemsa stain, 250 ×).

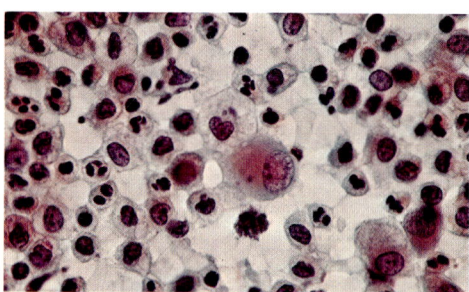

Plate 7.21. Ascites smear: pulmonary infarct: mesothelial cells; macrophages; polymorphonuclear leukocytes; lymphocytes; and histiocytes. This is the typical response to inflammation (250 ×).

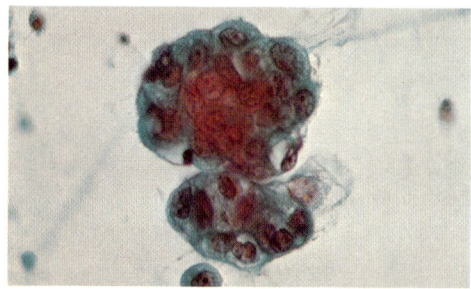

Plate 7.22. Ascites smear: adenocarcinoma of the breast: spherical aggregates of cells suggesting the structure of the embryonic blastula; cellular structures are poorly identifiable in these dense clusters (250 ×).

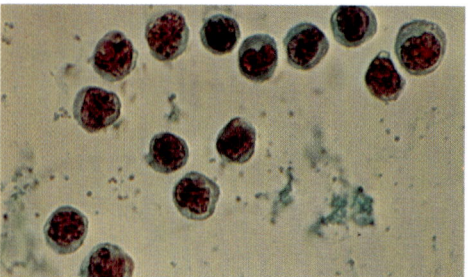

Plate 7.23. Ascites smear: malignant lymphoma, T-cell type: isolated, round cells with large hyperchromatic, sometimes indented nuclei (400 ×).

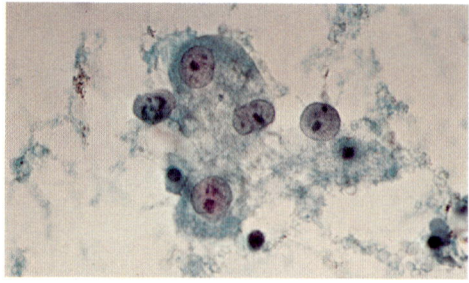

Plate 7.24. Pericardial fluid smear: malignant mesothelioma: rather regular nuclei with large nucleoli. Poorly preserved, vacuolated cytoplasm (250 ×).

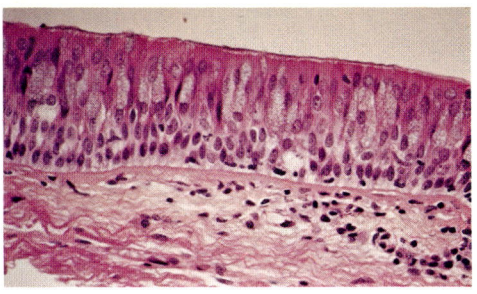

Plate 8.1. Biopsy: columnar stratified epithelium, bronchial mucosa: the multi-layered epithelium includes ciliated cells, mucous or goblet cells, basal-type cells (40 ×).

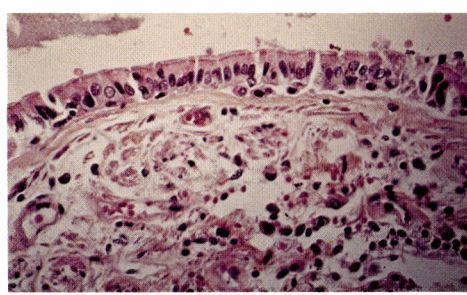

Plate 8.2. Biopsy: columnar stratified epithelium: single layer in alveoli (40 ×).

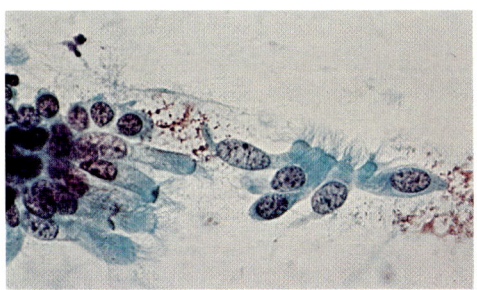

Plate 8.3. Bronchial aspiration smear: ciliated columnar cells (250 ×).

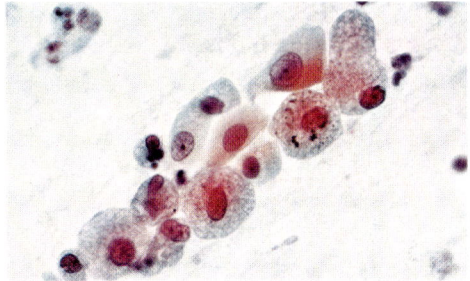

Plate 8.4. Bronchial aspiration smear: goblet cells and macrophages or pneumocytes (250 ×).

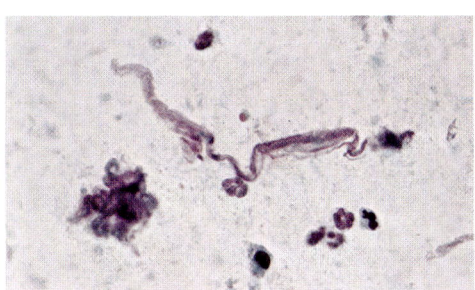

Plate 8.5. Sputum smear: Curschmann's spiral (250 ×).

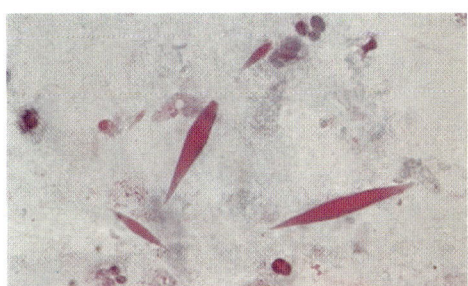

Plate 8.6. Sputum smear: Charcot-Leyden crystals (250 ×).

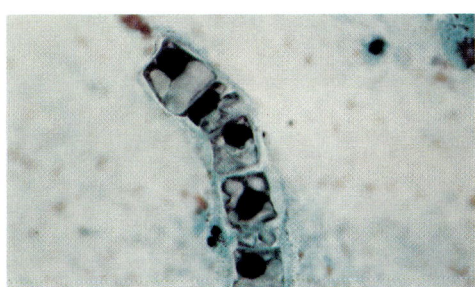

Plate 8.7. Sputum smear: vegetable cells: note the shape of the cells and the cellulose membranes (250 ×).

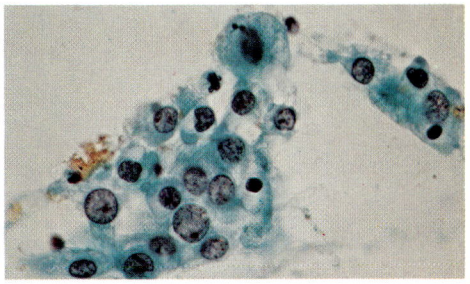

Plate 8.8. Paranasal sinus needle aspiration: columnar cells showing discrete inflammatory alterations (250 ×).

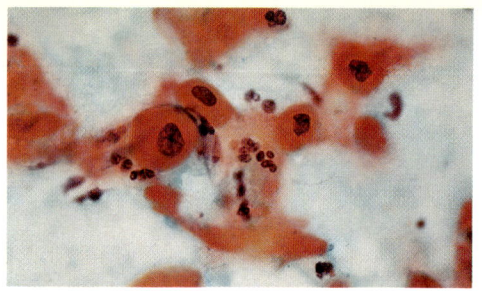

Plate 8.9. Mouth; direct scraping of the lesion: squamous carcinoma showing differentiated neoplastic cells; small, dense, irregular nuclei and keratinized cytoplasm (250 ×).

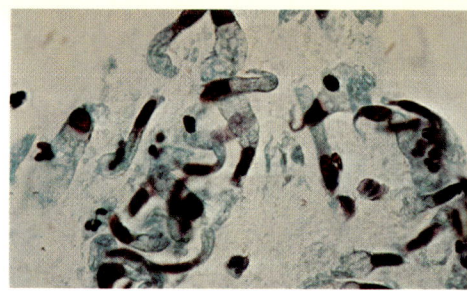

Plate 8.10. Bronchial aspiration smear: bronchiectasis: cluster of hyperplastic mucosecretory cells (250 ×).

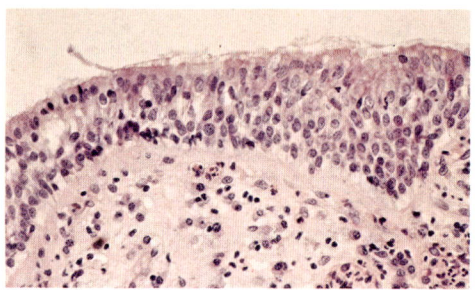

Plate 8.11. Biopsy: bronchial mucosa: hyperplasia of the basal cells and inflammatory infiltrate of the stroma (40 ×).

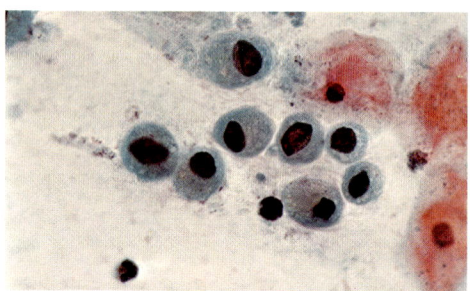

Plate 8.12. Sputum smear: squamous metaplasia: note the nuclear hyperchromatism (250 ×).

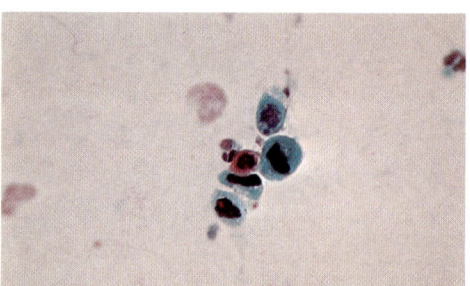

Plate 8.13. Sputum smear: bronchial dysplasia, epidermoid metaplasia: small, eosinophilic epidermoid cell (250 ×).

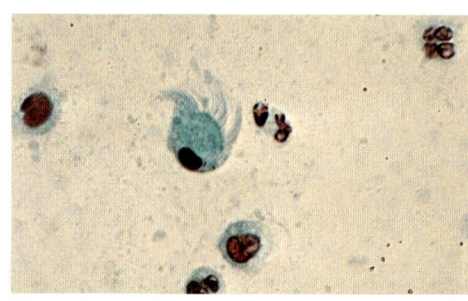

Plate 8.14. Sputum smear: ciliocytophtoria: partially degenerated cell with a ciliated border (250 ×).

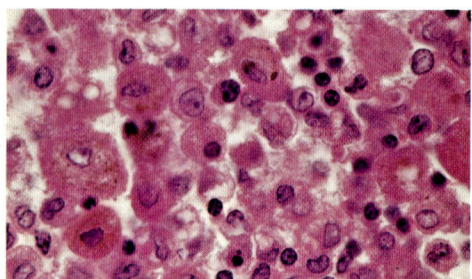

Plate 8.15. Sputum, paraffin-embedded inclusion: chronic inflammatory condition: numerous macrophages, histiocytes, and degenerated bronchial cell (250 ×).

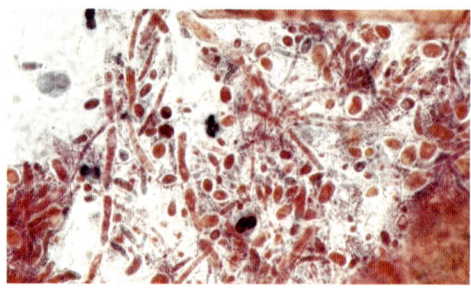

Plate 8.16. Sputum smear: mycotic infection (Candida): hyphae and spores (250 ×).

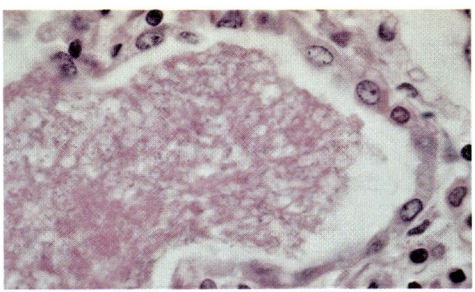

Plate 8.17. Biopsy: lung, *Pneumocystis carinii:* numerous elements present in the alveolar mucus (250 ×).

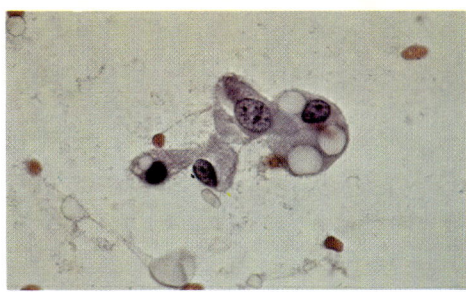

Plate 8.18. Sputum smear: lipid pneumonia: vacuolated macrophages containing the oil (continuing use of nose drops) (250 ×). Courtesy of Dr. Leduc.

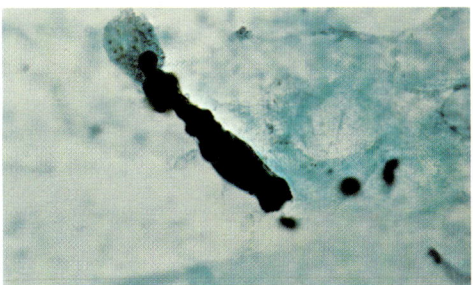

Plate 8.19. Sputum smear: asbestos body: deposition of protein and iron pigment on the surface of the asbestos fiber (250 ×). Courtesy of Dr. Leduc.

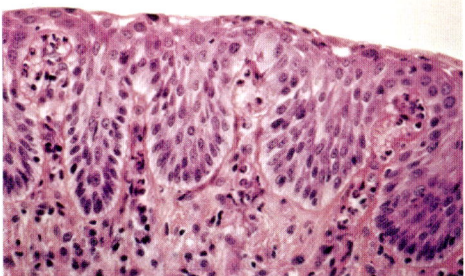

Plate 8.20. Biopsy: lung: simple dysplasia; thickening of the epithelium with acanthosis; moderate cellular atypias (history of heavy smoking) (40 ×).

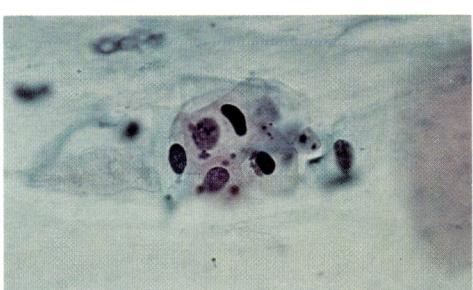

Plate 8.21. Sputum smear: atypical epidermoid cells in a case of bronchial carcinoma in situ (250 ×).

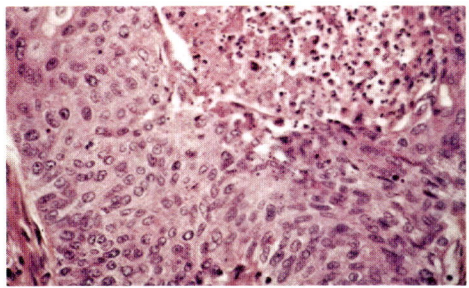

Plate 8.22. Biopsy: lung: poorly differentiated, invasive epidermoid carcinoma (40 ×).

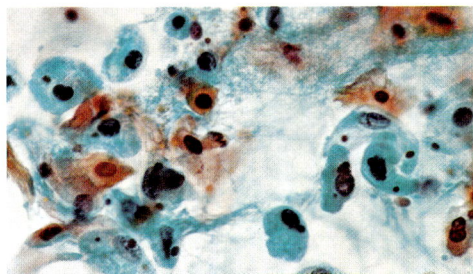

Plate 8.23. Sputum smear: differentiated epidermoid malignant cells: various nuclear and cytoplasmic abnormalities; marked tendency to keratin production (abundant eosinophilic cytoplasm) (250 ×).

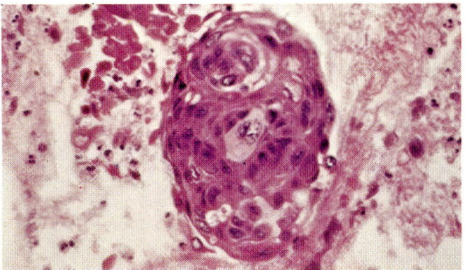

Plate 8.24. Sputum, paraffin embedding: presence of a cluster of malignant differentiated epidermoid cells showing pearly bodies (250 ×).

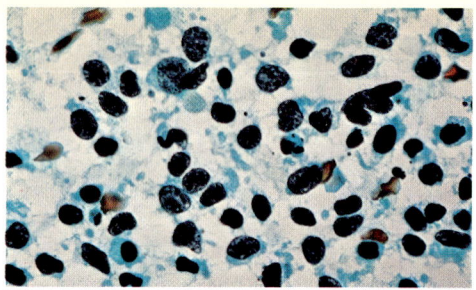

Plate 8.25. Blended sputum: small-cell anaplastic carcinoma: presence of isolated, small anaplastic cells. Note the very hyperchromatic nuclei and the poorly preserved cytoplasm (250×).

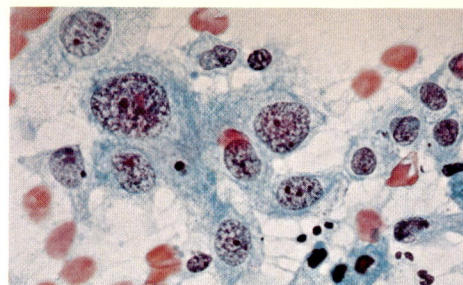

Plate 8.26. Sputum: large-cell anaplastic carcinoma: presence of large malignant elements with irregular, hyperchromatic nuclei and cyanophilic cytoplasm (250×).

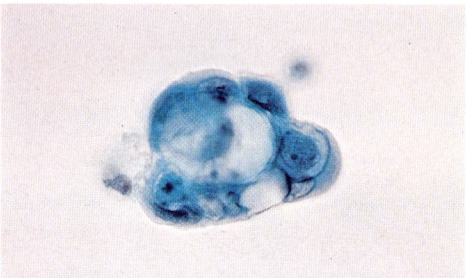

Plate 8.27. Sputum: bronchogenic adenocarcinoma: malignant cells arranged in clump which suggests glandular arrangement (250×).

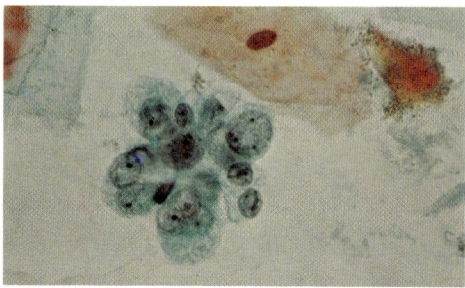

Plate 8.28. Sputum: bronchio-alveolar carcinoma: presence of a pseudospherical clump of cyanophilic cells with round nuclei and moderate hyperchromasia; large cytoplasmic vacuoles are present (250×).

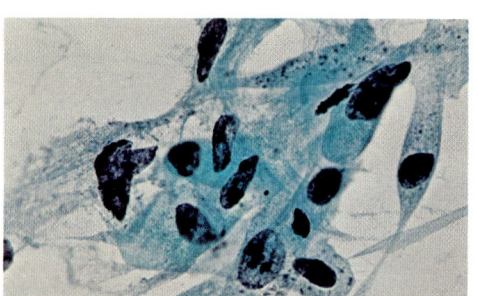

Plate 8.29. Sputum: large-cell undifferentiated carcinoma: group of large cells with considerable variation in nuclear and cellular size; marked hyperchromasia (250×).

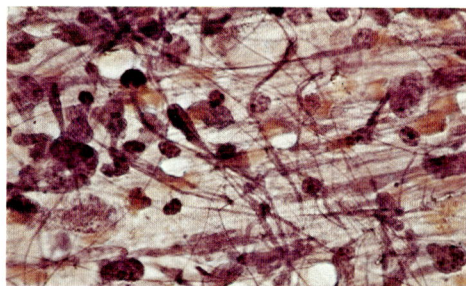

Plate 8.30. Bronchial aspiration: sarcoma: numerous elongated cells with fusiform hyperchromatic nuclei; presence of collagen fibers (250×).

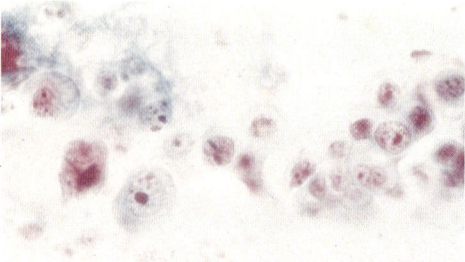

Plate 8.31. Sputum: pulmonary metastases of uterine choriocarcinoma: large malignant elements with round nuclei and prominent nucleoli (Papanicolaou stain, 250×). Courtesy of Dr. Pirson.

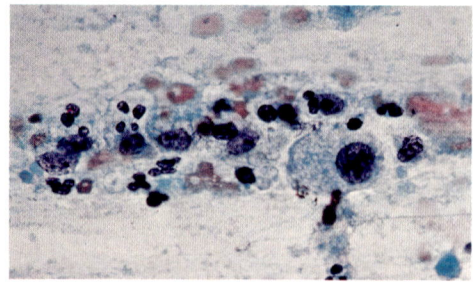

Plate 8.32. Sputum: pulmonary metastases of osteosarcoma: presence of poorly differentiated neoplastic elements with large, irregular, and hyperchromatic nuclei. The origin of the primary tumor cannot be suggested (250×).

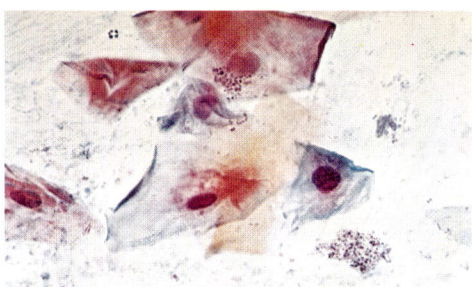

Plate 9.1. Aspiration: esophagus: mycotic esophagitis (250 ×).

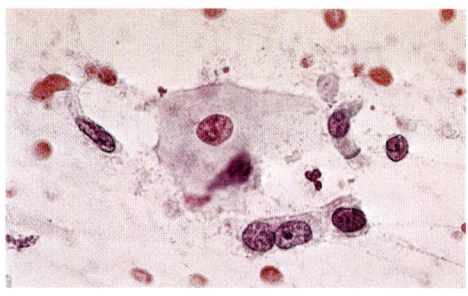

Plate 9.2. Aspiration of lower third of the esophagus (undifferentiated small-cell) squamous carcinoma: benign squamous cell; ciliated gastric cell, anaplastic small carcinoma cells (250 ×).

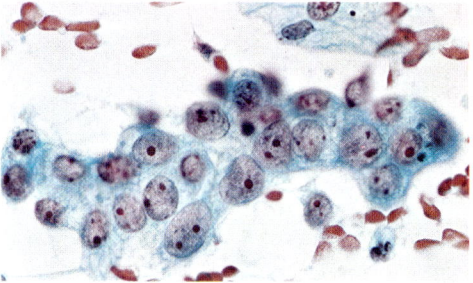

Plate 9.3. Brushing: esophagus, adenocarcinoma: cluster of neoplastic cells showing hyperchromatic nuclei and voluminous nucleoli (250 ×).

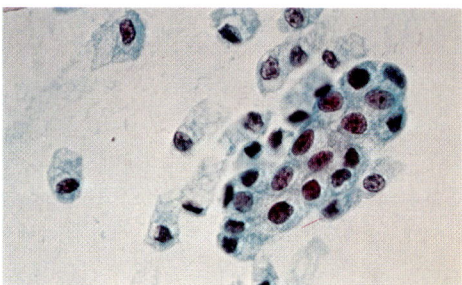

Plate 9.4. Aspiration smear: gastric washing: normal columnar gastric cells (250 ×).

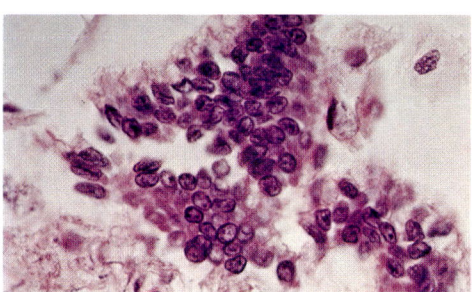

Plate 9.5. Paraffin-embedded specimen: gastric washing: normal columnar gastric cell (250 ×).

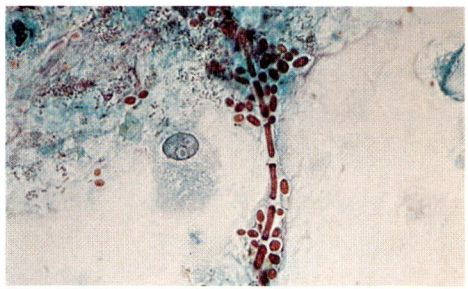

Plate 9.6. Gastric washing: presence of mycotic (Candida) hyphae and spores (250 ×).

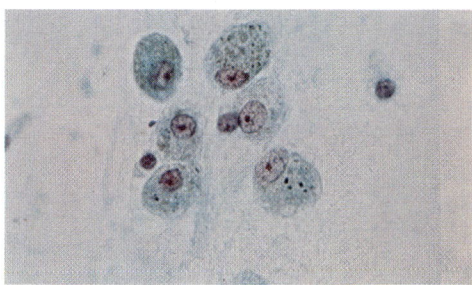

Plate 9.7. Gastric washing: presence of ingested pulmonary macrophages or pneumocytes containing hemosiderin particles in their cytoplasm (250 ×).

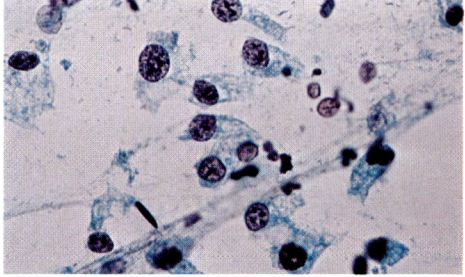

Plate 9.8. Gastric washing: chronic gastritis: abundant desquamation of altered cells; cytoplasmic vacuolation, presence of a pale nucleus in the upper left quadrant (250 ×).

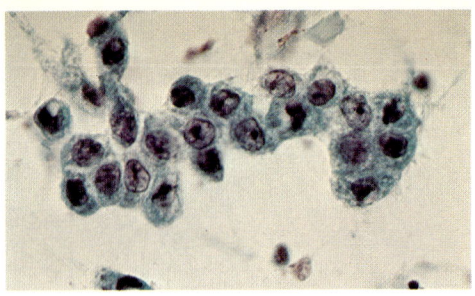

Plate 9.9. Aspiration, gastric washing: atrophic gastritis with intestinal metaplasia: clusters of columnar cells with dense, sometimes vacuolated cytoplasm (250 ×).

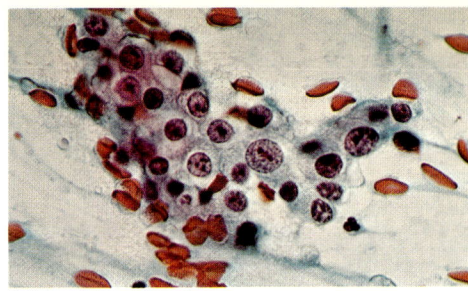

Plate 9.10. Gastric washing: differentiated adenocarcinoma: cluster of regular, columnar cells with round, hyperchromatic nuclei and well-preserved cytoplasm (250 ×).

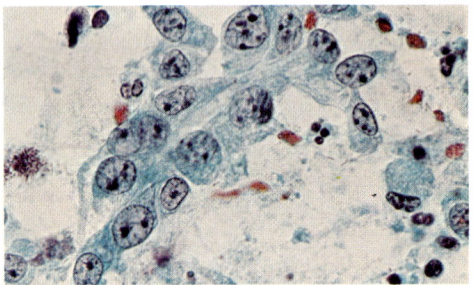

Plate 9.11. Aspiration, gastric washing: poorly differentiated adenocarcinoma: large anaplastic, adenocarcinoma cells. Note the chromatin, the nuclear membrane, and the nucleoli (250 ×).

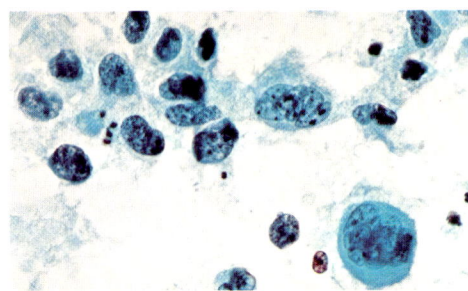

Plate 9.12. Aspiration, gastric washing: undifferentiated adenocarcinoma: anaplastic cells with lobular hyperchromatic nuclei (250 ×).

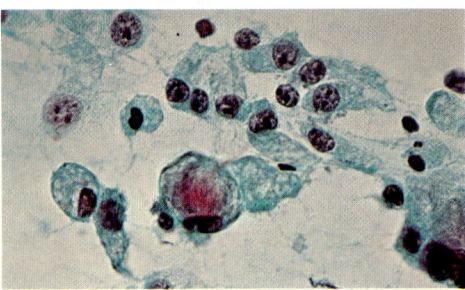

Plate 9.13. Aspiration, gastric washing: mucus secreting adenocarcinoma: note the abundant cytoplasm and the basal nuclei (250 ×).

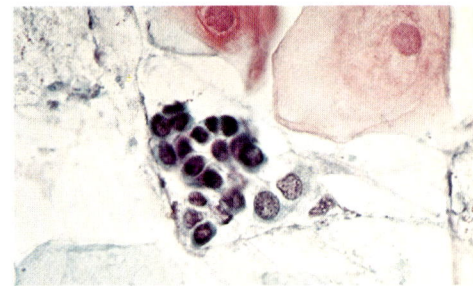

Plate 9.14. Gastric washing: undifferentiated small-cell adenocarcinoma: group of small malignant cells with hyperchromatic nuclei. These cells should not be confused with malignant lymphoma cells (250 ×).

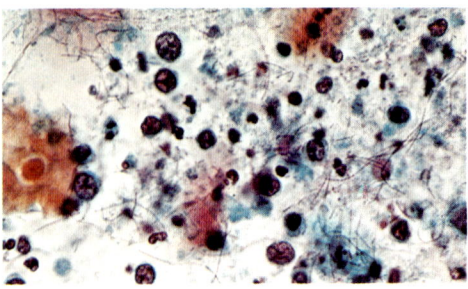

Plate 9.15. Gastric washing: poorly differentiated adenocarcinoma: abundance of dark, naked nuclei. Note the inflammatory infiltrate and the cellular debris (250 ×).

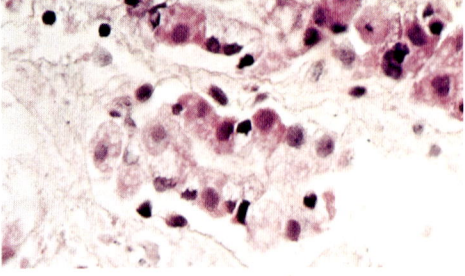

Plate 9.16. Gastric washing: chronic ulcer: cellular alterations due to chronic inflammation and intestinal metaplasia. These cells may be confused with elements of well-differentiated mucous carcinoma (250 ×).

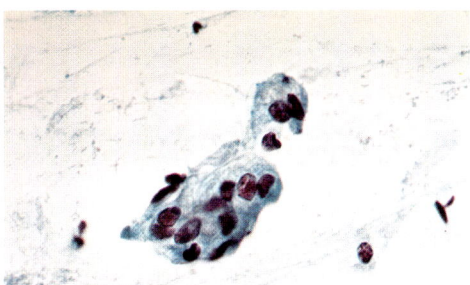

Plate 9.17. Gastric washing: mucous secreting adenocarcinoma: large group of malignant cells suggesting a papillary structure (250 ×).

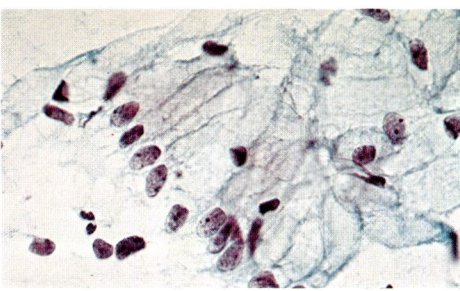

Plate 9.18. Duodenal washing: group of benign columnar cells probably originating from the biliary or pancreatic ducts (250 ×).

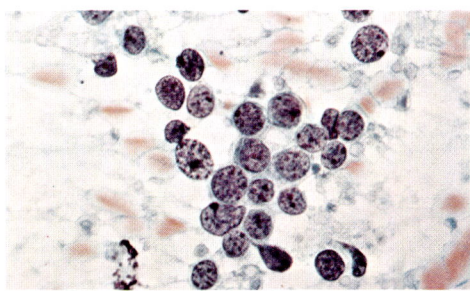

Plate 9.19. Colonic washing: benign adenomatous polyp: elongated columnar cells with an abundant, bulging, clear cytoplasm (250 ×).

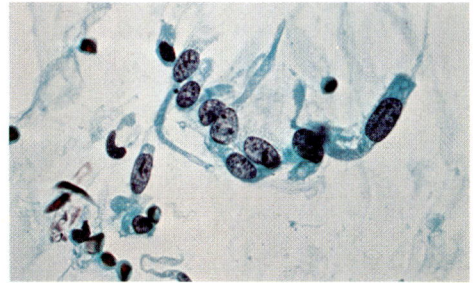

Plate 9.20. Colonic washing: undifferentiated adenocarcinoma: group of undifferentiated columnar cells; numerous naked hyperchromatic nuclei (250 ×).

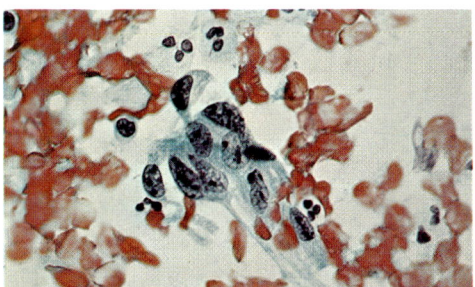

Plate 9.21. Smear: benign adenomatous polyp from the rectum; benign columnar cells (250 ×).

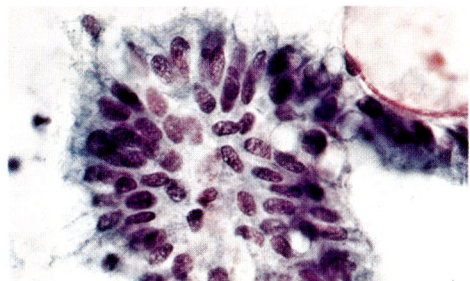

Plate 9.22. Rectal washing: differentiated adenocarcinoma of the rectum: group of elongated, columnar cells with hyperchromatic, irregular nuclei (250 ×).

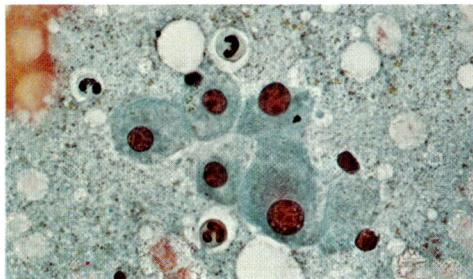

Plate 9.23. Needle aspiration: normal liver cells; note the round, regular nucleus and the homogeneous cytoplasm (250 ×).

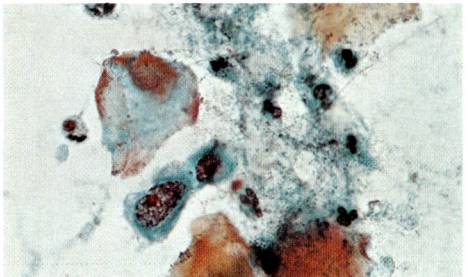

Plate 9.24. Needle aspiration: liver: metastatic poorly differentiated squamous carcinoma of the larynx. Note the presence of malignant cells with irregular hyperchromatic nuclei (100 ×).

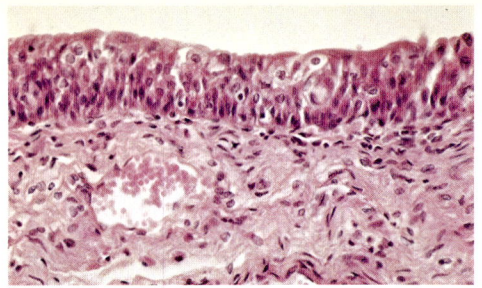

Plate 10.1. Biopsy: urinary bladder: normal urothelium (40 ×).

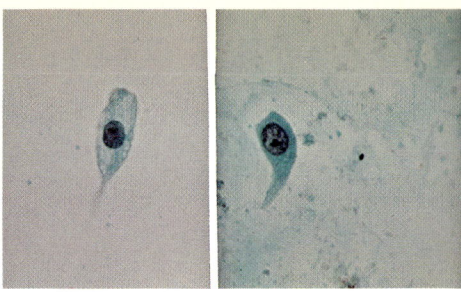

Plate 10.2. Smear: urinary sediment: urothelial cells (250 ×).

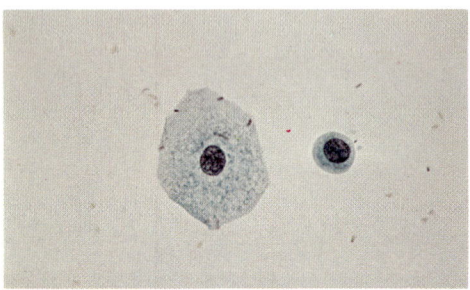

Plate 10.3. Smear: urinary sediment: superficial squamous cell and urothelial cell (250 ×).

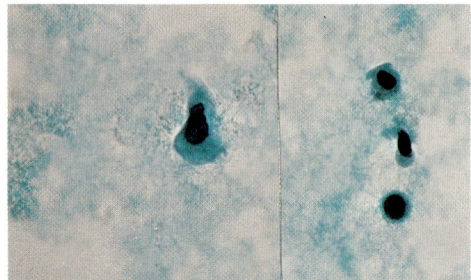

Plate 10.4. Smear: urinary sediment: degenerated epithelial cells (decoy cells) (250 ×).

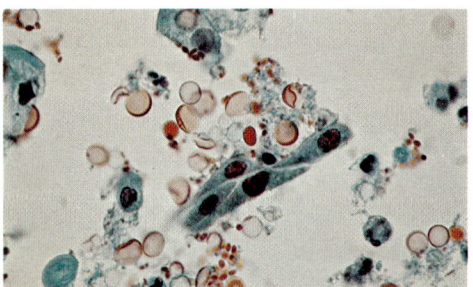

Plate 10.5. Smear: urinary sediment: chronic cystitis: umbrella cells: large, elongated, superficial urothelial cells (250 ×).

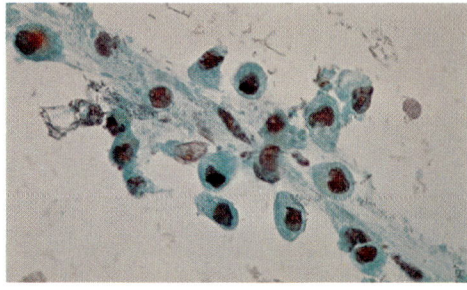

Plate 10.6. Urinary sediment: chronic cystitis: increased desquamation of altered urothelial cells; irregular, dense or hyperchromatic nuclei and vacuolated cytoplasm; abundant bacterial flora (250 ×).

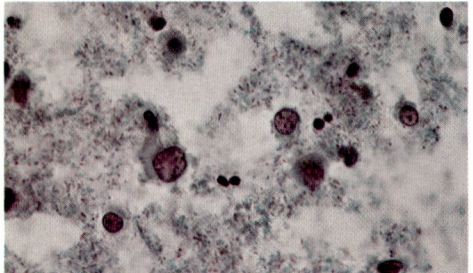

Plate 10.7. Urinary sediment: papilloma of the bladder mucosa: isolated urothelial cells (250 ×).

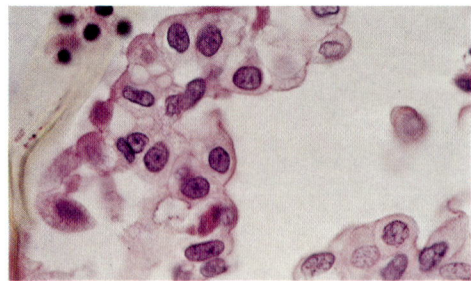

Plate 10.8. Urinary sediment: papillomas of the bladder: well-preserved papillary structures. Note the regular nuclei and the abundant, clear cytoplasm (250 ×).

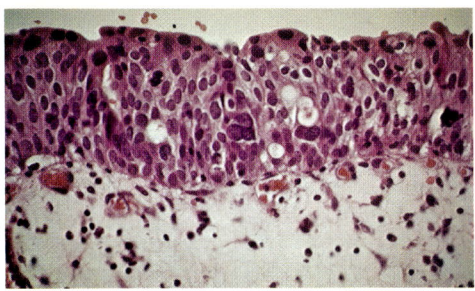

Plate 10.9. Biopsy: urinary bladder, in situ carcinoma (compare with Plate 10.1): thickened urothelium and moderate cellular anomalies; enlarged, regular nuclei and cytoplasmic vacuolation (40 ×).

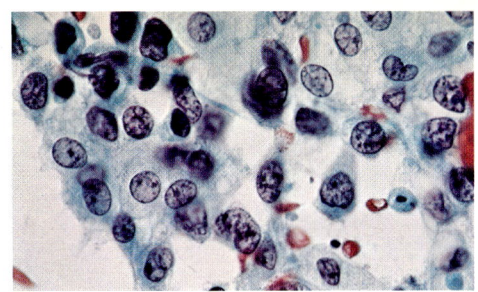

Plate 10.10. Urinary sediment: cluster of differentiated, malignant urothelial cells. Papillary carcinoma, grade I (250 ×).

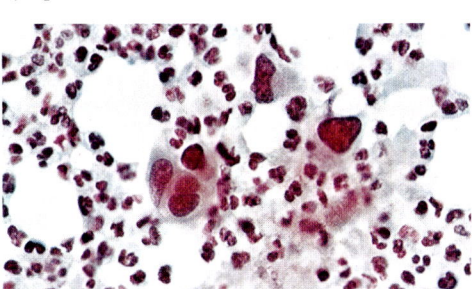

Plate 10.11. Urinary sediment: undifferentiated carcinoma of the urinary bladder: presence of anaplastic cells with dense, irregular, and hyperchromatic nuclei; incomplete cytoplasmic lysis (250 ×).

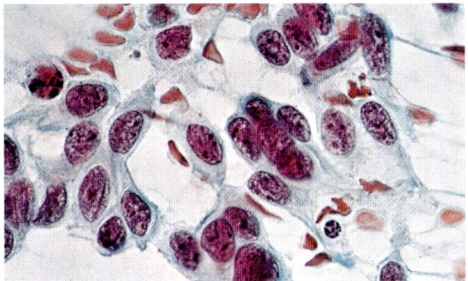

Plate 10.12. Urinary sediment: differentiated carcinomas of the urinary bladder: numerous malignant elongated cells with hyperchromatic nuclei; prominent nucleoli; well-preserved cyanophilic cytoplasm (250 ×).

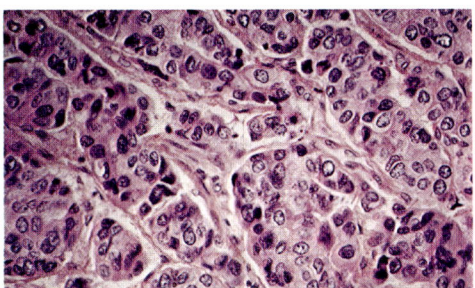

Plate 10.13. Biopsy: urinary bladder: invasive papillary carcinoma, grade III. Note the nuclear abnormalities (100 ×).

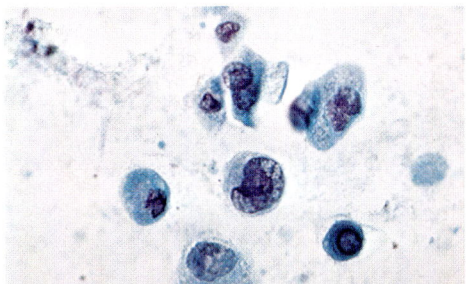

Plate 10.14. Urinary sediment: squamous cell carcinoma: note the presence of malignant, keratinized cells with a dense eosinophilic cytoplasm (250 ×).

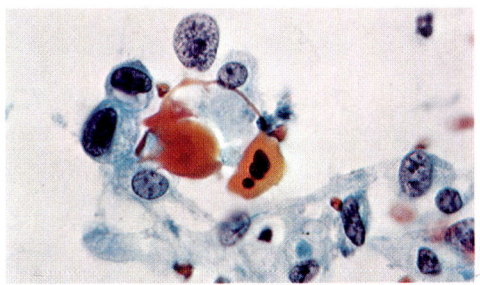

Plate 10.15. Urinary sediment: undifferentiated carcinoma of the bladder: isolated malignant cells with hyperchromatic nuclei and finely vacuolated, cyanophilic cytoplasm (250 ×).

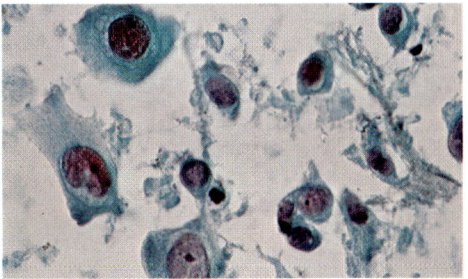

Plate 10.16. Urinary postradiation sediment: poorly differentiated carcinoma of urinary bladder. Note the persistence of malignant cells (250 ×).

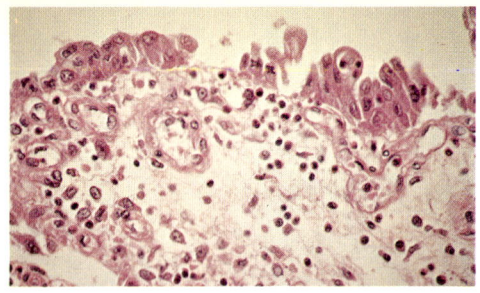

Plate 10.17. Biopsy: adenocarcinoma of the bladder: postradiation lesions of a benign area of the mucosa. Note the marked cellular atypias, the edema, and the inflammatory reaction (250 ×).

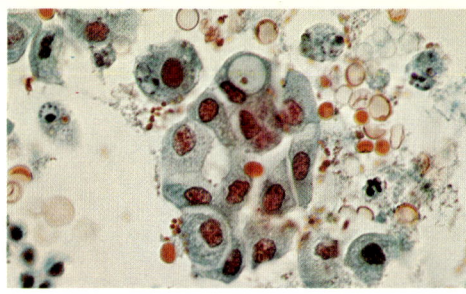

Plate 10.18. Urinary sediment: alkylating agent therapy (Busulfan). Note the cellular anomalies (anisonucleosis and cytoplasmic vacuolation) (250 ×).

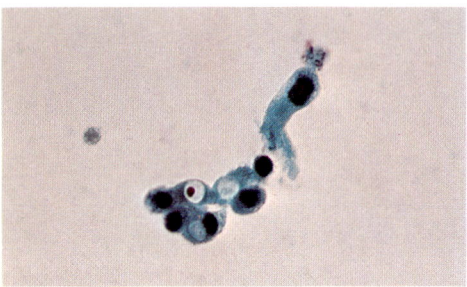

Plate 10.19. Urinary sediment: eosinophilic intracytoplasmic inclusion: a viral origin has been suggested but has not been proven (250 ×).

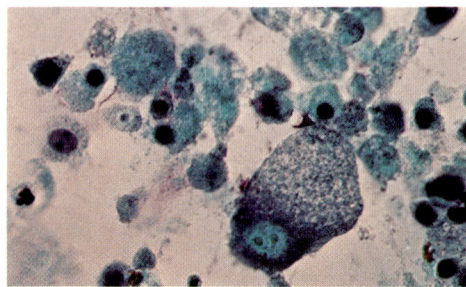

Plate 10.20. Urinary sediment: postradiation lesions of urothelial cells. Note the size of certain cells and the cytoplasmic alterations (250 ×).

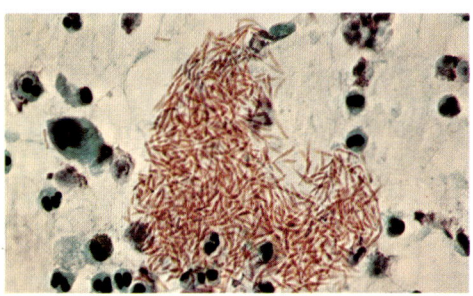

Plate 10.21. Urinary sediment: mycotic infection: note the inflammatory changes observed in epithelial cells. Similar changes may be developed after vesical or prostatic surgical procedures (250 ×).

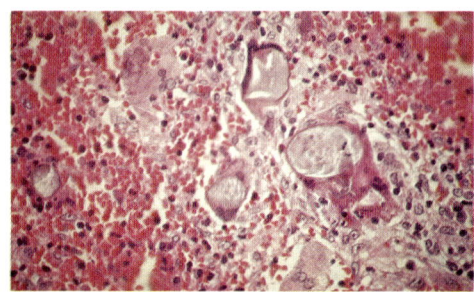

Plate 10.22. Urinary sediment: presence of *Schistosoma hematobium*. Note the presence of altered inflammatory urothelial cells and numerous red blood cells (100 ×).

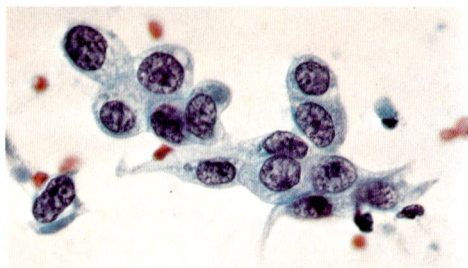

Plate 10.23. Urinary sediment: papillary carcinoma of ureter: urothelial differentiated neoplastic cells (250 ×).

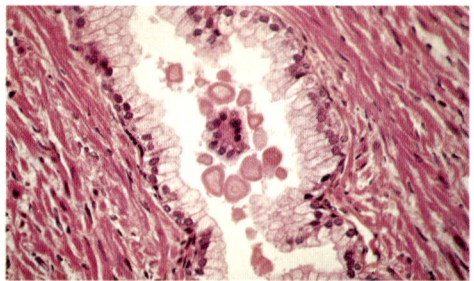

Plate 10.24. Biopsy: prostate: hyperplasia of both the glandular and stromal elements (100 ×).

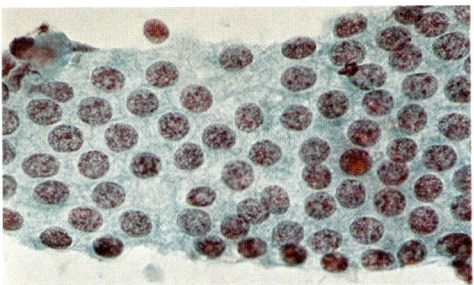

Plate 10.25. Needle aspiration: prostate: hyperplasia or benign hypertrophy; cluster of benign, regular acinar cells (250×).

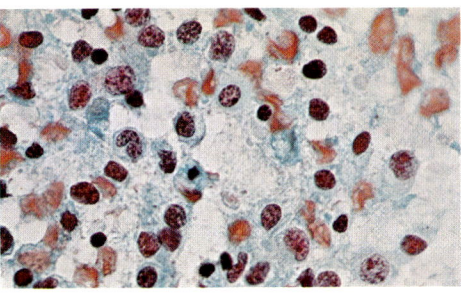

Plate 10.26. Needle aspiration, prostatic adenocarcinoma: isolated, differentiated, malignant cells showing hyperchromatic nuclei and blood elements (250×).

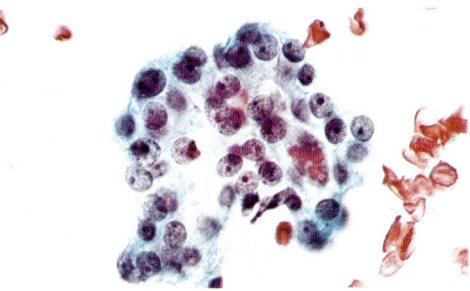

Plate 10.27. Needle aspiration: differentiated prostatic adenocarcinoma: discrete cellular criteria of malignancy. Note the prominent nucleoli (250×).

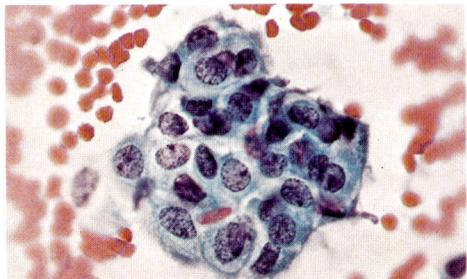

Plate 10.28. Preoperative needle aspiration of a solitary cyst of the kidney: presence of a dense cellular cluster. Note that the nuclei are regular and that the nucleocytoplasmic ratio is normal (250×).

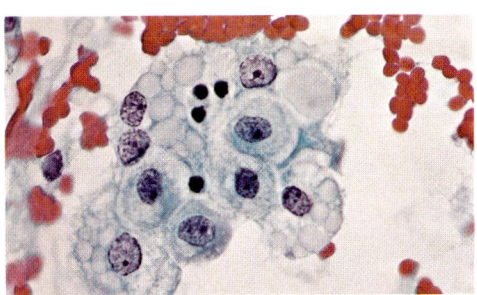

Plate 10.29. Urinary sediment: papilloma of the renal pelvis: presence of large cells with a regular nucleus and a clear cytoplasm. These cells were erroneously diagnosed malignant (250×).

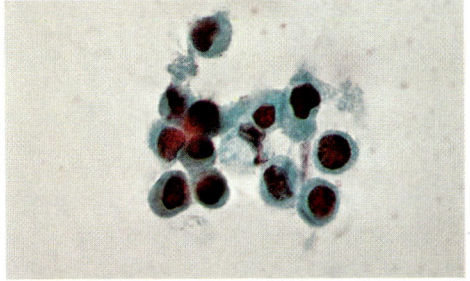

Plate 10.30. Urinary sediment: adenocarcinoma of the kidney: presence of rather poorly preserved malignant cells with dark, homogeneous, degenerating nuclei and cyanophilic cytoplasm (250×).

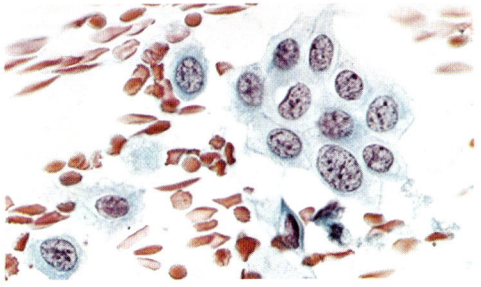

Plate 10.31. Ureteral catheterization: adenocarcinoma of the kidney: presence of differentiated malignant cells with regular nuclei and abundant cytoplasm (250×).

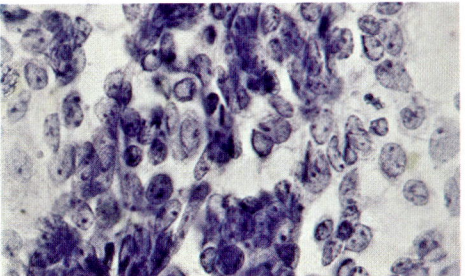

Plate 10.32. Imprint of tumor during operation: adenocarcinoma of the kidney. Note the numerous malignant cells; the cytoplasmic borders are poorly defined (Toluidine blue stain) (250×).

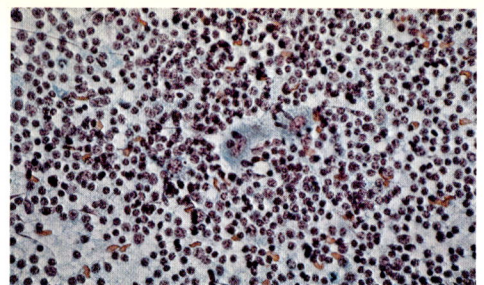

Plate 11.1. Lymph node aspiration: chronic lymphadenitis (Piringer-Kuschinka adenitis) (100×).

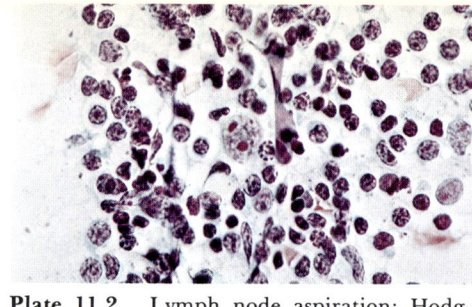

Plate 11.2. Lymph node aspiration: Hodgkin's disease: Hodgkin cells', lymphocytes. Hodgkin cells and Reed-Sternberg cells may be stages of activated lymphocytes, not stages of transformation from histiocytes (250×).

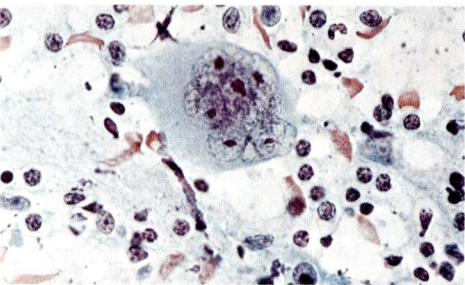

Plate 11.3. Lymph node aspiration: Hodgkin's disease: presence of a Reed-Sternberg cell with a multilobated nucleus, large nucleoli, and a dense cyanophilic cytoplasm (250×).

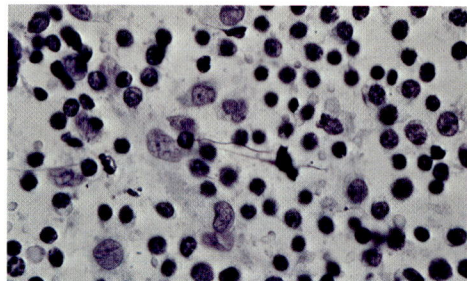

Plate 11.4. Lymph node aspiration: malignant lymphoma; lymphocytic poorly differentiated type (250×).

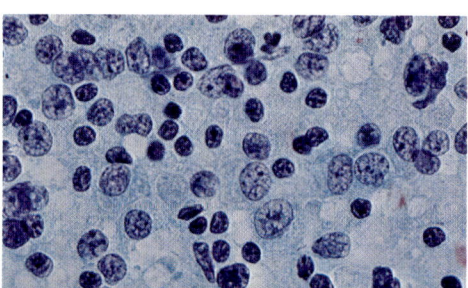

Plate 11.5. Lymph node aspiration: malignant lymphoma, histiocytic type: large noncleaved cells with prominent nucleoli (250×).

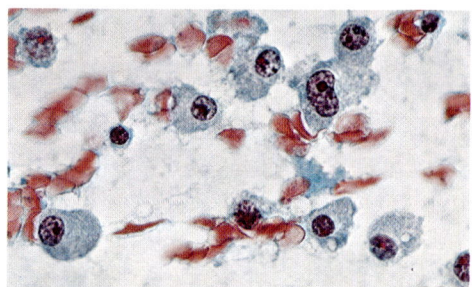

Plate 11.6. Lymph node aspiration: multiple myeloma: plasma cells some of which are immature and atypical. Note the eccentric nuclei and the chromatin structure (250×).

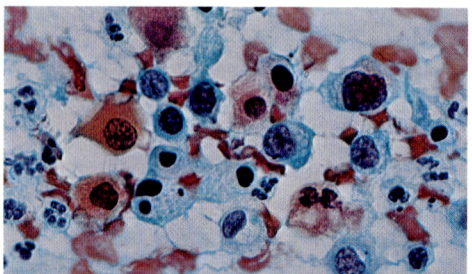

Plate 11.7. Lymph node aspiration: metastatic neoplasm, primary squamous carcinoma of the tongue: presence of keratinized, differentiated malignant cells (250×).

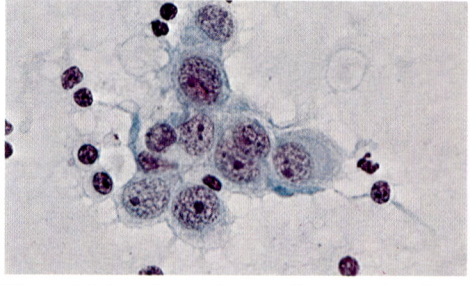

Plate 11.8. Lymph node aspiration: metastatic neoplasm, primary adenocarcinoma of the breast: aggregate of large cells with round, hyperchromatic nuclei and prominent nucleoli (250×).

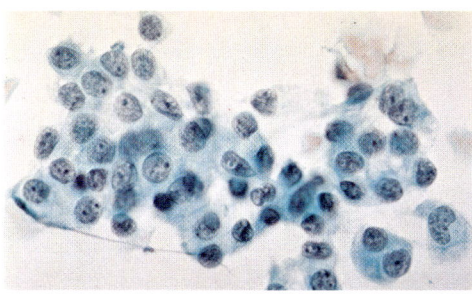

Plate 11.9. Thyroid needle aspiration: small regular cells of the follicle with round nuclei and apparent nucleoli (250 ×).

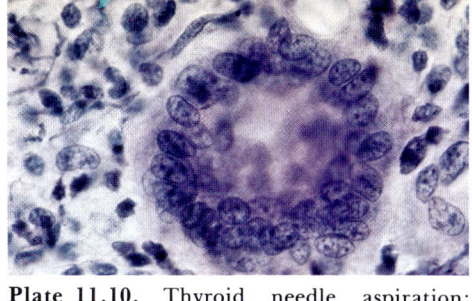

Plate 11.10. Thyroid needle aspiration: granulomatous thyroiditis (de Quervain's thyroiditis): presence of a typical giant cell suggesting a foreign body reaction to the colloid material (250 ×).

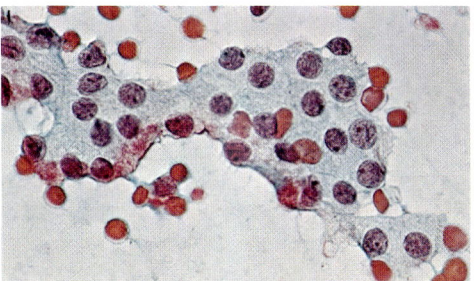

Plate 11.11. Ovarian needle aspiration: struma ovarii, particular type of ovarian teratoma constituted by benign thyroid parenchyma (250 ×).

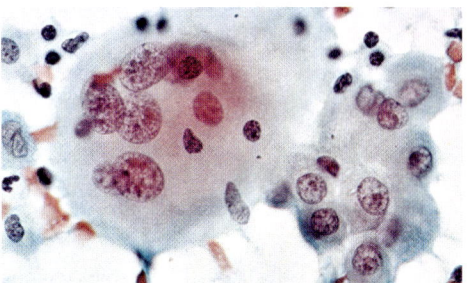

Plate 11.12. Thyroid needle aspiration: papillary carcinoma. Note the presence of slightly enlarged cells; nuclear hyperchromatism is moderate; a pseudo-giant-cell disposition is observed (250 ×).

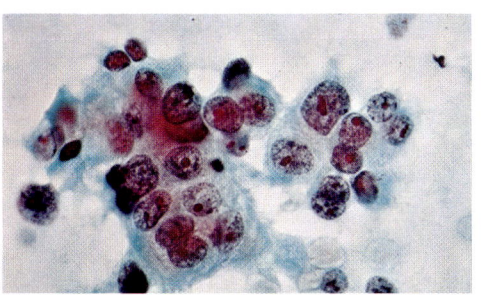

Plate 11.13. Thyroid needle aspiration: poorly differentiated carcinoma: the neoplastic elements have hyperchromatic nuclei and very prominent nucleoli (250 ×).

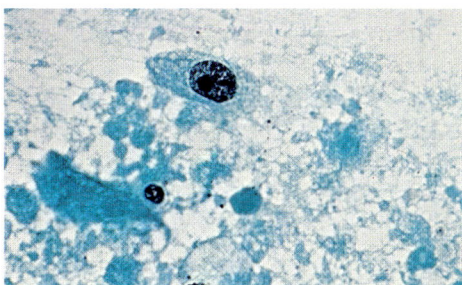

Plate 11.14. Lung needle aspiration: carcinoma, large cell, nonkeratinized type (250 ×).

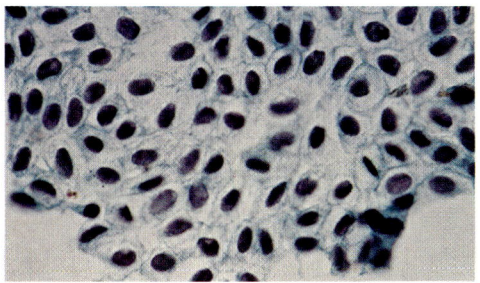

Plate 11.15. Liver needle aspiration: normal liver. Note the morphologic alterations of normal hepatocytes due to fine needle aspiration; the nuclei are markedly elongated (250 ×).

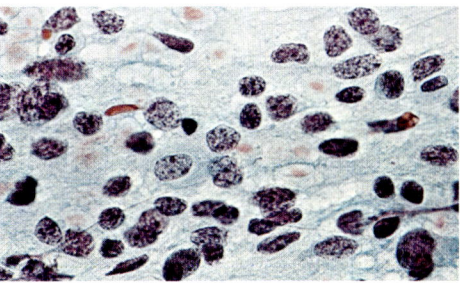

Plate 11.16. Adrenal needle aspiraiton: neuroblastoma: aggregate of small cells with elongated nuclei and finely dispersed chromatin; the cytoplasmic margins are poorly defined (250 ×).

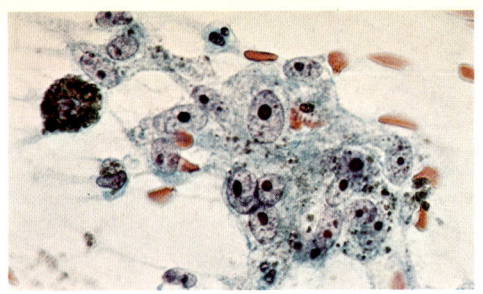

Plate 11.17. Skin nodule aspiration: malignant melanoma: large nucleoli; melanin granules are present in the cytoplasm (250 ×).

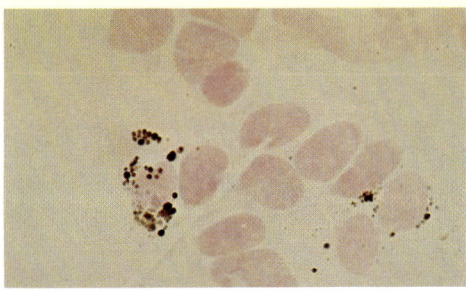

Plate 11.18. Skin nodule aspiration: malignant melanoma: demonstration of melanin with the Fontana-Masson silver method (250 ×).

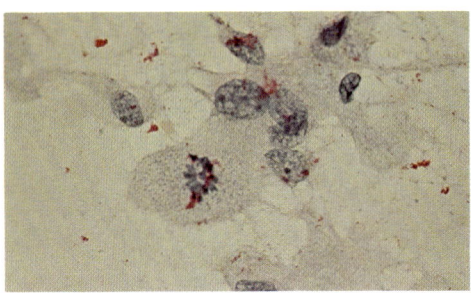

Plate 11.19. Retroperitoneal needle aspiration during operation: pheochromocytoma: ployhedral cells with fine granulations in the cytoplasm. Note the presence of a mitotic figure (250 ×).

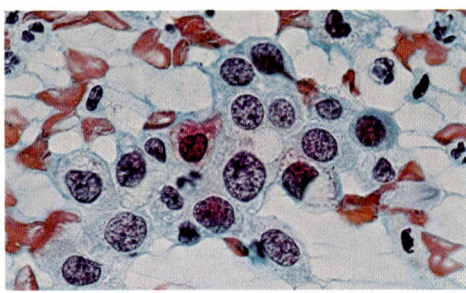

Plate 11.20. Nasal sinus needle aspiration: epidermoid carcinoma. Note the hyperchromatic nuclei and the abundant, well-defined cytoplasmic borders (250 ×).

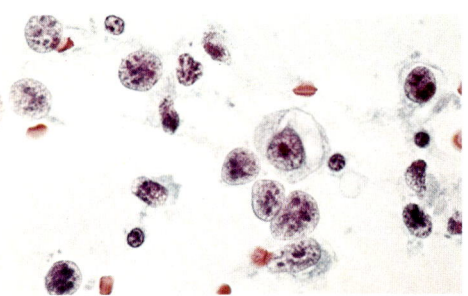

Plate 11.21. Testes needle aspiration: seminoma: isolated, neoplastic cells with poorly preserved cytoplasm; the chromatin is coarsely distributed and nucleoli are prominent; a few lymphocytes are present (250 ×).

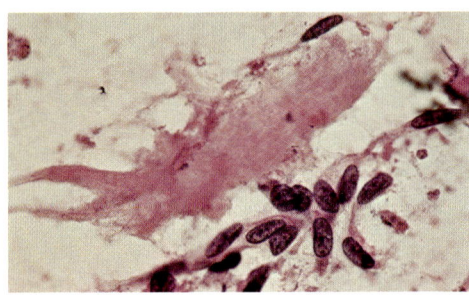

Plate 11.22. Femur needle aspiration: osteosarcoma: spindle-shaped malignant cells and bone tissue (250 ×).

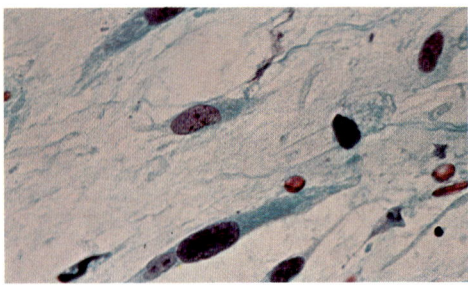

Plate 11.23. Upper extremity of the tibia, needle aspiration: differentiated fibrosarcoma: large elongated malignant cells with nuclear hyperchromatism (250 ×).

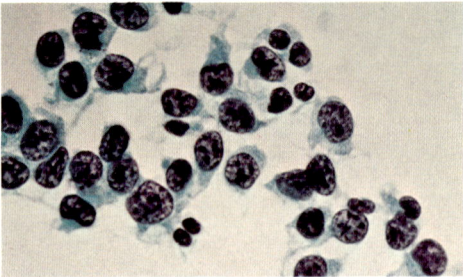

Plate 11.24. Femur needle aspiration: Ewing's sarcoma. Note the uniform cells with hyperchromatic nuclei and little cytoplasm (250 ×).

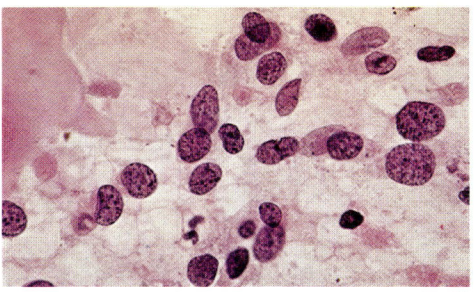

Plate 11.25. Pelvis bone needle aspiration: chondrosarcoma: large neoplastic cells with hyperchromatic nuclei and eosinophilic matrix of hyaline cartilage (250 ×).

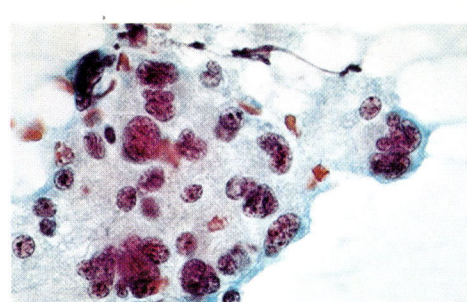

Plate 11.26. Thigh needle aspiration: rhabdomyosarcoma: undifferentiated rhabdomyoblasts, often multinucleated with abundant, finely granular cytoplasm. Cross- striations are not found in these cells (250 ×).

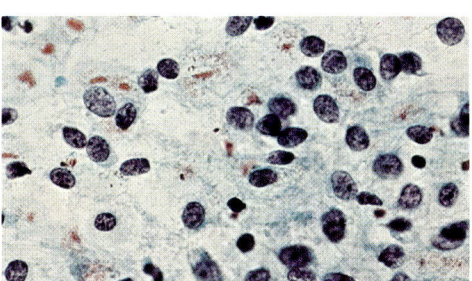

Plate 11.27. Thigh needle aspiration: liposarcoma: immature fibroblasts with hyperchromatic nuclei and poorly defined cytoplasm. Cytoplasmic fat is not always present in undifferentiated tumors (250 ×).

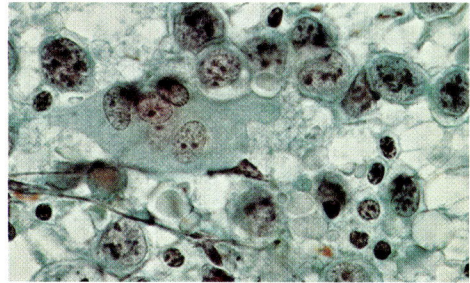

Plate 11.28. Mediastinoscopy: thymoma: presence of pale epithelial cells with large nuclei and rare lymphocytes (250 ×).

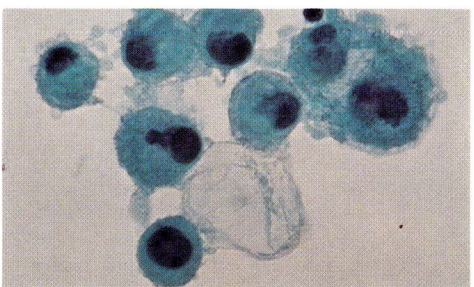

Plate 11.29. Cerebrospinal fluid: breast carcinoma: differentiated epithelial neoplastic cells (250 ×).

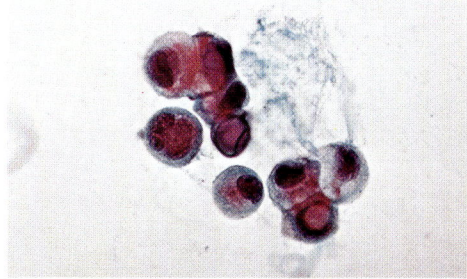

Plate 11.30. Cerebrospinal fluid: gastric carcinoma meningeal infiltrations. Note the presence of large mucus containing cytoplasmic vacuoles (250 ×).

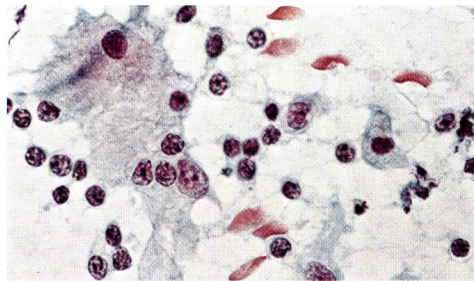

Plate 11.31. Parotid needle aspiration: Warthin's tumor (papillary cystadenoma lymphomatosum). Note the simultaneous presence of large epithelial benign-looking cells and of lymphocytes (250 ×).

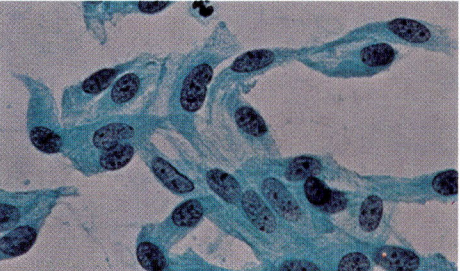

Plate 11.32. Parotid needle aspiration: mixed tumor: presence of large epithelial cells with regular round nuclei (250 ×).

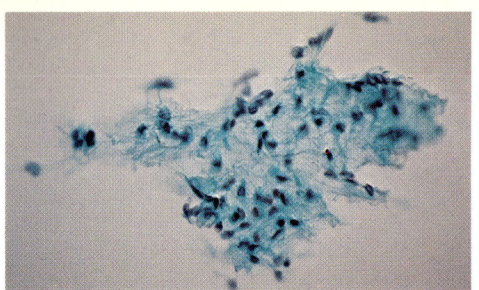

Plate 11.33. Parotid needle aspiration: mixed tumor: stromal elements retaining a mucoid character (100×).

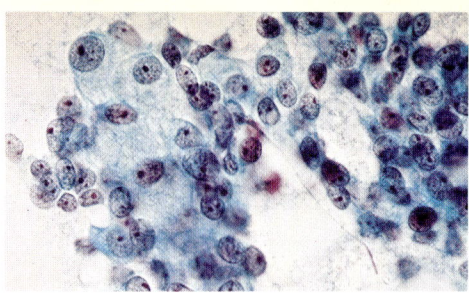

Plate 11.34. Parotid needle aspiration: adenocarcinoma: epithelial malignant-looking cells with hyperchromatic nuclei and prominent nucleoli (250×).

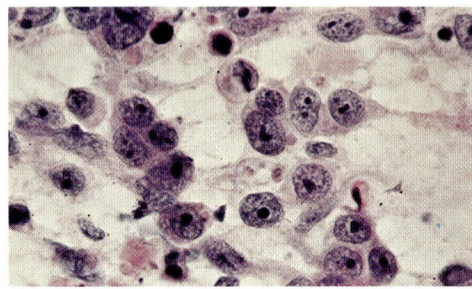

Plate 11.35. Submaxillary needle aspiration: adenocarcinoma: clusters of epithelial malignant cells with regular, hyperchromatic nuclei. Note the piling-up of malignant cells (250×).

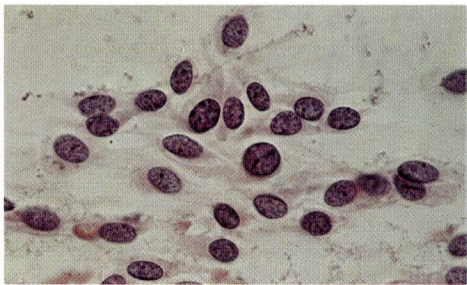

Plate 11.36. Orbital needle aspiration, presumably conjunctival differentiated carcinoma (250×). Courtesy of Dr. Malvi, Bombay.

5
Gynecologic Cytology

EMBRYOLOGY

The various relationships of the different constituents of the genitalia are best understood by considering their embryogenesis (Figure 5.1). While the vulva arises from ectodermal genital eminence, the inferior portion of the vagina derives from the urogenital sinus, which is endodermal in nature. The superior portion of the vagina, the cervix, the corpus of the uterus, and the fallopian tubes arise from the Mullerian ducts and surrounding mesenchyme.[204] These Mullerian ducts are formed from a groove in the coelomic mesothelium which is later closed off to form a tube. The ovaries arise from a thickening of the same coelomic epithelium as well as from the surrounding mesenchyme. Later, primitive sex cells migrate from the allantois to the coelomic fold.

HISTOLOGY OF THE VULVAR, VAGINAL AND EXOCERVICAL MUCOSAE

The vulvar, vaginal, and cervical mucosae (pars vaginalis) are covered with stratified squamous epithelium (Plate 5.1). As we shall see, these mucosae all react, to varying degrees, to hormonal stimulation by the steroids.[94, 97] The superior third of the vaginal mucosa is the best location for cell sampling and evaluation of such hormonal influence. However, it has been established that cervical and vulvar smears (medial face labium minus) can also be of value.

The influence of steroid hormones on the genital mucosae depends on their binding to intracellular receptors. The quality and quantity of hormone, the density of hormone receptors, and the stability of the hormone-receptor complexes are factors that explain differences in cellular response to stimulation. Once binding is accomplished, the hormone-receptor complex is transported into the target cell nucleus where it exerts its influence on gene activity. The cell responds by increased RNA and protein synthesis.

Stratified squamous epithelium includes many cellular layers: a basal layer formed by deep basal and parabasal cells, an intermediate layer, and a superficial layer, both several cells thick. The epithelium lies on a basement membrane which separates it from the connective tissue stroma. In reality, this basement membrane is composed of

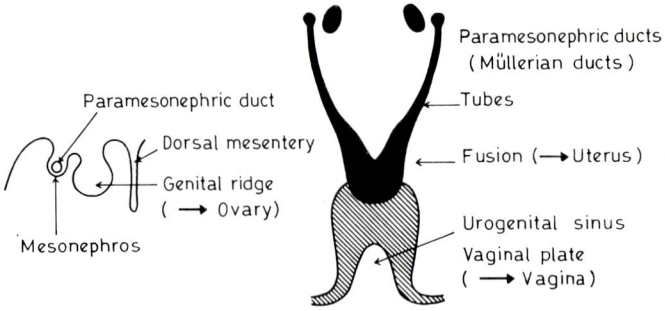

Figure 5.1. Embryology of the female genital tract.

three consecutive layers as shown by the electron microscope: the basal membrane of the epithelial cell, a 400 Å thick, less electron-dense zone and a thick basal plate of around 700 Å (basal lamina) formed by collagen filaments. The underlying loose connective tissue is formed by elongated cells and their intercellular secretions, collagen and reticulin fibers, and vascular and nervous elements.

The stratified squamous epithelium undergoes modifications in structure and function depending on the hormonal stimulation it receives (Figure 5.1). Thus, during the period of sexual activity, the histologic aspect of the epithelium will depend on the dominant hormonal activity.[129] Estrogen stimulation, for example, produces a thickened epithelium, all the layers of which are well developed, especially the superficial layer, which is rich in eosinophilic cells with pycnotic nuclei (Table 5.1). Progesterone stimulation favors the appearance and development of the intermediate, glycogen-rich layers.[61] Androgen stimulation also favors the growth of the intermediate layer but stimulates glycogen production to a lesser degree.

In the absence of genital hormonal stimulation (natural or artificial menopause, premenarchal period) the epithelium is reduced to a few basal layers and some deep intermediate cells.

While biopsy of the stratified squamous epithelium allows an evaluation of hormonal activity, the cytologic method is the technique of choice; it is simpler, faster, painless, and effective.

Table 5.1. Hormonal Administration in a Menopausal Patient with Atrophic Vaginal Smear

Estrogens	*Progesterone*	*Androgens*
nonspecific proliferation (intermediate cells)	nonspecific proliferation (intermediate cells)	nonspecific proliferation (intermediate cells)
maturation (superifical eosinophilic cells)	maturation glycogen rich (intermediate cells)	maturation glycogen rich (intermediate cells with vesicular nuclei)

HISTOLOGY OF THE ENDOCERVICAL MUCOSA

The endocervical mucosa is lined by a simple columnar mucosecretory epithelium (Plate 5.11). It is complemented by a system of glands formed by the same type of epithelium, branching into vascular connective tissue stroma. The junction of the cervical stratified squamous and columnar epithelia normally occurs at the external orifice of the cervix. This junction may constitute an abrupt or a gradual transition (Plate 5.10).

CYTOLOGY OF THE VULVAR, VAGINAL, AND EXOCERVICAL MUCOSAE

The stratified squamous epithelium of the mucosae exhibit three cell types: basal, intermediate, and superficial cells. (Figure 5.2).

Basal and Parabasal Cells

The deep or basal germ cell is rarely apparent on a smear unless a rather brutal scraping was performed on an atrophied mucosa or if erosion had occurred. Parabasal cells desquamate in sheets. They are round elements measuring 15 to 25 μ in diameter and have large nuclei with finely dispersed chromatin and prominent nucleoli. The cytoplasm is cyanophilic and has well-defined borders (Plates 5.4 and 5.5).

Intermediate Cells

The typical intermediate cell measures about 30 μ in diameter and, compared with the basal cell, has a smaller nucleus and a more voluminous cytoplasm because it possesses a high concentration of glycogen, especially during pregnancy; the cytoplasm assumes a characteristic boat shape—thus the term navicular cell. The navicular cell exhibits a clear cytoplasm delineated by a relatively dense border formed by organelles that have migrated toward the periphery. A progressive decrease is observed in the nucleocytoplasmic volume ratio as the cells approach the superficial layers (Plate 5.3).

Superficial Cells

The superficial cells represent the final step in the maturation of the stratified

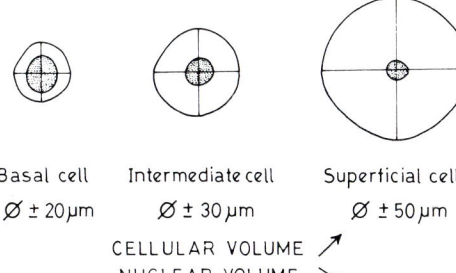

Basal cell   Intermediate cell   Superficial cell
$\emptyset \pm 20 \mu m$ $\emptyset \pm 30 \mu m$ $\emptyset \pm 50 \mu m$

CELLULAR VOLUME ↗
NUCLEAR VOLUME ↘

Figure 5.2. Mean cellular and nuclear volumes of squamous cells.

squamous epithelium. These are large cells, measuring from 40 to 50 μ in diameter, whose small nuclei progressively lose their internal structure to yield a pycnotic, homogeneous mass measuring 5 to 6 μ in diameter (Plate 5.2). The superficial squamous cell always conserves its pycnotic nucleus except in certain pathologic circumstances, such as prolapsus. The cytoplasm is voluminous compared with the size of the nucleus and is either cyanophilic or, at later stage, eosinophilic. The presence of keratin precursors, which are in effect basophilic proteins, explains the cytoplasmic cyanophilia. The transformation of these precursors into keratin, as well as the pH of the fixative agent, account for the change to eosinophilia. Basophilic keratohyalin granules are sometimes visible in the cytoplasm.

The superficial cells desquamate in sheets or as isolated cells, the latter case being restricted for the most mature elements. This mode of desquamation is explained by the disappearance of the desmosomes, whose role is to maintain cellular cohesion. The mode of desquamation is therefore an indication of the degree of cellular maturation. All squamous cells are not dead at time of exfoliation.[103]

CYTOLOGY OF THE ENDOCERVICAL MUCOSA

The columnar cells of the endocervical mucosa are elongated elements measuring from 25 to 30 μ in height. The nucleus, which is also elongated, has a greater diameter of about 7 μ and lies in the long axis of the cell. These cells desquamate in sheets or as isolated elements. In the former case, they assume a characteristic honeycomb pattern (Plate 5.13) or appear side by side in a row, depending on the incidence (Plate 5.12). The basophilic cytoplasm may contain a large mucous droplet which gives it a voluminous bloated appearance and the apical pole, when clearly visible, exhibits a dilated border. The nuclear chromatin appears abundant and disposed in more voluminous clumps than in the squamous cell.

The desquamated cells, which are often accompanied by mucous filaments, may be found in vaginal smears but are more numerous in cervical scrapings, particularly when a ectropion exists. They are more fragile than the squamous cells, and they often show signs of degeneration: cytoplasmic vacuolation, poorly delineated cellular borders, disappearance of cilia, and nuclear pycnosis. Their desquamation is increased during the postovulatory phase and is minimal during the follicular phase.

THE CONNECTIVE TISSUE CELLS
OF THE VAGINAL AND CERVICAL STROMA

The stratified squamous or columnar epithelium lies on a connective tissue stroma whose cellular elements may appear on smears in certain pathologic conditions: chronic cervicitis (Plate 5.15), erosion of the epithelium, destruction of the surface layers by tumors, or atrophy caused by radiation. Desquamation occurs either in sheets or isolated cells. These stromal elements, which are smaller than the epithelial cells, have oblong nuclei with dense chromatin. Their cyanophilic cytoplasm often has poorly defined contours.

OTHER CELLULAR ELEMENTS

The epithelial cells of genital smears may be accompanied by other cellular elements, whose presence may be of diagnostic significance. These elements include:

- Red blood cells appear as round or oval anucleate masses. They measure approximately 6 to 7 μ in diameter and are eosinophilic. Their appearance on smears taken during the menstrual phase is of course normal, and their presence also may be due to an excessively vigorous scraping. Other than in these circumstances, however, their presence must be considered pathologic.
- Neutrophilic polymorphonuclear leukocytes are always present in cervical smears and in endometrial aspirations; they may be rare or absent in vaginal smears. The number of these leukocytes increases considerably in cases of infections, erosions, and tumors. Eosinophils are also observed.
- Lymphocytes are normally observed less frequently than the other leukocytes.[41] Their numbers increase, however, in certain pathologic conditions.
- Histiocytes are mobile connective tissue cells having acquired the capacity of phagocytosis. They are slightly larger than leukocytes (12 to 15 μ in diameter).[87] The nucleus is often kidney-shaped, and its chromatin is arranged in fairly voluminous, irregular masses. A nucleolus is visible. The cytoplasm is basophilic and finely vacuolated; the cellular contours are more or less preserved (Plate 5.14). We frequently find these cells in small clusters, especially in inflammatory conditions, and the cytoplasm may contain cellular debris or pigments that are indicative of macrophagia. Some multinucleated and giant forms are observed in certain inflammations, particularly after menopause, and in erosive tumors. The nuclei are small in smears taken following radiation therapy.

CONTAMINATES

Among the contaminates that should be recognized are spermatozoa, polygonal crystals of talc, vegetable fibers, and pollen and fungi which may be present in staining solutions.

HORMONAL CYTOLOGY

Morphologic Criteria of Hormonal Evaluations

The different hormone-induced states of the individual's sexual evolution are reflected in the appearance of vaginal smears. Certain criteria permit an evaluation of such hormonal activity. These criteria are:

- The type of desquamate cells: superficial, intermediate, or parabasal. These cellular types have been defined previously. Examination of the smear at low magnification allows the determination of the dominant cell type.
- The mode of cellular desquamation: isolated cells or sheets. The mode varies,

depending on the cellular layers in question and on the presence or absence of desmosomes (cell cohesiveness). For example, the integrity of the desmosomal apparatus may be strongly altered during inflammation of the epithelium.

• The quantitative evaluation of nuclear pycnosis and cytoplasmic eosinophilia: karyopycnotic index, eosinophilic index, maturation index. The karyopycnotic index or pycnotic index (K.I.) is the percentage of superficial eosinophilic and cyanophilic cells having a nucleus whose diameter is less than 6μ. The eosinophilic or acidophilic index (E.I.) is the percentage of superficial eosinophilic cells. It is generally considered sufficient to count 200 cells to establish these indices. The eosinophilic index and the pycnotic index constitute a quantitative evaluation of estrogen stimulation. They do not encompass the percentage of parabasal cells, as do the maturation index and the estrogenic index. The maturation index (M.I.) simply represents the count of different cell types and their expression as percentages (for example, parabasal cells 20 percent; intermediate cells, 40 percent; superficial cells, 40 percent: M.I. 20/40/40).[81]

The maturation value is the count of different normal cells and their assortment into five types. The count of each type is multiplied by a determined index and the result is a number between 0 and 100.

eosinophilic superficial cells	index 1
cyanophilic superficial cells	index 0.8
large intermediate cells	index 0.6
small intermediate cells	index 0.5
parabasal cells	index 0

The maturation value indicates the proportions of these five cell types. Nonhormonal cellular alterations are not considered here, since only normal cells are counted.

To compare these values to results of other workers, Meisels has proposed a simplification of the maturation value considering only three cellular types: superficial cells (index 1), intermediate cells (index 0.5), and parabasal cells (index 0).[80]

While these different modes of quantitative evaluation can provide only relative indications, they nonetheless remain clinically useful. Such indices may vary widely however without being necessarily indicative of pathologic phenomena. Also, technical errors of fixation and staining may lead to mistakes in the evaluation of eosinophilia. Such evaluations are time consuming because the calculations must be based on at least 200 cells. We must point out, however, that associated characteristics such as the presence of leukocytes, red blood cells, histiocytes, bacterial flora, and mucus can strongly alter functional cytologic appearances. This is also the case for any inflammatory, degenerative, or tumoral phenomenon. Under such circumstances, therefore, one should not attempt evaluations of hormonal activity.

Once evaluated by cytologic means, the degree of hormonal activity is established by correlation with clinical facts such as age, the day of the cycle, hormonal therapy if in effect, and the state of ovarian function.

Quantitative evaluations of plasmatic steroid hormone levels, of course, can be more accurately performed by biochemical and radioimmunologic methods.[106]

Hormone-induced
Vaginal Cytologic Modifications

The administration of genital hormones to females with vaginal mucosal atrophy has afforded a better understanding of the specific effects of each type of hormone [129] (Figure 5.3). The estrogens provoke proliferation and maturation of the epithelium, characterized by the appearance on smears of isolated, eosinophilic superficial cells with pycnotic nuclei. Progesterone exerts a proliferative action that favors intense desquamation of intermediate cells including navicular cells. Thus, the essential difference between estrogenic and progesteronic action is that the latter provokes early desquamation of cells of the intermediate level, before they have reached the ultimate stage of maturation. Under normal circumstances, the estrogens and progesterone have a synergetic action on epithelial proliferation. Thus, the administration of progesterone in small doses does not produce an antiestrogenic effect; in large doses, however, the effects of the progesterone predominate over those of estrogen. This antagonistic effect can be explained by a reduction of the cytoplasmic estrogen receptors and, subsequently, a diminution of estrogen binding on the cellular receptors.

The administration of androgens induces an epithelial proliferation of the glycogenrich and intermediate layers. Under these conditions, a smear would contain clusters of cyanophilic parabasal and intermediate cells. One characteristic particularly is that under androgen stimulation, the nuclei become voluminous, and since the chromatin is still finely dispersed, such nuclei have a clear "hypochromatic" appearance (Plate 5.9). Much controversy has been raised about the existence of an "androgenic" cell. We think that this cell does indeed exist and is a large parabasal cyanophilic cell with a large, vesicular nucleus. The chromatin has a fine typical pattern. Active protein synthesis may explain this pattern. The massive and prolonged administration of androgens leads to a state of exhaustion of the mucosa; smears will then exhibit a relative increase in the percentage of parabasal cells. Most of the cells have a pale, glycogenrich, cyanophilic, vacuolated cytoplasm and a regular nucleus. Leukocytes are scarce.

The corticosteroids provoke a proliferation of the parabasal and intermediate cells. On smears, this is reflected by sheets of cyanophilic cells more or less resembling those

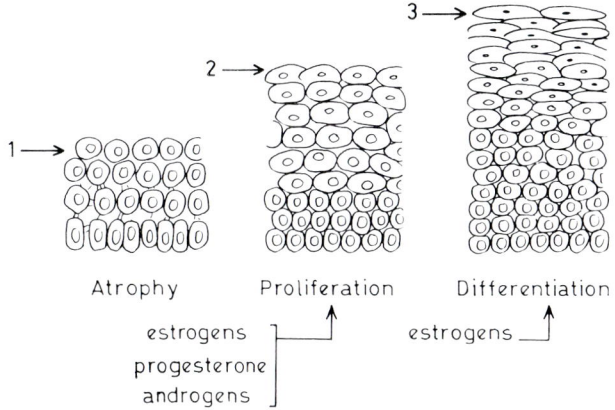

Figure 5.3. Specific effects of different hormone on the squamous mucosa.

appearing under androgenic stimulation. Again, one observes pale and voluminous nuclei.

As one would expect, the simultaneous administration of several hormones yields still different results. For example, small quantities of estrogens administrated after progesterone favor the appearance of the luteal aspect of the smear; the administration of testosterone produces an antiestrogenic effect characterized by a diminution of eosinophilic superficial cells with pycnotic nuclei. A summary of the results of such multiple hormonal stimulation is provided in Table 5.2.

We must point out that the administration of certain substances such as digitalis and tetracycline can falsify hormonal evaluations because of their effects on the vaginal mucosa. Digitalis induces maturation of the squamous epithelium while local administration of the tetracycline (suppositories) increases the desquamation of the epithelium. Finally, it must not be forgotten that inflammation disturbs the cellular ratios and invalidates any functional interpretation.

To summarize, cytology furnishes a rapid and inexpensive approach to functional genital hormone evaluations. Schneider et al.[108] have investigated the relationship between the plasma 17-β- estradiol and hormonal cytology: the maturation value and the karyopycnotic index are qualitative indicators of plasma 17-β- estradiol. Cytologic evaluation, however, does not match the precision of biochemical analyses. Even the utilization of the eosinophilic, pycnotic, and maturation indices does not appreciably increase the precision of the cytologic method in this respect. Therefore, cytology should be relied on only as a preliminary indicative tool that can tell the clinician when he should turn to biochemistry.

Cytology of the Menstrual Cycle

A cyclic evaluation of vaginal smears shows modifications in the number and type of

Table 5.2. Cumulative Effects of Hormonal Administration

Type of Smear	Estrogens	Progesterone	Androgens
Estrogenic type	accentuation of maturation then exhaustion phase with decrease in superficial eosinophilic pycnotic cells, decreased K.I. and E.I.	decreased maturation with diminution of the K.I. and E.I., increased desquamation of intermediate cells	antiestrogenic effect, decreased maturation of the K.I. and E.I., desquamation of intermediate cells but cluster formation is decreased compared with progesterone administration
Luteal type	estrogenic maturation, increased K.I. and E.I.	increase in intensity of progesterone-type image then exhaustion with decrease in intermediate clusters	increase in navicular cells with pseudo-pregnancy image
Androgenic type	increased androgenic maturation	increased proliferation and desquamation of clusters	reactional exhaustion

cells desquamating, the modes of desquamation, and the presence of nonepithelial elements (leukocytes, histiocytes, red blood cells). These modifications are the result of the fluctuating estrogenic and luteal stimuli of the normal menstrual cycle. We will consider six successive phases beginning at the arrest of menstrual bleeding.

Early Estrogenic Phase (Days 4 to 9)
A smear taken during this period contains sheets of cyanophilic intermediate and superficial cells with relatively voluminous nuclei. Eosinophilia and nuclear pycnosis, less prominent at first, increase progressively (30 percent in the case of eosinophilia; 50 to 60 percent for pycnosis). Leukocytes and histiocytes, first abundant, become rare, and red blood cells disappear. Mucus is scarce.

Advanced Estrogenic Phase (Days 10 to 14)
The percentage of isolated superficial cells increases with respect to superficial and intermediate cell sheets. Eosinophilic cells become more numerous (30 to 50 percent), and nuclear pycnosis increases to between 60 and 80 percent. Leukocytes are rare, and this gives the smears a "clean" look.

Ovulatory Phase (Days 14 and 15)
The eosinophilic superficial cells with pycnotic nuclei reach their highest proportion and constitute the majority of the cellular elements present on the smear. Leukocytes are very rare and mucus is absent. A few red blood cells may be found. These result from the transitory hemorrhage of the endometrial mucosa in midcycle.

Early Luteal Phase (Days 16 to 23)
One observes a diminution in the number of eosinophilic superficial cells with pycnotic nuclei and the reappearance of sheets of cyanophilic superficial and intermediate cells. Mucus and a few leukocytes reappear. In the sheets of intermediate cells, the presence of navicular-type elements (glycogen-rich cells) is observed.

Advanced Luteal Phase (Days 24 to 26)
Sheets of cyanophilic superficial and intermediate cells predominate, and the navicular cells are now relatively numerous. The proportion of eosinophilia and pycnosis tends to level off. The number of leukocytes increases and mucus is relatively abundant. At the end of the cycle, a few red blood cells may be present.

Menstrual Phase
The smear is rich in red blood cells, leukocytes, and histiocytes, while the epithelial elements are represented by sheets of cyanophilic superficial and intermediate cells. At the end of the menstrual phase, there is an increase in the eosinophilic and pycnotic indices. Endocervical columnar cells and endometrial epithelial and connective tissue cells are also present.

Physiologic Hormonal Insufficiency (Prepubertal and Postmenopausal Periods)

Prepubertal Period (Infancy and Adolescence)
For a few days following birth, scrapings of an infant's vagina contain superficial and intermediate cells that reflect maternal hormonal stimulation.[37,68] Towards the fifth

day, traces of such stimulation disappear and vaginal smears reveal only a discrete hormonal activity which continues until puberty. This is cytologically translated into sheets of parabasal cells and, to a certain degree, of intermediate cells. Superficial cells are practically absent.

One may observe the excessive cellular manifestations of hormonal activity, notably estrogenic, when hormones of this type are administered to the mother during pregnancy. Also, such images of mucosal hyperactivity have been found in newborns of mothers having received diethylstilbestrol.

As one would expect, the hormonal modifications provoked by an abnormally early puberty are reflected in the cytologic aspect of smears, for example, in cases of ovarian or suprarenal tumors and hypothalamic, pituitary, or ovarian lesions. The most frequent causes of early puberty are considered in Table 5.3.

Postmenopausal Period

The onset of postmenopausal atrophy is not immediate and may even take several years.[33, 80, 127] Hormonal stimulation of extraovarian (suprarenal) origin is characterized by the persistence of stratified squamous epithelial proliferation. Androgenic, cytolytic, and nonspecific smear types are common at the onset of menopause. Surgical removal or irradiation of the ovaries provokes the same modifications of the epithelium, but atrophy is generally more rapid. Before determining the postmenopausal hormonal level, it is of course necessary to eliminate external steroid sources and the administration of digitalis or tetracycline.

Severe atrophy of the vaginal mucosa may provoke senile vaginitis characterized by inflammation and various nuclear and cytoplasmic atypias. Administration of a small dose of estrogens helps to differentiate these senile atypias from dysplasia.[70] In pronounced epithelial atrophy, parabasal cells degenerate and form cyanophilic bodies.[369] These bodies are composed of granular material. They are round, oval, or poorly defined. They should not be misinterpreted as cancer cells or trichomonads· they represent degenerated parabasal cells.

Cytology of Amenorrhea

Different hormonal conditions provoke amenorrhea.[26] Hypoestrogenic and atrophic smears suggest primary ovarian failure or insufficient ovarian stimulation. Follicular cysts may produce hyperestrogenic smears; corpus luteal cysts produce a persistent and marked progesteronic image. Masculinizing tumors of the ovary (arrhenoblastoma, hilar cell tumor, etc.) will result in an androgenic smear.

Table 5.3. Causes of Early Puberty

Hypothalamic lesions: encephalitis, meningitis
Lesions of the tuber cinereum
Hypophysary lesions
Suprarenal lesions
Functional tumors of the gonads
Albright's syndrome
Hepatoma
Juvenile hypothyroidism

The Functional Cytology of Pregnancy

Vaginal cytology during pregnancy is characterized by the disappearance of cyclic modifications and the progressive accentuation of progesteronic stimulation. Three periods are distinguished in the course of a normal pregnancy.[64, 78, 98]

Cytology of the First Trimester
(Activity of the Corpus Luteum)

Activity of the luteal type persists and is accentuated. Thus the intermediate cells are the best represented. They desquamate as isolated elements more often than during the postovulatory phase, and when they desquamate in sheets, these are small (Plate 5.6). The folding of cells is marked, and the increased quantity of cytoplasmic glycogen produces numerous elements of the navicular type. The nuclei may appear rounded and vesicular or flattened and deformed as the stored glycogen claims more and more space. Superficial cells still represent 20 to 30 percent of the squamous elements during the first 2 months, but later become more rare and represent only 10 percent on the average. The parabasal cells are normally absent. In about 10 percent of the cases, smears exhibit cytolysis. This is provoked by Döderlein's bacillus (lactobacillus), whose presence is favored by the abundance of glycogen. This type of smear can persist for an entire pregnancy, in which case hormonal evaluation is impossible because the smears consist essentially of bare nuclei. In about 5 percent of the cases, smears preserve a premenstrual aspect with no pathologic cause. Thus, three types of smears may be found at the beginning of pregnancy: the navicular, the cytolic, and the premenstrual type.[64, 110, 119, 131]

Cytology of the Last Two Trimesters
(Placental Activity)

The typical aspect of the smear during pregnancy is established by the fourth month and persists until term.[28, 103] Desquamation of the intermediate cells in broad sheets predominates; the cells, principally the navicular type, present a pronounced folding of the cytoplasm, and the nuclei are flattened by the abundance of glycogen. The eosinophilic and pycnotic indices are less than 10 percent. Smears of the cytolytic type are found in 10 percent of the cases, and a menstrual type smear is rare. Transfer of the progesteronic activity from the corpus luteum to the placenta is sometimes expressed by a transitory increase in the eosinophilic and pycnotic indices at the end of the third month.

Cytology of the Last 2 Weeks of Pregnancy

During this period, comparison of smears demonstrates the changes announcing the end of the pregnancy. One normally observes a diminution in the number and size of sheets of intermediate cells, with a concomitant increase in the number of superficial cells. The eosinophilic and pycnotic indices increase moderately (to about 15 to 20 percent). In many cases, it can be predicted that birth will occur within 5 days.

Bercovici et al.[28] have proposed a classification of pregnancy scrapings obtained at term by taking into account only two factors: the abundance of desquamation in sheets and the karyopycnotic index.

Group I	normal pregnancy—abundant clusters	CPI	<	10 percent

Group II	moderate progesteronic insufficiency		
	—clusters fairly abundant	CPI	5–10 percent
Group III	marked progesteronic insufficiency		
	—clusters rare or absent	CPI >	15 percent
Group IV	Marked progesteronic insufficiency		
	—moderate or marked estrogenic		
	insufficiency, clusters absent	CPI <	5 percent

The application of this classification can be useful in certain pathological circumstances that require that labor be induced, such as cardiac insufficiency, toxemia, and placental hypermaturation.[103]

Cytologic Diagnosis of Ruptured Fetal Membranes

Rupture of the fetal membranes is an important step in the process of delivery. Cytologic features may help to recognize this event: presence of anucleated squamous cells of the fetal skin and of fetal lanugo hair and the increase of the vaginal pH due to the presence of the alkaline amniotic fluid.[97]

Postpartum Cytology

After birth, smears change rapidly: the desquamated sheets rich in navicular elements disappear, as does cytolysis.[37,43,119] The smear assumes a "dirty" appearance occasioned by the presence of leukocytes, histiocytes, and mucus. About the fifth postpartum day, numerous parabasal or intermediate cells of a particular type are found: they have oval or rounded contours delineated by a thickening of their peripheries (Plate 5.7). The nuclei are rounded and the cyanophilic cytoplasm contains some vacuoles and glycogen. This postpartum cytologic image, however, may vary from one case to another and depends on the intensity of cellular desquamation due to the trauma of birth. Certain authors report it in only 25 percent of the cases.

During lactation, smears are essentially comprised of cyanophilic parabasal and intermediate cells. If there is no lactation, atrophy is followed by resumption of the normal menstrual cycle around the sixth week.

Cytology of the Pathologic
Conditions of Pregnancy

If hormonally induced premature interruption of pregnancy is about to occur, cytologic examination will show an increase in the cytologic signs of estrogen activity and an occasional regression in the cytologic manifestations of luteal activity.[79, 97] Cytologic modifications appear before clinical manifestations. Naturally, only hormonally induced abortions can be detected cytologically. A spontaneous abortion is preceded by an increased number of superficial cells. Smears taken after show the usual postpartum cytology and then a return to normal. If the abortion is provoked, the cytology is normal and followed by a postabortum smear and a return to normal.

The administration of estrogens in the case of imminent spontaneous abortion provokes a return of the normal cytologic aspects of pregnancy if the risk disappears, and it provokes an accentuation of the signs of estrogenic activity if the menace of abortion becomes definite. The administration of progesterone in the case of imminent abortion provokes an accentuation of the luteal-type aspect whether the risk of abortion disappears or becomes definite.

Diagnosis of Pregnancy by Cytology

The typical pregnancy smear appears only after the third month: the cytologic method is therefore not a diagnostic test of pregnancy. Furthermore, the typical aspect of the pregnancy smear is not highly specific: both menopause and amenorrhea of excess hormonal stimulation will provoke the same type of smear.

Value of Cytology
in the Surveillance of Pregnancy

A good correlation exists between hormonal activity and the aspect of the vaginal smear during pregnancy. If the smear is normal, it can be deduced that the hormonal equilibrium is normal. If an augmentation of the eosinophilic and pycnotic indices is noticed, a hormonal disequilibrium whose nature must be determined by biochemical analyses must be suspected. It must be kept in mind that fetal disorders have no repercussion on vaginal smears.

Summary: Necessary Steps
of Cytologic Hormonal Evaluation

What should be noted:

* The nature of the cells and their mode of desquamation
* The indices of hormonal activity
* the cellular diathesis (leukocytes, histiocytes, necrosis, the bacterial flora)

What should be reported:

* integration and evaluation of observations (note: hormonal evaluations based on vaginal cytology are approximations: their reliability is improved by the examination of several smears during the course of the functional cycle; the observed cytologic tableau is the result of different hormonal activities, modified by variations of sensitivity and local receptivity of the mucosa.)

Pitfalls to avoid:

* hormonal evaluation in presence of inflammatory alteration
* insufficient sampling, defective fixation, and staining techniques
* misinterpretation of hormonal status due to poor correlation with clinical facts
* forgetting that the method has only an indicative value

VULVAR CYTOPATHOLOGY

Malignant tumors of the vulva are generally seen in older women; they consist of epidermoid carcinoma[5,7,10,13] (in situ[2,5,23] and invasive[10]), Paget's disease,[1,9,16] adenocarcinoma of Bartholin's gland, malignant melanoma,[9,15] and sarcomas.[19] Vulvar metastases are rare; the most frequent are of uterine origin.[1,6] The majority of vulvar tumors are of the squamous type and the anomalies encountered are identical to those of the cervix or vagina.

The clinical accessibility of tumorous vulvar lesions makes direct scraping the obvious sampling technique. It is particularly useful in cases where biopsy is contraindicated or to obtain a rapid diagnosis for orientation purposes. Direct imprints of the visible lesions may provide rapid valuable information on the nature of the vulvar alteration (Plate 5.61). The presence of malignant cells of epidermoid nature will confirm the clinical diagnosis before the result of the biopsy is available.[4,6,8]

The surveillance of dysplastic conditions (e.g., lichen sclerosis and atrophicus, leukoplakia) is a good example of the application of imprint cytology. An intraepithelial carcinoma whose cytologic aspect is similar to that of vaginal or cervical in situ carcinoma can be recognized by vulvar scrapings. Urethral caruncles and carcinomas of the urethra also can be diagnosed by the scraping or imprint method.

In postradiation follow-up, cytology allows multiple harmless morphologic controls which could not be effected by repeated relatively brutal biopsies of fragile tissues altered by radiation.

Great care should be taken when scraping vulvar lesions since drying of smears is very frequent. Tips moistened by saline solution should be used to prevent cell alterations.

NON-NEOPLASTIC VAGINAL AND CERVICAL CYTOPATHOLOGY

Inflammatory Cytopathology

Structural Modifications of the Epithelium

Among various modifications, inflammation provokes the phenomena of hyperplasia (via division of the germinative basal cells) and metaplasia and the repair of a damaged region (chronic cervicitis) (Plates 5.15 and 5.16). These epithelial transformations are expressed on smears by:

- the presence of an unusually large number of cells of a definite type (e.g., hyperplasia of the columnar epithelium in chronic cervicitis)
- a marked cytoplasmic basophilia indicating increased protein synthesis
- foci of squamous metaplasia

Cervical Metaplasia

Epithelial metaplasia is a phenomenon of tissular defense. It constitutes a proliferation of undifferentiated cells (reserve cells) located at the base of the epithelium (Plate 5.56). These reserve cells have retained their unspecialized nature, which allows them to participate in the defense mechanisms and to undergo maturation and differentiation. This cellular differentiation is directed either in the squamous or columnar direction. Microscopically, it is reflected in squamous foci or islands of columnar cells.

Squamous metaplasia is more frequent and easier to recognize (Plate 5.57). The reserve cells situated at the basal part of the epithelium (and usually in the zone of squamocolumnar junction) give birth to epidermoid cells that repel the neighboring columnar cells either toward the surface or toward the periphery. Columnar metaplasia is the differentiation of reserve cells into cells of the endocervical mucoid type. It manifests itself in the form of small columnar islands of papillarylike structures (Plate 5.28).

Metaplastic cells may be undifferentiated (immature metaplasia) or differentiated (mature metaplasia). Metaplasia is present in most of the cervices submitted to an active genital life (sexual relations, miscarriages, etc.).

Cytology of Inflammatory Lesions

Generalities

Inflammation is the local tissue response to any aggression, no matter what its nature. It may be caused by bacteria, [18,173, 223, 227] viruses, [242, 271,273, 348] fungi, [221] parasites,[151, 171,225] or the result of chemical, physical, or traumatic lesions. Some inflammatory modifications are specific to certain causes and others are not.[194, 293]

In the vaginocervical tract, inflammation produces the following modifications:

- morphologic modifications of the squamous and columnar epithelial cells and the connective tissue cells
- histologic modifications of the epithelium (metaplasia and hyperplasia)
- appearance of inflammatory diathesis (leukocytes, histiocytes, red blood cells, and plasma cells). Immature and mature lymphocytes and plasma cells have been described in follicular cervicitis [194] and in chronic plasma cell cervicitis.[292]

These lesions are more frequent in the cervical mucosa than in the vaginal.

Cellular Morphologic Alterations

The following are the cellular morphologic alterations observed in inflammatory conditions:

- nucleus: pycnosis, karyorrhexis, homogeneous aspect of the chromatin (ground-glass appearance), loss of tinctorial affinities, anisonucleosis, karyolysis
- cytoplasm: disappearance of well-defined cellular borders, vacuolation, autolysis, alterations of tinctorial affinities, necrosis, anomalies of size and form

Most of these alterations are common to all inflammations; we will see, however, that certain modifications are associated with definite etiologic agents. Plates 5.17 to 5.60 illustrate the various aspects of cervical inflammation.

Metaplasia must not be confused with the repair mechanism of an eroded zone, which occurs by epithelization from the margins of the erosion. Squamous metaplasia is characterized by the presence of elongated cells with nuclei resembling those of columnar endocervical cells. The cytoplasm is densely stained and often vacuolated (Plates 5.18 and 5.60).

Appearance of Inflammatory Diathesis

Neutrophilic leukocytes are numerous in cases of acute inflammations, eosinophilic leukocytes accompany lesions of parasitic or allergic origin, and lymphocytes are numerous in chronic inflammations. An abundance of lymphocytes and younger cells of lymphoid lineage is found especially in follicular cervicitis (Plate 5.26); the presence of simple or multinucleated histiocytes, which may or may not exhibit phagocytosis, is also a constant characteristic of this condition. Some cases have been associated with endometritis and salpingitis.[238]

The existence of chronic cervicitis accompanied by a large number of plasma cells has been described.

The appearance of connective tissue cells reflects phenomena of necrosis with erosion of epithelial linings.

Cytolytic Smears

Despite its bacterial origin, the cytolytic smear should not be considered as indicative of a pathologic state (Plate 5.8). The presence of the Döderlein's bacillus provokes cytoplasmic destruction in the intermediate cells. This phenomenon, called cytolysis, develops in the simultaneous presence of lactobacillus (among which is the acidophilic Döderlein's bacillus) and of vaginal cell layers rich in glycogen. The bacillus transforms glycogen into lactic acid and aids in the maintenance of the acid pH (4) of the vaginal medium. This acidity itself discourages the multiplication of pathogenic flora. The cytolysis is generally not accompanied by inflammation.

Human vaginal flora includes various bacteria that are generally nonpathogenic. Lactobacillaceae, for example, are nonspore producing, rod-shaped, gram-positive bacteria. *Leptotrichia vaginalis* is a poorly identified bacterium, probably a member of the genus Fusobacterium. It is a rod, 1 to 1.5 md wide and 5 to 15 md long. It is not pathogenic. Included in this family are *L. acidophilus, L. casei, L. fermentum,* and *L. cellobiosus.* Döderlein's bacillus is a general term used for all these bacteria.

Smears of vaginal flora can be classified into different types depending on the nature of the identified species:

Type 1	lactobacilli (Döderlein's
Type 2	lactobacilli, some cocci and gram-positive bacilli, leukocytes
Type 3	few lactobacilli, numerous gram-positive bacilli and cocci, numerous leukocytes
Type 4	lactobacilli absent, very numerous gram-positive bacilli and cocci.
Types 5 and 6	specific staphylococcal pathogenic flora, streptococcal flora, enterobacteria (*Neisseria gonorrheae*),[223] Hemophilus, etc.

The following information can be drawn from cytolytic smears:

• the presence of glycogen-rich intermediate cells eliminates the possibility of an atrophic-type smear or the intense estrogenic type (parakeratotic squamous cells protecting the intermediate cells)

• the cytolytic smear is found in many circumstances: moderate or insufficient estrogenic, progesteronic, or androgenic activities, pregnancy.

Cytolysis must not be confused with autolysis, a cellular degenerative process that is a stage in cellular necrosis. Autolysis often occurs among endocervical and endometrial columnar cells, resulting in progressive destruction of the entire cell.

Whatever the path of infection (vascular, extension of a neighboring infection, or direct invasion of the genital tract), the modifications of the smear are fairly similar. Certain pathogenic agents such as Trichomonas, fungi, and herpes provoke more specific cellular modifications.

Nonspecific Infections

The presence of bacterial infection may be observed on smears stained by the Shorr or the Papanicolaou method. These methods however generally do not allow precise determination of the bacteria. This is a matter for microbiologists.[329]

The inflammatory cellular anomalies are the same as those already described (page 67).

On an evenly spread smear, the presence of a localized zone of necrotic cellular debris and a high concentration of leukocytes and histiocytes should suggest a cervical erosion (Plate 5.49). Epidermoid metaplasia is frequent.

Actinomycetes,[43] an anaerobic, gram-positive, branching filamentous bacterium, has been described in cervicovaginal smears of I.U.D. users. The organisms form compact, irregular masses in which filaments, thought for many years to be fungi, are recognized. Branching is characteristically present.

Parasitic Infections

Cases of inflammatory smears resulting from *Enterobius vermicularis* (Plate 5.45), Schistosoma[49, 151, 324] and *Entamoeba histolytica*[35, 171] have been reported. While such organisms are rare, they are easily identified, since their morphology is quite characteristic. For example, chitinous shells of schistosomal ova are refractile, and they have a terminal or a lateral spine. Toxoplasma cysts containing isolated toxoplasmas are round or irregular, dark structures measuring approximately 15 to 25 μ in size. Vorticella,[225, 317] a ciliated protozoan measuring from 15 to 50μ, has been found in cervical smears.

Trichomonas Infections

Trichomonas vaginalis is a frequent protozoan infection easily recognized by the cytologic method. It is most rapidly identified in vaginal smears examined by phase contrast microscopy. Fixed and stained smears also allow an efficient diagnosis but take more time.[260]

Recognition of Trichomonas is important for more than one reason. It causes vaginitis and cervicitis and sometimes cellular anomalies which, to an inexperienced cytologist, resemble dysplastic and even neoplastic lesions. Generally speaking, the inflammatory nature of the smear and of course the presence of the protozoan itself make diagnosis easier.[150]

The parasite may be oval, rounded, or oblong and assumes a pale green tint with Shorr's or Papanicolaou's stain. When well preserved, its small nucleus is characterized by a finely dispersed chromatin, but its flagellae are rarely preserved (Plates 5.33 and 5.34).

Cytologic Modifications of Trichomonas Infection
The following modifications are specific to Trichomonas infection:

- a marked cytoplasmic eosinophilia not attributable to estrogen-induced maturation (i.e., without concomitant rise in the pycnotic index); this eosinophilia is observed in intermediate as well as in many superficial cells.
- the presence of clear, perinuclear halos, particularly in the eosinophilic cells
- some nuclear anomalies: anomalies of form, apparent thickening of the nuclear membrane due to condensation of the chromatin on its internal surface, blurring of nuclear structure, multinucleation, voluminous but irregular nuclei, finely granular hyperchromatism, necrosis
- anomalies of columnar endocervical cells

Mycotic Infections

Among the mycoses, that due to *Candida albicans* is the most frequently encountered, notably in pregnant women and diabetics. Mycotic smears exhibit an increase in cytoplasmic eosinophilia (without the perinuclear halo of trichomonas smears) and a leukohistiocytic infiltration of variable degree.[224]

Diagnosis is accomplished by detection of mycotic elements such as mycelial spores and filaments. The spores appear as small masses measuring 3 to 6 μm and showing a clear central zone limited by a well-delineated membrane. The presence of macroconidia is rarer. Yeast cells may be observed as extracellular or intracellular elements (Plates 5.42, 5.43, and 5.44).

Viral Infections

Certain viral infections provoke mucosal modifications which show up on smears. Among these are herpes simplex,[24,46,237,273,331] condyloma acuminatum,[266,272,291] venerian vegetations, benign lymphogranulomatosis (Nicolas-Favre disease), and inclusion vaginitis.

While *herpes simplex* is the most frequent of these infections, published statistics indicate that it is rarely diagnosed (cytologically only 62 out of 40,000 cases) (Plates 5.35, 5.36–38). This frequency of the infection is demonstrated by systematic antibody detection.

The acute period during which cytologic anomalies are visible is brief. The typical herpes cervicovaginal smear is of the inflammatory type, i.e., rich in leukocytes and histiocytes. The epithelium manifests the following characteristics:

- cells with voluminous ground-glass nuclei exhibiting a homogeneous, poorly stained central zone that contrasts with the prominent nuclear membrane (This aspect is due to chromatin modifications.)
- eosinophilic or amphophilic intranuclear inclusions. These may be small, and dispersed throughout the nucleus, or large and single, occupying much of the nuclear volume and surrounded by a clear halo (Cowdry's type A inclusion). The formation of these inclusions represents the last stage of the infection's evolution.
- the presence of large, multinucleated squamous cells of irregular form. There may be as many as 15 nuclei per cell; they appear tightly clustered and their chromatin may be either visible or homogeneously blurred forming a thin peripheral margin. The cytoplasm is abundant and vacuolated and its borders are sometimes poorly delineated.

These large, multinucleated cells must not be confused with trophoblastic cells. It must also be noted that these anomalies can be found in smears of the oral mucosa.

As is the case with Trichomonas, the efficient diagnosis of herpes is important for more than one reason: should it go undetected in a pregnant woman, its consequences can be rather serious for newborn (hepatic or surrenal necrosis). Intranuclear inclusions and abnormal mitotic figures have been induced in vitro by adenovirus.[348]

In condyloma acuminata, Naib[272] has described some multinucleation and basophilic cytoplasmic inclusions among intermediate cells as well as the presence of hyperplastic basal cells. Papanicolaou reported nuclear enlargement and hyperchromasia.[287] Cytologic descriptions of these conditions by Meisels and Forton[266] include cellular enlargement, multinucleation, hyperchromasia, perinuclear clearing, amphophilia, and dyskeratotic changes. These authors believe that a certain number of mild dysplasias represent various stages of condylomatous lesions. If this notion is accepted, these cases should no longer be classified as dysplasia; the remaining cases of dysplasia should acquire a more serious prognosis as potential precancerous lesions.

Benign lymphogranulomatosis is described by Naib as producing inflammatory smears containing macrophagic histiocytes with azurophilic cytoplasmic granulations.[271]

Differential Diagnosis

Cellular modifications of an inflammatory nature may be mistaken for simple or aggravated dysplasia and sometimes even carcinoma. In cases of inflammation, hyperplastic endocervical cells can exhibit highly atypical modifications, particularly their enlarged nuclei. However, the regular, rounded nuclear shape and the normal chromatin distribution rule out malignancy.

In the absence of unquestionable morphologic proof of malignancy, diagnosis becomes a question of somewhat subjective interpretation. Recognition of a pathogenic factor (parasitic or bacterial), the general inflammatory aspect of the smear, and relatively unpronounced nuclear and cytoplasmic anomalies plead in favor of an inflammatory origin. The presence of the more specific modifications of certain inflammations (Trichomonas, herpes and other viruses) is also an important aid. Finally, a nucleocytoplasmic ratio close to normal for a given cellular type also helps rule out cancer.

If doubt persists in the mind of the cytopathologist, a control smear should be performed after the administration of adequate antiinflammatory therapy. In this case, the persistence of abundant lymphocytic or plasmocytic infiltration could possibly inculpate malignant lymphoma or myeloma. More frequently, however, such persistent infiltrations will arise from chronic follicular cervicitis or extended granulomatous cervicitis. The absence of nuclear hypochromasia points to the latter diagnosis.

TUMORS OF THE VAGINA

Tumors of the vagina are rare; they represent about 1 percent of all genital tumors. They include benign lesions such as adenosis and carcinomas, malignant melanomas, and sarcomas. Among the metastatic tumors are in situ or invasive carcinoma of cervical origin, endometrial carcinoma, and chorioepithelioma. The most frequent histologic type is epidermoid carcinoma. Cytologically, epidermoid carcinoma of vaginal origin cannot be distinguished from cervical carcinoma; we will therefore include it in the section on the cervix (Plates 5.62 and 5.63). As in the cervix, in situ carcinomas, microinvasive forms, and invasive forms are described.

Vaginal adenosis is a lesion caused by the persistence of embryonic Mullerian glandular derivations in the superficial vaginal wall. These glandular masses of mucinous type usually occur in the vaginal mucosa. However, one in four cases occurs in the submucosal tissue and is accompanied by foci of extensive squamous metaplasia, factors that appreciably complicate cytologic diagnosis. It results from an anormalous localization of the squamocolumnar junction in the vagina.

Although relatively infrequent (not more than 8 percent of all vaginal lesions), adenosis is more prevalent in patients whose mothers received diethylstilbestrol during pregnancy. The vaginal smear contains columnar and metaplastic squamous cells in various quantities and proportions. Inflammatory phenomena are discrete or absent. The development of squamous dysplasia has been reported in some cases, again suggesting a possible relationship with use of the hormone.[115]

Adenocarcinoma of the vaginal wall is rare.[93] It occurs in the elderly woman, and it has been reported that the administration of diethylstilbestrol to pregnant women is a predisposing factor for adolescent offspring (Plate 5.64). In this eventuality the incidence has been estimated to be lower than 4 per 1000.[121] These tumors are accompanied in more than half the cases by vaginal adenosis.[106] The carcinomas observed are of the clear cell type or, rarely, of the endometrial type.[32, 40, 47, 60, 62, 63, 112]

When the tumor breaks the mucosal surface, cytologic examination reveals sheets of cells arranged in cords or in papillary formations. The cells exhibit rounded or oval excentric nuclei, prominent nucleoli, and finely distributed chromatin (Plate 5.65). The cytoplasm is clear, discretely eosinophilic, and abundant. Overlapping of cells within the sheets is characteristic. The presence of ciliated cells is also noted.

It is by direct scraping of the vaginal lesion that we obtain the most indicative smears.[83] The cells will be less numerous or even absent in vaginal or cervical smears.

One lesion requires special attention: vaginal endometriosis. Endometrial cell types may arise from foci of vaginal endometriosis. These should not be confused with cells from a differentiated endometrial adenocarcinoma. This diagnosis should be made in presence of benign-looking clusters of endometrial cells in the vaginal smear and of a clinical lesion of the vaginal wall.

Botryoid sarcomas,[337] melanomas,[52,72,76,236] malignant lymphomas,[36,39] and metastatic cancers can be detected by vaginal cytology (Plates 5.66 and 5.67). The nature of these tumors can sometimes be determined by the cell's aspect on smears (for example, melanin-containing cells in melanoma).

TUMORS OF THE CERVIX

Generalities

Cervical cancer was formerly the most frequently found tumor of the genitals. It is now on the decline, particularly in North America, and endometrial cancer is presently found about as frequently.[168,169] Different data are proposed to explain this decline; better sexual hygiene and systemic cytologic screening of cervical lesions are among the factors most often mentioned.[192, 254, 255, 258, 352] However, this decline is not reported by all authors; Robinson[306] has noted that in Nova Scotia the disease has doubled in frequency in 15 years. Even with active cervical cancer detection programs, cytologic screening is not reaching a large high-risk group of population.[201] The most common histologic form is squamous carcinoma. The incidence of cervical cancer is explained by factors such as multiple pregnancies and abortions, low socioeconomic status involving an early and active sex life, and deficient hygiene leading to chronic infections.[138,182,241,294,295] A positive correlation exists between the frequency of type 2 herpetic infection and cervical cancer.[145] It is relatively infrequent in the nulliparous female.

Cancer of the cervix arises most often at the junction of the squamous and columnar epithelia. Its evolution may be described by the following histologic classification.

Stage 0 is a localized intraepithelial lesion that may last up to several years. Once the tumor invades the underlying stroma, it becomes a stage 1 cancer. Stage 2 cancers go beyond the cervix but not as far as the pelvic wall; the superior portion of the vagina is involved but not the lower third. In stage 3, the tumor has reached both the pelvic wall and the lower third of the vagina. In stage 4, the tumor has involved the bladder and/or the rectum or has otherwise extended beyond the previously described limits.

Some authors recognize a microinvasive type: this amounts to an epithelial penetration of the stroma not exceeding a depth of 5 mm from the basement membrane.[140,279,281,319,346] A large biopsy including the whole thickness of the lesion is needed to confirm this diagnosis.

The interest of this classification resides in the direct relationship between the clinical stage and the prognosis. Correctly treated, stage 0 cancers are cured in practically 100 percent of the cases. From there on, the prognosis becomes worse as a function of the stage.

Epidermoid carcinoma represents around 85 percent of all cancers of the cervix; adenocarcinoma, around 5 percent; the mixed form (adenosquamous carcinoma), about 5 percent; and all the rare tumors (sarcomas, malignant lymphomas, malignant melanomas, and metastatic tumors) account for the remaining 5 percent.

Numerous histologic classifications have been proposed to characterize the differentiation of epidermoid cancers, with the intention of revealing a correlation between morphology and the clinical behavior of the lesion. These attempts have not been very successful and have resulted in highly elaborate systems ill-adapted to practical routine.[160]

A simple and easily applicable classification distinguishes among the keratinizing epidermoid carcinomas, the large-cell nonkeratinizing carcinomas, and the small-cell nonkeratinizing epitheliomas.[289a]

Before elaborating on cervical cancer, we shall discuss the cytologic manifestations of precancerous states.

Precancerous and Cancerous States

The biologic stages in the cancerous transformation of a normal cell are not yet well defined, and it is not our intention to detail the knowledge acquired thus far. Fixation and staining are empirical techniques that at best reproduce a modified image of the living cell. In the large sense, therefore, everything the microscopist observes is an artifact.

When confronted with the morphologic aspect of the cell as observed under light microscopy, the cytopathologist observes the progressive changes of cellular structure that finally lead to what experience tells him are alterations of a cancerous nature.[146,205,233,249] These morphologic alterations have been verified by different methods, for example, scanning electron microscopic studies have shown the modifications in the structure of microvilli in dysplasia and carcinoma.[310]

The study of these structural modifications involves a certain degree of subjectivity, which explains the existence of numerous classifications and interpretations, particularly concerning borderline lesions. In the absence of objective and undeniable morphologic criteria of malignancy, we think a dogmatic attitude is not justified. The classification that we use does not pretend to resolve the problem, and we intend it to be the simplest possible.

Dystrophia

By dystrophia we mean minimal modifications in cellular structure. Dystrophia may be provoked by tissular atrophy, inflammation, and physical or chemical irritation. The lesions that we have described in the preceding paragraphs on inflammation therefore can be qualified as dystrophic.

The aspect of these modifications makes for no hesitation in the mind of the cytopathologist: they are of benign nature and are most often reversible. However, experience shows that, in certain cases, dystrophia may constitute the first step toward dysplasia and malignancy. Accordingly, dystrophia requires clinical and cytologic surveillance but not the immediate institution of untimely and perhaps useless medical therapy. Should it persist, however, adequate therapy is necessary.

The term leukoplakia is used to define a clinical lesion: a white spot (Plates 5.68 and 5.69). For the cytopathologist, it corresponds to a hyperplasia and a hyperkeratosis accompanied or not by a parakeratosis of the squamous epithelium. Hyperkeratosis is a benign lesion, but in rare instances it may cover a dysplasia of the underlying epithelium.

Dysplasia

Dysplasia is defined as a disorder of development and maturation arising in a zone of squamous epithelium or of squamous metaplasia (Plates 5.70 to 5.89). It is manifested by:

- an atypical maturation of different cellular layers
- abnormal and often precocious keratinization
- nuclear and cytoplasmic alterations the degree of which varies from one case to another
- hyperactivity of the basal and parabasal layers, constituting a major portion of the epithelium

Dysplasia often constitutes a major diagnostic difficulty, since it is characterized by cellular modifications somewhat intermediate between the benign and the malignant. Such modifications may be so discrete as to result in differing diagnosis from different pathologists. While large-scale karyotype examinations can be quite helpful in this respect (cf. Chapter 4), the technique involved is a long one and is therefore not applicable in routine procedure.

On the average, dysplasia occurs in a younger age group than intraepithelial carcinoma and appears most frequently at squamocolumnar epithelial junctions.[205] The important influence of the socioeconomic status in the genesis of cervical lesions has been reported by several authors.[184,269,282,283,342] Cases of dysplasia have been observed in patients on immunodepression therapy for renal transplantation.[240]

The existence of dysplastic lesions on the peripheries of intraepithelial or invasive carcinomas, and the higher frequency of intraepithelial carcinomas among females with dysplasia, further illustrate the relationship between these lesions.

Diagnostic problems are further complicated by the reversible nature of certain dysplasias. Every cytopathology laboratory has had the same experience: initial, undeniable cellular anomalies fail to show up on control examinations. Whether these lesions undergo spontaneous regression or are somehow affected by the biopsy itself is uncertain. Both hypotheses seem credible (Figure 5.4). In the series studied by Nasiell, more than half the cases of mild or moderate dysplasia show regression of the lesion and 12 to 21 percent persist. On the other hand, 25 percent of the moderate dysplasias will develop into severe dysplasia. In the series of Yajima et al.,[366] 18 percent of severe dysplasia progressed to carcinoma in situ or microinvasive carcinoma. In the series of dysplasia of Rummel et al.,[313] a carcinoma in situ developed in 21 percent; a microinvasive carcinoma, in 7 percent, and an invasive carcinoma, in 2.8 percent. In any case, experimental proof of lesional reversibility has been provided.[351]

Several authors have performed analytic studies of the cytologic modifications in dysplasia and have compared the latter with those of intraepithelial carcinoma as well as with those of normal cells: these findings are summarized in Table 5.3.[215,289,298,301] We repeat that the value of these findings is at best relative, and no definite diagnostic criteria have yet been proposed. Automated analysis of cell structure may in the future offer a valuable assistance in solving this problem.

The cellular anomalies in question occur to widely differing degrees, and therefore the terminologies utilized to describe them are complex. In an attempt to simplify matters, we shall retain only three types of dysplasia: simple, moderate, and severe (Plates 5.70, 5.75, 5.76, and 5.84). We must keep in mind however that the correlation between cellular morphology and clinical manifestations is not always evident.

Histologically, dysplastic epithelium exhibits a disorganization of the normal cellular stratification, a certain degree of acanthosis, hyperkeratosis, parakeratosis, and dyskeratosis. Different degrees of cellular anomalies are present depending on the type of dysplasia.

Cytology of Dysplasia

Exfoliative cytology, like histologic examination, cannot reveal any strictly definitive morphologic criteria of dysplasia. Dysplasia and intraepithelial carcinoma present the same cellular modifications and, as we have seen previously, the same epithelial architectural anomalies; it is the lesser intensity of the lesions and the persistence of superficial maturation that identifies dysplasia.[156,215] The structural modifications of the nucleus include anisonucleosis, increase in nuclear volume, atypical chromatin

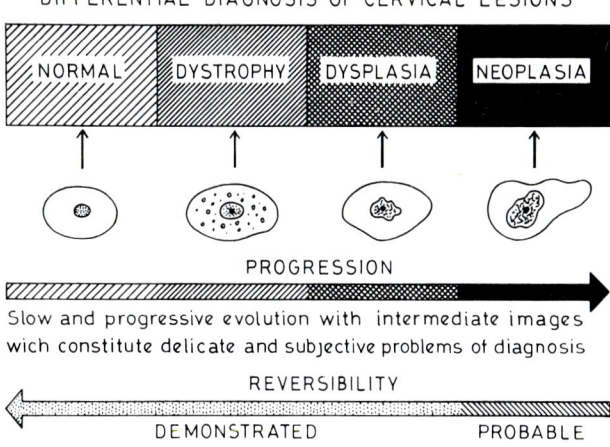

Figure 5.4. Differential diagnosis of cervical lesions.

distribution, hyperchromasia, and abnormal mitoses. The nucleolus also may be enlarged. The cytoplasm exhibits abnormal maturation (precocious or atypical keratinization) as well as changes in size and shape. Desquamation occurs more often as isolated elements than as clusters, and the general appearance of the slide is a "clean" one, indicating the absence of inflammatory phenomena and secondary necrosis.

The degree of cellular differentiation permits differentiation of three types: keratinized, nonkeratinized, and immature or metaplastic. The degree of cellular alterations permits three degrees of differentiation: slight, moderate, and severe (Plates 5.71, 5.74, 5.77, 5.82, 5.85, 5.88). In keratinized types, the cytoplasm is often eosinophilic, and the nucleocytoplasmic ratio is closer to normal. In nonkeratinized and immature types, the cells are smaller, the cytoplasm is cyanophilic, and the nucleocytoplasmic ratio is higher than normal.

Reagan's measurements of nuclear and cellular size deviations show a correlation between the degree of cellular volumetric variation and the gravity of the lesions.[298] These observations, however, are of a statistical value and must be considered with reservation when dealing with individual cases.[156]

Koilocytotic atypia was described by Koss and Durfee;[247] it consists of nuclear abnormalities associated with cytoplasmic vacuolation and ballooning of the upper layers of the squamous epithelium. Butler et al.[162] report a prevalence of koilocytosis of 1.25 percent in women under 20 years of age compared with 0.22 percent in the women in the 40–49 age group.

Transmission electron micrographs confirm that the halo of these cells contains glycogen and free ribosomes.

The association of koilocytosis with dysplasia is evident, but the etiology and the significance of these alterations remains unclear. The association of koilocytosis with condylomata has been reported.[266,291]

An evaluation of DNA content in simple dysplasias reveals diploid or polyploid nuclei, whereas in severe dysplasia and intraepithelial carcinoma, aneuploid cells are frequent. This indicates an increase in cellular anomalies from mild dysplasia to in situ carcinoma.

Dysplastic evolution may take different courses: dysplasias may regress, persist for variable periods of time, or undergo intraepithelial or invasive carcinomatous transformation. Unfortunately, it is impossible to foresee the exact evolution of dysplasia by morphologic examination of the lesion. As we have pointed out regression may be spontaneous or secondary to an intentional trauma (biopsy), or it may result from antiinflammatory therapy (antibiotics).

In a series of cases with severe dysplasia, less than half the cases show signs of regression, one-third progress to carcinoma in situ, and the remaining patients reveal persistence of dysplasia.[295] These data show that there is a definite difference in evolution between mild and moderate dysplasia on one hand and severe dysplasia on the other. However, the risks of developing severe dysplasia or in situ carcinoma in patients with moderate dysplasia is 2000 times greater than in patients with a normal cervix.

Since the risk of developing invasive carcinoma is low, mild and moderate dysplasia can be treated conservatively provided that regular controls are obtained. Conization should be advised if controls performed over a period of a year show a persistence or aggravation of the lesion. Severe dysplasia should be treated as in situ carcinoma, i.e., conization with careful histologic control of the surgical specimen to rule out an eventual microinvasion of the stroma.

Dysplasias that develop during pregnancy may regress after childbirth only to reappear during a subsequent pregnancy. While this reversible nature has been observed by every cytopathologist, the mechanism of this regression is still a mystery.[353]

Differential Diagnosis of Dysplasia

Cellular anomalies encountered in simple or severe dysplasia must not be confused with the following:

- Severe inflammatory type modifications
  To be considered as dysplastic, alterations must be quite severe; the cytologic inflammatory context will facilitate the differential diagnosis: abundant leukocytic infiltration, cellular necrosis, and possibly the presence of a pathogenic agent (fungi, Trichomonas, bacterial flora, viruses).

- Postionizing radiation modifications
  The modifications of initially normal cells can be easily identified: cellular and nuclear gigantism with a normal nucleocytoplasmic ratio indicates the benign nature and the physical origin of the lesions. Nuclear anomalies, which sometimes are quite marked, should not be misconstrued as being of a neoplastic nature.[212,335]

- Anomalies provoked by folic acid deficiency
  These anomalies are comparable to those brought on by ionizing radiation. They are characterized by an increase in nuclear volume, by irregular distribution of the chromatin, and by cytoplasmic vacuolation. (A deficiency of the coenzymes of folic acid brings on disruptions in DNA synthesis and consequently in mitoses.)[347]

- In situ or invasive carcinoma
  The number and intensity of cellular modifications as well as necrosis and inflammatory reactions are more pronounced in cancerous lesions. Cases of mild and moderate dysplasia contain fewer abnormal cells than cell samples from cases of carcinoma in situ or invasive carcinoma.[156]

Intraepithelial Carcinoma In Situ Carcinoma, Noninvasive Carcinoma of the Cervix

Generalities

Squamous intraepithelial carcinoma is a lesion of the cervical epithelial lining which exhibits the classic cellular modifications of malignancy.[178,200,208,300] The epithelium shows abnormal squamous differentiation and various nucleocytoplasmic anomalies, but invasion of the underlying stroma does not occur (Plate 5.90). This histologic form of beginning cancer is found in the cervix as well as in numerous other locations (bronchi, esophagus, etc.). There is also an endocervical glandular form. Its incidence varies from 1.5 to 2 percent depending on the type of population examined.

The histologic tableau of intraepithelial carcinoma was first described near the end of the nineteenth century; however it was only 20 years ago that the relationship between the cytologic tableau and the clinical conception of beginning cancer was established. The literature on this subject is rather abundant and we shall try to outline the essentials. Let us add, however, that there are certain questions still to be answered: which cases will persist, which will regress, and which will become invasive. In other words, does there exist an adequate correlation between lesional morphology and biologic behavior?

In conclusion, we wish to stress three facts. (1) The histologic image of in situ carcinoma cannot be defined without ambiguity: none of the morphologic criteria used to describe the lesion is definitive, and thus a diagnosis still has some degree of subjectivity. The experience of the pathologist can reduce this margin of uncertainty but not eliminate it. (2) The extension of the lesion into endocervical glands is not considered to be a criterion of invasion. (3) Purely mechanical phenomena frequently can cause a flattening of the superficial epithelial layers; therefore, this should not be interpreted as a sign of cellular differentiation.

Origin of In Situ Carcinoma

The cells responsible for early neoplastic transformation have not yet been clearly identified. Several hypotheses have been proposed:

- proliferation of the reserve cells of squamous or columnar epithelium
- zones of squamous metaplasia undergoing malignant transformation
- proliferation of undifferentiated cells (of the deep layers of squamous epithelia) that replace the normal reserve cells

The transformation process therefore may originate in undifferentiated cells or in cells that already have acquired a certain degree of differentiation (metaplastic squamous differentiation).

The process has been experimentally obtained in rats by administration of carcinogens such as methylcholantrene.[351]

Localization

Intraepithelial carcinoma originates:

- in most cases in the transition zone between squamous and columnar endocervical epithelia. The surface area occupied by such a lesion is highly variable.
- more rarely in the exocervix or in the endocervix. The later site should be suspected when the exocervix is healthy but smears are atypical.
- in multiple localizations.

Intraepithelial carcinoma can appear in young women and is initially asymptomatic. Its clinical evolution is variable, since it may persist unchanged for years, develop quickly, disappear after minor therapy, or reach the point of no return and develop into invasive carcinoma. The cytologic or histologic appearances of the lesion—for lack of precise malignant morphologic criteria—do not help predict its evolution. Therefore, in view of our partial ignorance of the nature of this lesion, it must be considered as potentially malignant and, as such, treated conservatively.

The clinical discovery of intraepithelial carcinoma is often fortuitous. It may be detected by colposcopy, rarely by macroscopic appearance and particularly by cytology and histology.

The literature confirms that the average age of appearance is lower than for invasive cancer and higher than for dysplasia.[158,182,185,205,315.] The comparison, in different age groups, of the cases of in situ carcinoma and invasive carcinoma 10 years later shows a much higher number of in situ carcinoma particularly in the younger age group. The interpretation of this is that a large number of these lesions would disappear spontaneously. Two arguments are not in favor of this interpretation: first, a spontaneous regression of in situ carcinoma anomalies on smears is rarely observed; secondly, we know that the percentage of false-negative reports is not negligible and represents approximately 20 percent of positive cases.

If the histologic diagnosis of intraepithelial carcinoma is partially subjective, what are the morphologic and clinical criteria that nonetheless may suggest its neoplastic nature?

- Transitional images have been described intermediate between normal squamous epithelium and invasive squamous carcinoma. These images correspond to the definition of in situ carcinoma.
- Histologic and clinical evidences exist of the transformation of cases of intraepithelial carcinoma into invasive carcinoma; similar morphologic and biochemical cellular anomalies have been described in invasive and noninvasive carcinoma.
- Intraepithelial and invasive lesions similar to those described clinically have been reproduced experimentally using chemical carcinogens.

We must point out that certain clinical characteristics distinguish in situ carcinoma and dysplasia from invasive tumors: local trauma, such as biopsy, as well as certain drugs can cause their regression.

Histology

Histologically, intraepithelial carcinoma is defined as a zone of epithelium that does not exhibit the normal characteristics of maturation and differentiation (Plate 5.90). Three types are generally distinguished:

- the small-cell undifferentiated type
- the large-cell undifferentiated type
- the differentiated keratinized type

The entire epithelium gives way to cells that exhibit nucleocytoplasmic anomalies and abnormal mitoses; the normal orderly stratification of the epithelium disappears. In most cases, atypical cells are found as high as the superficial layers. Rarely, there is a superficial, rather poor differentiation formed by partially keratinized pathologic elements. The nuclei are hyperchromatic and the chromatin irregularly distributed. Mitoses are found throughout the epithelium and not just in the basal layer. These mitoses may have a normal or a pathologic aspect. Cytoplasmic maturation is also modified and cellular eosinophilia is indicative of precocious keratinization.

In biopsy samples, the zones invaded by intraepithelial carcinoma may appear as separate parcels: this image is merely a result of the incidence of sectioning. They should not be interpreted as foci of invasive carcinoma.

The underlying stroma, which is separated from the epithelium by the basal lamina, exhibits a leukohistiocytic infiltration of variable intensity. The endocervical glands may be invaded by the lesion which migrates between the basal lamina and the columnar cells. The columnar cells are progressively replaced by islands of squamous cells and fill the glandular lumen.

At the peripheries of intraepithelial carcinomas, one may find areas of simple or severe dysplasia, foci of squamous metaplasia, and zones of erosion. The coexistence of such lesions emphasizes their interrelations.

Cytologic Description of In Situ Carcinoma

The number of atypical elements found depends on both the localization of the lesion and the type of smear examined. Smears taken from the vaginal cul-de-sac, for example, exhibit fewer neoplastic elements than do those made from direct scrapings of the cervix. If the lesion is located in the endocervix, an endocervical sampling will be the most productive.

A control examination performed shortly after a thorough cervical scraping will be relatively poor in atypical cells, and it is therefore preferable to wait at least 2 weeks before repeating such an examination. The number of atypical elements is generally higher in smears of intraepithelial carcinoma than in dysplastic smears. This difference results from the abundant desquamation typical of tumoral cells having suffered damage to their desmosomal apparatus.

In most cases, the diagnosis of intraepithelial carcinoma can be facilitated by the "clean" appearance of the smear: tissular necrosis and marked inflammatory reactions are lacking. Neoplastic squamous elements desquamate separately or in sheets and constitute cellular populations easily distinguished from normal cells.

Because, in most cases, dysplastic lesions accompany in situ carcinoma, smears may exhibit all the varying degrees of cellular anomaly, from the most innocuous to the most severe. The estimation of mean cellular and nuclear areas may help in the establishment of the differential diagnosis (Table 5.3).

The most frequently encountered malignant cells are those of the small-cell undifferentiated carcinoma. The are relatively small (roughly 23 μ in diameter) and have hyperchromatic nuclei which are more voluminous than those of parabasal cells (Plates 5.95 and 5.96). The chromatin is distributed in larger clumps than in dysplastic cells, but the nucleolar volume is relatively unchanged. The cytoplasm constitutes a reduced cyanophilic border. Syncitialike arrangements (clusters of cells with poorly defined borders) may be found. (Plate 5.94).

In the large-cell undifferentiated type, the volumes of both nuclei and cytoplasm are greater and the chromatin more finely distributed. Cell diameter is about 25 μ. Desquamated sheets of these elements exhibit disorganization of the cellular stratification (Plates 5.96, 5.97, and 5.98).

In the differentiated keratinized type, the nuclei are hyperchromatic and irregular in size and shape. The cytoplasm is polymorphous, eosinophilic, and often quite abundant. These elements therefore recall the normal differentiated squamous cell. Here again, marked cellular disorganization is apparent within the desquamated clusters.

Table 5.3.   Comparison of Cytologic Manifestations of Dysplasia and Carcinoma In Situ[a]

	Mean Cell Area (μ^2)	Mean Nuclear Area (μ^2)
Normal squamous cell	1550	36–39
Dysplasia	995–1081	167–174
Carcinoma in situ	238–381	109–116

[a]From Patten S.F., Reagan J.W., Wied G.L.[289,297,361]

Generally speaking, the variations of cellular size and shape are more discrete than those found in invasive carcinoma (Plates 5.92 and 5.93).

During pregnancy, the biologic comportment and cytologic alterations of intraepithelial carcinoma remain unchanged. This fact, until recently highly contested, is now established.

Some authors [140, 281] have pointed out the possibility of cytologic differentiation between microcarcinoma and in situ or invasive carcinoma. An analysis of a large number of cells (5000) shows that certain cellular features are related to the extent of tissue infiltration. Isolated cells, irregular configuration, coarsely granular pattern of the chromatin, and increased volume of nucleolus are factors that become quantitatively greater as lesions become more infiltrating. Obtaining such data is time consuming and requires the accumulation of much information. Such information however is only of statistical value. Biopsy of the lesion will give a quicker answer.

Differential Diagnosis
Intraepithelial carcinoma must be distinguished from:

Dysplasia
This distinction is a delicate one. It is quantitative nuances and not qualitative facts that separate the diagnosis of severe dysplasia from that of intraepithelial carcinoma. The presence of marked nuclear anomalies, small desquamated sheets exhibiting disorganization of the normal epithelial layers, blatant changes in the nucleocytoplasmic ratio, precocious and abnormal eosinophilia, and atypical mitoses all point to intraepithelial carcinoma. Atypical cells with irregularly shaped eosinophilic cytoplasm frequently are found. There is, however, no clear dividing line between the two types of lesions, and in borderline cases it is presumptuous to try to categorize.

A comparison based on the systematic examination of many cases shows that the frequency of atypical cells is generally higher in intraepithelial carcinoma than in dysplasia, and that the atypical cancerous cells desquamate more often as isolated elements than in sheets. Also, inflammatory infiltration of the stroma is more frequent in carcinoma than in dysplasia. Multinucleation on the other hand, is more frequent in dysplastic cells. Let us repeat, that these notions are statistical ones and as such should be interpreted with caution when studying a particular case.

Invasive Carcinoma
In cervical invasive carcinoma, inflammation and necrosis are the rule, and their modifications are reflected in the smear. Both the nucleus and the cytoplasm are more voluminous in invasive types and, more particularly so, in differentiated keratinized types.

The invasive lesion whose morphology is closest to that of in situ carcinoma is the small-cell undifferentiated type.

Postradiation Lesions
Postradiation modifications in nontumoral cells bear certain resemblances to cancerous elements. However, these nontumoral cells maintain their normal nucleocytoplasmic ratio and have less pronounced cytoplasmic anomalies. It is interesting to note that modifications resulting from ionizing radiations may persist for years.

Administration of alkylating agents (e.g., cyclophosphamide or busulfan) may cause cellular changes that mimic dysplasia or carcinoma in situ or resemble postradiation lesions.

Inflammatory Lesions

Certain inflammatory disorders such as bacterial, viral, or Trichomonas infection provoke the appearance of perplexing cellular anomalies. The disappearance or regression of these anomalies after antiinflammatory therapy or administration of small doses of estrogen indicates their benign nature. Vaginal scarring following hysterectomy (necessitated by cancer or other reasons) may provoke inflammatory anomalies that must not be confused with tumoral lesions.[207] Such granular tissue, especially if the trauma is recent, includes highly atypical cells characteristic of the repair process.[157] The cells and their nuclei are irregular in size, and multinucleation may occur. Mitoses are sometimes present. Large nucleoli are observed and are a manifestation of active cellular protein synthesis accompanying tissue repair. Cytoplasmic vacuolation is common, and bizarre cells may be encountered. The presence of fibroblasts and histiocytes indicates the predominantly inflammatory nature of the smear. Leukocytes and erythrocytes are usually abundant. This diagnostic problem becomes all the more complicated if radiotherapy has been instituted in addition to surgery. The radiation cancer cells changes coexist with inflammatory cellular alterations. A knowledge of the clinical facts naturally will help in the differential diagnosis.

Diagnostic Errors and Their Causes

Diagnostic errors can result from one or several of the following causes:

False positive

- interpretational errors due to negligence or incompetence
- smear contaminations by cells from other samples (The cells may be detached from the slide while in the staining bath and adhere to subsequent slides.)
- mismatched slides (right cells, wrong patient)

False negative

- errors in reading the slide: negligence or incompetence
- disinfection and washing of the cervix prior to sampling
- sampling errors: smears containing no cells from the lesion itself (the most frequent occurrence)
- mismatched slides

We have included these remarks here, but they are valid for all types of cytology.

Invasive Cancer of the Cervix

Histology

Invasive cancer of the cervix is so termed because it breaks through the cervical epithelium and the basal lamina to infiltrate and destroy the underlying tissue. This lesion thus is defined by its topographic characteristics. It arises most often at the junction of squamous and columnar epithelia found normally at the external orifice. The lesion is localized and isolated, but at times there may be several foci (Plates 5.99 and 5.108).

Table 5.4 summarizes the classification of the malignant tumors proposed by the WHO.[289a]

Table 5.4. WHO Classification of Malignant Tumors of the Cervix Uteri

Epithelial Tumors

   Squamous cell carcinoma (epidermoid carcinoma)
      keratinizing
      large-cell nonkeratinizing
      small-cell nonkeratinizing
   Adenocarcinoma, endocervical type
   Endometrioid adenocarcinoma
   Clear-cell adenocarcinoma
   Adenoid cystic carcinoma
   Adenosquamous carcinoma
   Undifferentiated carcinoma

Nonepithelial Tumors

   Leiomyosarcoma
   Embryonal rhabdomyosarcoma (sarcoma botryoides)

Miscellaneous Tumors

   Müllerian mixed tumor

The squamous type is the most frequent (more than 80 percent of all cervical cancers); adenocarcinoma and the mixed (adenosquamous) type are much less frequent (15 percent and 5 percent respectively). Other malignant cervical tumors occur but they are rare. Let us mention unusual types of malignant tumors: sarcomas, malignant lymphomas, malignant melanomas, epitheliosarcomas, and Mullerian mixed tumors.

Macroscopically, invasive cancer of the cervix manifests itself as a fungating (the exophytic type) or as a deep and infiltrating lesion (endophytic type).

Many systems of staging have been proposed in hopes of revealing a correlation between the histologic form and its clinical evolution.[160, 230, 354] Unfortunately, none of these systems is easily applicable to the day-to-day routine, and the results they provide are generally dubious ones. It does seem, however, that the nonkeratinizing large-cell forms, which are the most frequent, have a better prognosis than keratinizing forms and finally nonkeratinizing small-cell forms.[251, 356]

We find that the division of invasive squamous carcinoma into three histologic stages constitutes a simple classification that respects the histologic realities, allows valid comparisons with cytologic findings, and yields information on the sensitivity of the lesion in question to radiation therapy.

Cytology of Invasive Carcinoma of Cervix Uteri

On smears, we encounter the three types of squamous carcinoma defined histologically: classification is facilitated when one cellular type predominates.

The large-cell nonkeratinizing type, which is the most frequent variety, contains neoplastic cells whose average size is smaller than that of normal cells. The nuclei present the classic signs of malignancy: the chromatin is dense and abundant or irregularly clumped with clear areas of parachromatin. Nucleoli are apparent. At times, the nuclei appear uniformly dense and darkened: this homogenization of the chromatin represents the beginning stage of cellular senescence. The nuclear envelope may appear thickened due to chromatin adherence to the inner membrane. Cell boundaries are visible or indistinct and vacuolation is frequent. Cyanophilia is the rule. Exfoliated cells are isolated or arranged in clumps (Plates 5.105 and 5.106).

The small-cell nonkeratinizing type shows clusters of small irregular cells with enlarged, hyperchromatic, round or irregular nuclei. The nucleocytoplasmic ratio is high and most often the cyanophilic cytoplasm is indistinct. Nucleoli are quite visible. Cells are isolated or arranged in sheets and clumps (Plate 5.107).

The keratinizing type is characterized by the appearance of larger, irregular cells with cyanophilic or eosinophilic cytoplasm. Oddly shaped elements such as fiber cells and tadpole cells are frequent. Disturbed keratinization with the eventual information of pearls is common. Nuclei are irregular and often large; modifications in the chromatin structure are very marked. A variety of nuclei is seen, from the large nucleus with an apparent chromatin to the dense, homogeneous, very dark mass. The large squamous elements usually desquamate in clusters (Plates 5.100–5.104, and 5.109).

A comparison of the mean cellular area of the different types gives the following data: 275 μ^2 for the keratinizing type, 256μ^2 for the nonkeratinizing large-cell type, and 169μ^2 for the nonkeratinizing small-cell type.[298]

Tumoral elements of all grades exfoliate separately or in sheets and are generally quite numerous. The desquamation of neoplastic cells consistently occurs in a necrotic and inflammatory context: leukocytes, histiocytes, erythrocytes, cellular debris, hemosiderin and protein precipitates (inflammatory diathesis). This aspect may be interpretated as the patient's reaction to the tumor and as the consequence of superficial necrosis of the lesion itself. When a pronounced inflammatory image is observed, which does not at the same time exhibit its pathogenesis (e.g., bacteria or Trichomonas), one must consider the possibility of invasive carcinoma with superficial necrosis. Inflammatory reactions are less pronounced in keratinized exophytic and exocervical carcinomas.

Postradiation Cervical Cytology

Radiation changes may occur both in benign and malignant cells. These changes are acute or may become chronic. Non-neoplastic cells show marked cytoplasmic vacuolization, nuclear enlargement without hyperchromatism, multinucleation, and cellular gigantism. Bizarre, polymorphous elements result from mitotic alterations. Neoplastic cells exhibit severe injuries including nuclear fragmentation, vacuolization, and necrosis. The majority of the malignant cells disappear within a week. If there is persistence of neoplastic cells one month after irradiation, one must fear that the tumor mass has not been totally destroyed. There is also a difference in survival rate between patients whose neoplastic cells disappear at 4000 rads and those whose malignant elements persist.[335]

Attempts have been made to establish a correlation between the intensity of these cellular changes and the clinical prognosis of the tumor.[212,217,257,264] The cellular radiation response (RR) studied by Graham may be defined as the cellular changes observed in benign cells under the influence of ionizing radiation.[216] The intensity of the radiation response does not seem to represent a reliable method to evaluate the survival rate of patients with cancer of the cervix. Many prognostic factors, other than the local modifications in the structure of the benign cells, influence this survival rate: among them are the local extension of the tumor, the type of radiation technique used, the existence of metastases, and the quality of the individual immunologic response.

To conclude, regular postradiation cytology is a good technique to detect superficial local recurrence of cervical tumors or to diagnose postradiation cellular alterations. It is needless to point out that deep malignant infiltration of the uterus by the tumor will not be detected by superficial scraping.

Differential Diagnosis of Invasive Carcinoma

Invasive carcinoma must be distinguished from:

Dysplasia

See page 77.

In Situ Carcinoma

See page 81.

Sarcoma

In the presence of isolated pleomorphic, neoplastic elements, one should think of sarcoma first. Anaplastic carcinoma cells, which resemble sarcomatous elements, more frequently desquamate in groups.

Reparative or Regenerative Process

In the regenerative process, cells are found primarly in aggregates. Chromatin structure alterations and nucleoli must be prominent. Generally, anisonucleosis and hyperchromasia are less marked in this benign condition.

Endometrial Carcinoma

Endometrial cells are smaller and their concentration, in a cervicovaginal smear, is lower. Nucleoli are apparent and anisonucleosis is never so pronounced as in epidermoid tumors.

Cells of Seminal Vesicles

Cells of seminal vesicles may represent contaminants of cervicovaginal smears after recent coitus.[265] They are small elements with a vesicular and hyperchromatic or pycnotic nucleus and a foamy cytoplasm sometimes containing a yellowish brown pigment (lipofuscin). Anisokaryosis and giant polypoid nuclei may be present.

Adenocarcinoma of the Cervix

Adenocarcinoma constitutes less than 10 percent of all cancers of the cervix and occurs predominantly in the over 50 age group.[338] A few rare cases have been described in young women and children.[228] These tumors arise in the endocervix and either infiltrate deeply or bulge into the cervical cavity. They must be differentiated from endometrial carcinoma that has secondarily invaded the cervix. Cases of in situ adenocarcinomas have been described.[355]

Diagnosis is performed cytologically, particularly by direct scraping of the endocervical region. Histologically, we distinguish different types of glandular carcinoma of the cervix (see Table 5.4). The most frequent ones arise from the Mullerian epithelium, others from the Wolffian epithelium (Gartner duct type) (Figure 5.5).

Adenocarcinomas of the endocervical epithelium exhibit highly varying degrees of maturation, from differentiated forms recalling normal columnar and mucosecretory

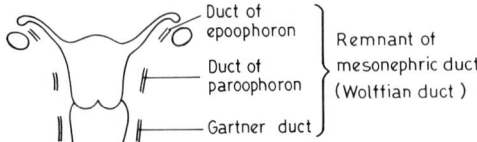

Figure 5.5. Remnant of mesonephric duct.

cells, to undifferentiated and anaplastic types (Plate 5.111). Foci of squamous metaplasia are found. The differentiated forms seem to have a more optimistic clinical prognosis. The endometrioid type closely resembles adenocarcinoma originating in the endometrium. Adenoid cystic carcinoma is very rare.[210]

The Gartner duct type originates deep within the cervical tissue and is not connected with the epithelial endocervical lining. Its tubular formations are lined with a cuboidal or columnar nonmucrosecretory epithelium. Certain varieties are rare, such as the clear-cell and/or hob nail carcinoma.

The association of squamous intraepithelial carcinoma and glandular carcinoma of the cervix has been described. Extragenital adenocarcinomas metastatic to the cervix are rare.[183]

Cytology

Tumor cells are of the columnar cuboidal type and are rare and easily recognized in differentiated forms. They are elongated, have a peripherally located nucleus, and a highly visible cytoplasmic apical pole (Plates 5.112–5.114). The rounded or oval nucleus exhibits small distinct clumps of chromatin and a prominent nucleolus which is often enlarged. The nuclear substance appears clear between the chromatin clumps, and the nuclear envelope is prominent. Anomalies of nuclear size and shape are frequent, the average nuclear diameter being greater than in normal cells. Multinucleation is encountered. The cytoplasm is prominent in differentiated forms and often appears vacuolated.

Desquamation occurs more often in clusters than as isolated cells, and a palisadelike arrangement is found in differentiated forms. At times a smear may show incomplete papillary structures.

In undifferentiated types, cellular anomalies are marked, the nucleocytoplasmic ratio is higher and glandular or tubular structures do not appear. In mixed types (adenosquamous carcinoma), the lucky examiner may observe both squamous and glandular elements (Plate 5.115). Columnar cells often exhibit a mucous secretion. Malignant epidermoid cells should not be confused with foci of benign squamous metaplasia: the abundance and the intensity of cellular anomalies point to the diagnosis of malignancy.

Differential Diagnosis

Neoplastic endocervical cells may be confused with malignant or hyperplastic endometrial cells, with benign hyperplastic endocervical columnar and stromal cells (Plate 5.117), and with the cells of large-cell nonkeratinized carcinoma of the cervix (Plate 5.116).

Distinction between the neoplastic endocervical cell and the neoplastic endometrial cell is sometimes impossible, especially in poorly differentiated forms where cervical mucous secretion is absent. Generally the endocervical cell is larger and has a more voluminous nucleus than the endometrial cell. Direct scraping of the endocervical canal results in a very high concentration of malignant cells on the slide; this is rarely obtained when the tumor is primarily localized in the endometrial cavity.

In cervicitis with erosion, one may find aggregations of voluminous bare nuclei piling up in large clusters. This hypertrophied nuclear aspect may cause some doubt about the benign nature of such elements; the chromatin distribution, however, presents more discrete clumps than in neoplastic nuclei and the nucleolus is smaller. In any case, when such doubt exists, a histologic examination is required before a definitive diagnosis can be made.

Much more rarely, glandular neoplastic elements may arise in the ovaries or the fallopian tubes. The clinical history will suggest such a possibility, since a cytologic examination alone cannot pinpoint the anatomic sources of undifferentiated or poorly preserved columnar cells.

Value of the Method

The detection of cancer of the cervix is extremely important since it is the only common localization of cancer that can be easily detected at a preinvasive stage; moreover, adequate treatment of these lesions results in practically certain eradication.

However, large-scale screening programs conducted for more than 10 years have not brought about the expected dramatic fall in the death rate from cervical cancer.[170, 201] Such programs reveal an apparent increase in survival time. This does not prove, however, that early diagnosis has been particularly useful, but only that the diagnosis has been made earlier in the evolution of the cancer. The period of time between the biologic beginning of the lesion and death may have remained the same.

Different facts may partially explain these results: the majority of women who take advantage of cancer detection programs belong to the low-risk population (young women from middle or high socioeconomic classes), and in any case, this type of populations seeks treatment at early stages of the disease. High-risk populations (old women and those from low socioeconomic classes) are not as sensitive to public health propaganda and do not regularly attend screening programs.

Moreover, causes with a long clinical evolution have a better chance of being diagnosed and have the best prognosis. Cases with rapid evolution and a brief in situ period have a lesser chance of being diagnosed and have worse prognoses. Luckily, these cases are rare.

The accuracy of the method can be estimated by the following criteria. The *sensitivity* of the method is the ratio of the number of true positive cases to the sum of positive tests and false negatives.

$$S = \frac{a}{a + c}$$

a = number of positive cases
c = number of false negative tests

The *specificity* of the method is the ratio of the number of true negative cases to the sum of negative tests and false positives.

$$SP = \frac{d}{b + d}$$

d = number of negative cases
b = number of false positive tests

In cervical cytology, the sensitivity should be between 90 and 100 percent and the specificity should approach 95 percent.[209, 303]

The frequency of the disease will affect the percentage of false-positive and false-negative cases. For example, if the disease is not frequent, the percentage of false-positive cases may become very high. False negatives have been estimated in cervical cytology to range from 1 to 20 percent. False negative rates are underestimated in most laboratories because the follow-up of large populations with negative smears is impossible. If the cause of the negativity is the absence of suspicious elements on the smear, a repeat of the test within 1 year may correct the mistake. This is why we recommend taking two smears, with a 1-year interval, for every woman who has never had a Pap smear before. Afterward a 3-year interval between smears is adequate to detect the slow-growing lesions, which are the most frequent.

Summary

Indications for the method:

- detection of asymptomatic and symptomatic tumors
- recognition of specific and non-specific inflammatory lesions
- follow-up of irradiated cases

What should be noted:

- the type, origin, and quantity of cells desquamated, and their configuration
- the presence of specific inflammatory agents and the intensity of the inflammatory reaction

What the lesions are that may be diagnosed by cytology:

- inflammations: chronic and acute, specific and non specific
- precancerous conditions
- neoplastic diseases: primary or metastatic

What should be reported:

- definite malignancy
- cellular atypias ranging from dystrophia and mild dysplasia to severe dysplasia
- presence of inflammation and possibly its origin
- necessity of cytologic or histologic control

HISTOLOGY OF THE ENDOMETRIAL MUCOSA

At the internal orifice of the cervix, the endocervical mucosa becomes continuous with the endometrial mucosa. The latter is formed by a columnar lining that itself is deeply continuous with a ramified glandular system. These glands are also lined by columnar epithelium and exhibit a highly variable morphology resulting from differing hormonal activities (Figure 5.6). Surrounding the glands, there is a connective tissue stroma which is also subject to hormone-induced structural modifications.

The intercalated cells, which are quite prominent in histologic preparations, are flattened columnar elements with a dense cytoplasm. Compression by neighboring columnar cells explains this "structured out" appearance. These are probably degenerative secretory cells found after ovulation. Other smaller intercalated elements situated just above the basement membrane and packed tightly between neighboring

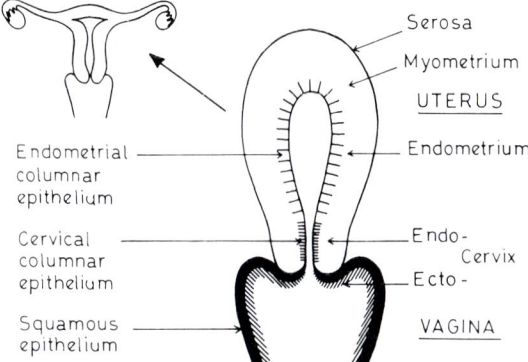

Figure 5.6. Schematic representation of the female genital tract.

columnar cells are probably leukocytes having migrated from the underlying stroma (Plates 5.118 and 5.122).

Four types of major glandular activity are encountered:

• proliferative activity corresponding to normal estrogenic stimulation occurring during the first half of the menstrual cycle
• secretory activity such as that occurring during the second half of the cycle
• atrophy of the glandular system characterizing a lack of hormonal stimulation
• hypertrophy of the glandular system characterizing excess hormonal stimulation

Glands in the proliferative phase are straight, of moderate size, and lined with a highly mitotic columnar epithelium. Secretory-phase glands exhibit enlarged cells whose abundant cytoplasm manifests glycogen secretory activity. The nuclei are rounded and centrally situated. These morphologic characteristics may be modified in pathologic conditions. The estrogenic phase stroma is filled by cells that at first resemble fibroblasts. These subsequently become rounded and dispersed in an edematous interstitial fluid. Here also, mitoses are observed.

During the secretory phase, the highly edematous stroma exhibits rounded cells that will become loaded with glycogen (decidual reaction); another type of cell is also present, one containing mucopolysaccharides important in heparin synthesis. These are the oncocytic cells of Hamperln or K cells.

The endometrial arterial system proliferates and becomes most developed near the end of the menstrual cycle (spiral arteries).

CYTOLOGY OF THE ENDOMETRIAL MUCOSA

Cellular Types

Two cellular types result from desquamation of the endometrial mucosa: columnar cells and stromal cells (Plates 5.119–5.121). The columnar cells of the glandular systems may be subdivided into four categories:

• cells of the endometrial epithelial lining
• functional cells of the uterine glands

- ciliated cells
- intercalated cells

The stromal cells include fibroblasts and the oncocytic or K cells. On smears of endometrial aspirations, superficial columnar cells may be distinguished from their glandular counterparts by the former's larger size, rounder nuclei, and homogeneous or vacuolated lightly stained cytoplasm.

The glandular columnar cells are elongated elements measuring 15 to 20 μ in height; they are smaller than endocervical columnar cells. Their nuclei may be rounded or elongated and may exhibit finely dispersed chromatin and a prominent nucleolus. When observed in isolated elements, borders of the basophilic cytoplasm are often poorly preserved and ill defined. Depending on the incidence, fixed cells may appear in clusters of small, regularly rounded elements that have well-defined borders (honeycomb appearance) or as elongated elements forming a palisade. When the cytoplasm has degenerated one observes aggregates of naked nuclei. Certain cells have a ciliated border which is visible on smears. At the base of this border is a row of fine granules, the basal or ciliary apparatus. The nuclei of ciliated cells are elongated or round.

Intercalated cells are rarely recognized on smears except in certain clusters having partially preserved their glandular structure.

Stromal cells are observed as clusters of small cells with elongated or rounded nuclei depending on their functional state. The size of the nuclei is similar to that of glandular cells and the cytoplasm is cyanophilic. These cells increase slightly in size during the proliferative and secretory phases. During the second half of the cycle, the stromal cells become loaded with glycogen and appear as rounded elements with dense, finely vacuolated cytoplasm.

The appearance of endometrial cells varies with the mode of sampling. In endometrial aspirations, clusters are more abundant than isolated elements, and if the aspiration was particularly vigorous, small tissue fragments may appear which for all practical purposes, can, be treated as biopsies and embedded in paraffin.

Endometrial Cells During the Menstrual Cycle

In addition to cancer diagnosis, the cytologic method may offer a means of hormonal evaluation when the endometrial cells are well preserved. Aspiration methods will provide better cellular material than cervical or vaginal smears. The nature of desquamated endometrial cells may be correlated with the vaginal cytohormonal pattern. The nuclear size and shape as well as the abundance of the cytoplasm will reflect the type of activity of the endometrium: proliferation or secretion.

Proliferative cells are elongated elements with long nuclei and a dense chromatin; secretory cells have a round, pale nuclei and a clear abundant, often vacuolized cytoplasm (Plates 5.123 and 5.124). However, these functional modifications are difficult to interpret since the state of cell preservation often leaves much to be desired.

The presence of endometrial cells during the first 10 days of the menstrual cycle is no cause for concern: they generally result from a physiologic desquamation. The nuclei are clumped together, and the cytoplasm exhibits varying degrees of cytolysis.

Endometrial cells arranged in a syncitium should not be confused with multinucleated histiocytes (Plate 5.125). Stromal cells may sometimes be very numerous at the end of the menstrual flow ("exodus period" of Papanicolaou).[287] They provide phagocytic elements during the period of endometrial desquamation and repair of the epithelium. Similar cells are observed after abortion and parturition.

The presence of normal glandular and/or stromal endometrial cells during the luteal phase or after menopause constitutes a sign of possible pathologic conditions of the corpus, such as mucosal polyps, various types of hyperplasia, or adenocarcinoma.

Presence of endometrial cells has also been reported at the time of ovulation in women using intrauterine contraceptive devices and in postmenopausal women receiving estrogen therapy.

When the centrifugation method is used, cell preservation of the aspiration fluid in saline may be enhanced by the technique. Embedding the sediment in paraffin allows preservation and recognition of epithelial fragments suitable for histologic evaluations.

To summarize, although cellular functional evaluations sometimes may be based on well-preserved endometrial cell clusters, such favorable instances are rare.

INFLAMMATORY LESIONS OF THE ENDOMETRIUM

Endometrial cytology can be quite informative with regard to infections. Thus, the cytopathologist will observe the following cytologic inflammatory manifestations in the aspiration fluid: cytolysis, nuclear necrosis, polymorphonuclear leukocytes, and histiocytes. Should such a smear occur other than during the menstrual cycle, further microbiologic or pathologic examinations are indicated. A diagnosis of endometritis can be evoked in these cases and should be confirmed by clinical examination.

Cytological Modifications Caused by I.U.D.

The presence of an I.U.D. can provoke perplexing cytologic anomalies [371, 388, 421, 424, 429] (Plate 5.136). These occur in endometrial and endocervical cells and are characterized by modifications in nuclear size and shape, by hypertrophy of nucleoli, by the distribution of chromatin in relatively large clumps, and by an extensive cytoplasmic vacuolation.[380] Epidermoid metaplasia has been observed in the endocervical clusters.

The presence of calcified bodies and foreign giant cells has been found in the cervical smear in a patient wearing an intrauterine device for more than a year.[403]

Loopal smears reveal the presence of macrophages and foreign body giant cells. They indicate the usually focal endometrial inflammatory reaction caused by the polyethylene device. The association of pelvic actinomycosis and the use of an I.U.D. has been reported.[430]

These modifications, which mimic malignant anomalies, do not constitute a source of confusion if the cytopathologist is aware of the clinical history.

TUMORS OF THE ENDOMETRIAL MUCOSA

Histology of Endometrial Tumors

Benign Tumors

The majority of benign tumors of the endometrium are polyps. They consist of a mass of endometrial tissue appended to the mucosa. The erosion of the superficial epithelium is the source of inflammatory lesions.

Cyclic changes of the glandular system are not always present in polyps and it is not rare to find in the endometrial biopsy the simultaneous presence of nonactive proliferative tissue and secretory tissue.

A submucosal leiomyoma may protude into the uterine cavity and suggest an endometrial tumor.

Malignant Tumors

The majority of the malignant epithelial tumors of the endometrial mucosa are adenocarcinomas.

Tables 5.5 and 5.6 summarize the different forms that can be recognized according to the WHO classification.

Glandular carcinomas arise among the columnar cells and exhibit various morphologic aspects depending on their degree of differentiation. Broders' classification[160] uses four grades according to the percentage of differentiated and undifferentiated cells. Grade I consists mostly of differentiated cells; in grade IV 75 to 100 percent of cells are undifferentiated. Grades II and III are intermediate. We classify the different types of adenocarcinoma of the endometrium into three groups depending on the degree of differentiation: well-differentiated, poorly differentiated, and undifferentiated.

The differentiated types resemble normal glandular structures; the undifferentiated anaplastic types bear no resemblance whatsoever to glands; adenoacanthoma is a glandular carcinoma showing foci of benign squamous metaplasia, and mixed forms contain glandular and squamous neoplastic elements.

The clear-cell carcinoma shows typical cells with abundant clear cytoplasm forming

Table 5.5. WHO Classification of Epithelial Tumors of the Corpus Uteri

Benign
    endometrial polyp
    endometrial hyperplasia

Atypical endometrial hyperplasia

Malignant
    adenocarcinoma
    clear-cell adenocarcinoma
    squamous cell carcinoma
    adenosquamous (mucoepidermoid) carcinoma
    undifferentiated carcinoma

Table 5.6. WHO Classification of Nonepithelial Tumors of the Corpus Uteri

Benign
    leiomyoma

Malignant
    leiomyosarcoma
    endometrial stromal sarcoma

glandular, tubular, or papillary structures. Identical tumors (of Mullerian origin) are described in the vagina, cervix, and ovary. Squamous carcinomas are rare.

Sarcomas of the endometrial stroma are formed by clusters of small cells with pronounced nuclear anomalies and frequent mitoses. Certain sarcomas of the internal muscularis may compress and ulcerate the overlying endometrium, thus provoking the appearance of tumoral elements in the uterine cavity. These may be leiomyosarcomas, undifferentiated sarcomas, or mixed mesodermal tumors.

Metastatic tumors are extremely rare.

Cytology of Endometrial Tumors

The principal indication of endometrial cytology is the screening for tumoral cells.[372,374,389,425] This screening is performed either with exfoliated cells in vaginal (Plate 5.131) or cervical smears or with endometrial aspirations.[409-412] The latter technique has the advantage of furnishing greater numbers of relatively well-preserved cells. The techniques of negative pressure endometrial washing[394] (as described by Gravlee) also offer certain advantages.[377, 416, 432, 434] We shall examine the respective qualities of these methods in Chapter 13.

Benign Epithelial tumors
Endometrial polyps sometimes may give an abundant cellular desquamation when an endometrial aspiration is performed. However, there is no specific cytologic criterion suggesting the nature of the lesion. The presence of smooth muscle cells can be observed when a mucosal erosion takes place, for example, in the case of a submucosal leiomyoma (Plate 5.126).

Malignant Epithelial Tumors
Neoplastic cells arising from adenocarcinoma of the endometrium may desquamate in groups or, less frequently, as isolated cells. Determination of malignancy in an isolated endometrial cell is a delicate matter and requires much experience and caution, particularly in well-differentiated lesions. The extent of this spontaneous desquamation depends on the histologic type of the tumor: it is abundant with exophytic, papillomatous, and undifferentiated tumors and less so with undifferentiated or infiltrating tumors. The concentration of cells is larger in endometrial aspiration than in cervical scrapings and is minimal in vaginal smears.

The malignant endometrial cell is small and round, oval, or columnar when its configuration is not altered by compression or clumping. The mean cell area is about 148 μ[427] (Table 5.7). The nucleus is round, oval, or deformed and pushed aside by cytoplasmic vacuolation. The chromatin is irregularly distributed and hyperchromasia is usually moderate.

Table 5.7. Mean Values of Surface of Different Endometrial Cell Types[a]

Functional glandular cell	35–70μm²
Stromal cell	40–70μm²
Cells of hyperplasia	95–120μm²
Cells of adenocarcinoma	130–200μm²

[a]From REAGAN J.W., Ng A.B.P.[427]

Nucleolar hypertrophy is a fairly constant characteristic of the adenocarcinoma cell, and this distinguishes it from the malignant squamous cell. Multiple nucleoli are less frequent. At times, the nucleus is so peripherally located that the nuclear and plasma membranes are in contact. In the majority of cells, the cytoplasmic borders are indistinct. When intact, the cytoplasm is pale and cyanophilic and exhibits one or several large vacuoles, or it may be foamy due to numerous microvacuoles. In the case of active mucous secretion, the vacuole occupies most of the cytoplasm (signet ring cell) (Plate 5.132). A leukocyte or cellular debris may be present in the vacuoles (cellular cannibalism) (Plate 5.132). The mean size of adenocarcinoma cells increases from grade I (differentiated type) to grade IV. (undifferentiated type). The presence of malignant ciliated cells may be observed[392] (Plate 5.133).

When nuclei are naked and exhibit signs of degeneration (karyorrhexis, ground-glass appearance) their recognition is difficult. The presence of anisonucleosis in these nuclei suggest their malignant nature. An inflammatory diathesis is often present and contains cellular debris leukocytes, erythrocytes, histiocytes, and fibrin, Small necrotic cells can be incorrectly assessed as histiocytes or leukocytes.[376]

On atrophic smears, the detection of neoplastic endometrial elements is a more delicate matter, since these elements may be mistaken for cervical cells modified by chronic inflammatory reactions. The presence of voluminous nucleoli indicates the glandular nature of such cells. In these cases, bare nuclei are observed frequently. These are generally less voluminous than in hyperplastic inflammatory endocervical cells.

Morphologic signs of inflammation without malignant cells are sometimes present in the vaginal smear of patients with endometrial adenocarcinoma. In absence of cervicovaginal anomalies explaining this inflammation, the cytopathologist should consider an endometrial lesion.

Undifferentiated tumor cells are markedly irregular in both size and form (Plate 5.134). They have a peripheral nucleus which may appear rounded, elongated, or irregular. The chromatin is finely distributed in small, clearly defined strands, which stand out on a relatively clear nuclear matrix. The chromatin material sometimes may appear granular. The nuclear envelope is delineated by the adherence of chromatin to its inner membrane giving an impression of thickening. Multinucleation is rare. The presence of cell aggregates is greater in poorly differentiated lesions.

Differentiated tumor cells are small and regular and may resemble benign ones (Plate 5.135). For this reason, the presence of benign-looking endometrial epithelial clusters should always draw the cytopathologist's attention, particularly in postmenopausal women.

Papillary tumors exfoliate numerous malignant cells that sometimes suggest their origin by the presence of structures resembling bunches of grapes.

In those rare well-preserved smears we have had the opportunity to study, cells of clear-cell carcinoma look like differentiated, regular elements with round nuclei and abundant cytoplasm.

In endometrial aspiration, the simultaneous presence of columnar malignant cells and benign squamous elements may suggest the diagnosis of adenoacanthoma (Plate 5.139). Keratin bodies, representing small typical hyalin masses of keratin, accompanied by leukocytes and histiocytes are observed in adenoacanthoma.[381]

In mixed types, one finds in the majority of cases both squamous and glandular malignant elements. Squamous elements are more frequently nonkeratinized.

Sarcomatous Tumors

Endometrial sarcomas liberate highly disorganized clusters whose cells exhibit voluminous polymorphous nuclei (Plate 5.149). The diameter of the latter may be 25μ; they are hyperchromatic and sometimes possess a large nucleolus. The irregularly shaped cytoplasm is foamy and its borders are ill defined.

The polymorphism of these cells may suggest a diagnosis of sarcoma. However, if cytolysis is pronounced, the cytopathologist must limit his report to the possibility of malignancy and wait for the results of a pathologic examination before defining the tumor.

Sarcoma botryoides is a rare mixed mesodermal tumor occurring in children or young adults.[379] It reveals bizarre, irregular cells of various sizes with hyperchromatic nuclei. With great luck, one may find muscular cross-striations in some elements.

Choriocarcinoma originates from the trophoblast, the fetal portion of the placenta. Proliferating, often monstruous, and large trophoblastic elements with multiple or giant nuclei as well as small hyperchromatic cells are found in hemorrhagic uterine secretions after abortion or delivery. Clinical data and a positive pregnancy blood test confirm the diagnosis.

Clusters of trophoblastic cells may be rarely seen in hydatidiform mole after abortion. They consist of grapelike cysts. The cells show round, uniform, moderately hyperchromatic nuclei surrounded by pale, scarce cytoplasm.

Recognition of the malignant nature of sarcomatous cells is not a difficult diagnostic problem, but definition of the type of tumor by cytology only is sometimes hazardous.

Metastatic Tumors

Malignant columnar cells may originate from the ovaries or the fallopian tubes and migrate into the uterine cavity[442, 445-447] (Plates 5.150-5.152). Psammoma bodies have been described and their identification is associated with ovarian neoplasms.[149]

We have observed neoplastic cells arising in a rectal carcinoma that have migrated into the vaginal cavity through a rectovaginal fistula (Plate 5.153).

The Hormonal Cytologic Context
of Endometrial Cancers

Endometrial cancer, whose frequency is highest after menopause, results either in atrophic smears or in smears indicative of hormonal insufficiency. However, the persistence of an estrogenic hyperactivity may be observed. The cytopathologist should report this anomaly since it may reveal an abnormal hormonal stimulation related to an endometrial cancer or a hyperplasia or to the existence of an extraovarian source of steroid hormones (suprarenal gland).

The role of excessive and long estrogenic activity in the genesis of endometrial cancer has been widely discussed. This matter remains controversial.[414, 433]

Postradiation Endometrial Cytology

Treatment of endometrial cancer by ionizing radiation leads to rapid destruction of tumoral elements and to their disappearance from smears. The cellular anomalies which are observed arise predominantly from benign cells modified by radiation. However, the differential diagnosis between benign cellular anomalies and cancer cells is a delicate one. Postradiation endometrial biopsies show that the rare benign endometrial glands that reappear bear marked morphologic anomalies. The preservation of normal glandular structure pleads in favor of the benign nature of the proliferation. Necrosis, inflammatory infiltrates, and hemorrhage may hide the existence of scattered abnormal cells.

Endometrial aspiration is a valid screening technique for recurring superficial endometrial cancers after irradiation. It does not reflect the eventual invasion of the deep layers of the mucosa and the myometrium.

Value of Cytology in the Diagnosis of Asymptomatic Endometrial Cancer

Several authors [442] report a recent rise in the proportion of endometrial cancers with respect to cervical cancers. Thus augmentation indicates the necessity of effective diagnosis of both early and asymptomatic forms. The sampling should be performed in the endometrial cavity but practical considerations make routine, large-scale endometrial aspirations impossible. Endometrial aspirations must be restricted to symptomatic cases or high-risk individuals (e.g., postmenopausal hyperestrogenesis).

Cervicovaginal smears may pick up some cases. Our observations of a female asymptomatic population of 60,000 cases with no age restrictions revealed 6 cases of endometrial cancer. The cytologic diagnosis was positive in 4 cases and suspicious in 2; biopsies were used to confirm the malignancy in all 6 cases.

Studies performed on series of proved endometrial cancers show that vaginal cytology is positive in approximately 25 percent of the cases. An acceptable accuracy of about 80 percent is obtained with cervical scraping and endocervical aspiration. Endometrial aspirations and washings should give unsatisfactory samples in approximately 10 to 20 percent of the cases. In any case, endometrial biopsy is the adequate procedure to obtain the final diagnosis of malignancy.[370]

Differential Diagnosis

The size of the cells and their poor state of preservation create the following problems in differential diagnosis:

- Recognition of different endometrial cell types: cell preservation and fixation must both be favorable to allow recognition of the different types of glandular cells; functional cytology cannot replace endometrial biopsy.

- The distinction between endocervical and endometrial cells: in cervicovaginal samples, distinction is difficult, especially when one observes only bare nuclei. The average size of endocervical nuclei is slightly greater than that of endometrial nuclei. The mucosecretory or ciliated aspect of the cervical cell makes diagnosis easier. Similarly, cells derived from endocervical adenocarcinomas are larger than cells originating in endometrial carcinomas, and they retain a more typical columnar configuration. Also, nuclear chromatin distribution is coarser, nucleoli are larger, and cytoplasmic vacuolation is less common in endocervical carcinomas.

- The distinction between the endocervical cell and the histiocyte: histiocytes are generally recognized by the following characteristics—irregular or kidney-shaped nucleus, absence of a voluminous nucleolus, foamy or microvacuolated cytoplasm, multinucleation, macrophagia, and, notably, lipophagia.

- The distinction between the normal and the hyperplastic cell: this distinction is possible when the cellular material is abundant and when clusters show the increased cellular volume, anisonucleosis, pseudostratification, and cellular aggregation characteristic of endometrial hyperplasia.

- Malignant endometrial cells may be confused with the cells of endometrial polyps and hyperplastic lesions; abundant desquamation and cellular aggregation may lead to false positive results. The absence of definite malignant cellular changes is a pertinent criterion in favor of benignity.

- Neoplastic glandular cells may arise in the fallopian tubes or in the ovaries. These cells can migrate in the tubes and may be found in the endometrial cavity or as far as the cervix or vagina. This migration is reported to occur in between 5 and 40 percent of all ovarian cancers. The cell clusters which result are generally few in number. They contain cells with rounded nuclei and vacuolated cytoplasm. Cytoplasmic vacuoles may be extremely large. These elements are aggregated in globular masses which appear tridimensional. Distinction between these cells and neoplastic glandular endometrial cells is difficult, if not impossible (Plates 5.150–5.152). The presence of psammoma bodies has been described in a very small number of ovarian cancers and in still fewer endometrial cancers.[437,438,444] The presence of ovarian cells in samples collected from the vagina or the cervix has been described in one case of a wearer of an intrauterine device, as well as in certain benign papillary lesions of the ovary.

- The presence of highly altered ciliated cells from the ovaries or fallopian tubes may sometimes be mistaken for protozoa, as described by Gaudefroy and Coliche,[54] in genital tract smears or ovarian needle punctures. These cells are generally characterized by a well-preserved ciliary border surrounding an undefinable structure. The latter is none other than the debris of a highly degenerated cell. Such images have been experimentally reproduced. Cilia should not be misinterpreted as elongated, enlarged microvilli.

- Large cells measuring up to 75 μ in diameter with large homogeneous nuclei have been described in late pregnancy during amnioscopy. These benign elements arise in the lower segment of the decidua and should not be mistaken for anaplastic uterine cancer cells.

Summary

Indications for the method:

- study of endometrial cells obtained by aspiration or present in the cervical vaginal pool

What should be noted:

- the presence of endometrial cells: both epithelial and stromal
- the configuration and abundance of endometrial cells depending on the type of smear (vaginal, cervical, or endometrial)
- the existence of an inflammatory diathesis

What the lesions are that may be diagnosed by cytology:

- hormonal evaluation of ovarian function, possible in technically favorable cases

- recognition of inflammatory alterations
- detection of precancerous lesions and benign and malignant tumor cells

What should be reported:

- description of the uterine elements and analysis of their configuration
- interpretation of the inflammatory diathesis
- presence of tumoral elements and tentative diagnosis of their origin

BIBLIOGRAPHY

Vulva

1. Bennington J.L., et al. Detection of cells from extramammary Paget's disease of the vulva in a vaginal smear. Report of a case. *Obstet. Gynec.* 37:772, 1966.
2. Boutselis, J.G. Intraepithelial carcinoma of the vulva. *Amer. J. Obstet. Gynec.* 113:733, 1972.
3. Broen E.M., Ostergard D.R. Toluidine blue and colposcopy for screening and delineating vulvar neoplasia. *Obstet. Gynec.* 38:775, 1971.
4. Carter B., Kaufmann L.A., Cuyler K. Smear preparations in the diagnosis of vulvar cancer. *Surg. Gyn. Obst.* 91:600, 1950.
5. Collins C.G., Roman-Lopez J.J., Lee F.Y.L. Intraepithelial carcinoma of the vulva. *Amer. J. Obstet. Gynec.* 108:1187, 1970.
6. Cuyler W.K., et al. Cytologic studies in malignant lesions of the vulva. *Surg. Gyn. Obst.* 96:115, 1953.
7. Daly J.W., et al. Carcinoma of the vulva. *JAMA* 59:38, 1972.
8. Dennerstein G.J. The cytology of the vulva. *J. Obst. Gyn. Brit. Commw.* 75:603, 1968.
9. Fenn M.E., et al. Paget's disease of vulva. *Obstet. Gynec.* 38:660, 1971.
10. Figge D.C., Gaudenz R. Invasive carcinoma of the vulva. *Amer. J. Obstet. Gynec.* 119:382, 1974.
11. Franklin E.W. Clinical staging of carcinoma of the vulva. *Obstet. Gynec.* 40:277, 1972.
12. Janovski N.A. Classification of dysplastic and premalignant lesions of the vulva with histologic and histochemical considerations. *Int. J. Gynaec. Obstet.* 8:581, 1970.
13. Katayma K.P., Woodruff J.D., Jones F.H.W., Preston E. Chromosomes of condyloma acuminatum, Paget's disease, in situ carcinoma, invasive squamous cell carcinoma and malignant melanoma of the human vulva. *Obstet. Gynec.* 39:346, 1972.
14. Kaufman R.H., Gardner H.L., Brown D., Beyth Y. Vulvar dystrophies: an evaluation. *Amer. J. Obstet. Gynec.* 120:363, 1974.
15. Morrow C.P., et al. Melanoma of the vulva. *Obstet. Gynec.* 39:745, 1972.
16. Neilson D., Woodruff J.D. Electron microscopy in in situ and invasive vulvar Paget's disease. *Amer. J. Obstet. Gynec.* 113:719, 1972.
17. Ridley C.M. A review of the recent literature on diseases of the vulva. Carcinogenesis and tumors. *Brit. J. Derm.* 87:163, 1972.
18. Ridley C.M. A review of the recent literature on diseases of the vulva. Part II: Vulvitis: infections. *Brit. J. Derm.* 87:58, 1972.
19. Sara P.J., Rutledge F., Smith J.P. Sarcoma of the vulva. Report of 12 patients. *Obstet. Gynec.* 38:180, 1971.
20. Woodruff J.D., Davis H.J., Jones Jr. J.H. Correlated investigate technics of multiple anaplasias in the lower genital canal. *Obstet. Gynec.* 33:609, 1969.
21. Woodruff J.D., Julian C., Puray T. The contemporary challenge of carcinoma in situ of the vulva. *Amer. J. Obstet. Gynec.* 115:677, 1972.
22. Woodworth H., Dockerty M.B., Wilson R.B., Pratt J.H. Papillary hidradenoma of the vulva: a clinicopathologic study of 69 cases. *Amer. J. Obstet. Gynec.* 110:501, 1971.

23. Woyke S., Domagala W. Olszewski W. Hidradenoma papilliferum vulvae. Evaluation of ultrastructure. *Nowotory* 22:109, 1972.

Vagina

24. An S.H. Herpes simplex virus infection detected on routine gynecologic cell specimens. *Acta Cytol.* 13:354, 1969.

25. Anthonioz P. Premières observations continues de corps ciliés dans les frottis vaginaux. (A propos de 57 cas). *Rev. Cytol. Clin.* 8:9, 1975.

26. Batrinos M.L., Eustratiades M.G. Vaginal cytology in primary amenorrhea. *Acta Cytol.* 16:376, 1972.

27. Batrinos M.L., Eustratiades M.G. The diagnostic significance of parabasal cells. I. Correlation with the clinical diagnosis in 209 patients. *Acta Cytol.* 19:97, 1975.

28. Bercovici B., Diamant Y., Polishuk W.Z. A simplified evaluation of vaginal cytology in third trimester pregnancy complications. *Acta Cytol.* 17:67 , 1973.

29. Berget A., Gyldensted M., Skaarup P., Szczepanski K. Clinical consequences in patients with suspect cell findings on initial vaginal cytology. *Danish Med. Bull.* 19:131, 1972.

30. Bhagavan B.S., Weinberg T. Cytologic diagnosis of metastatic cancer by cervical and vaginal smears with report of a case. *Acta Cytol.* 13:377, 1969.

31. Bibbo M., Ali I., Al-Naqeeb M., Baccarini I., Climaco L.A., Gill W., Sonek M., Wied G.L. Cytologic findings in female and male offspring of DES treated mothers. *Acta Cytol.* 19:568, 1975.

32. Blaikley J.B., et al. Vaginal adenosis: clinical and pathological features with special reference to malignant change. *J. Obstet. Gynaec. Brit. Commonw.* 78:1115, 1971.

33. Botella Llusia J., and van Keep P.A. Vaginal cytology in the postmenopause. A study into some correlates. *Acta Cytol.* 21:18, 1977.

34. Bourg R. Phase constrast microscopic study of vaginal exfoliative cytology. Technic and indications. *Brux. Med.* 52:189, 1972.

35. Braga C.A., Teoh T.B. Amoebiasis of the cervix and vagina. *J. Obstet. Gynaec. Brit. Commonw.* 71:289, 1965.

36. Buchler D.A., et al. Primary lymphoma of the vagina. *Obstet. Gynec.* 40:235, 1972.

37. Butler E.B., Taylor D.S. The postnatal smear. *Acta Cytol.* 17:237, 1973.

38. Cabanne F., Michiels C.R., Mottot C., Bastien H. Ciliated bodies in gynecologic cytopathology: parasite or cellular debris? *Acta Cytol.* 19:407, 1975.

39. Ceelen G.H., Sakurai M. Vaginal cytology in leukemia. *Acta Cytol.* 6:370, 1962.

40. Chiaffitelli H.S. de, Dominguez S. Vaginal cytology in patients treated with steroid combinations. *Acta Cytol.* 14:344, 1970.

41. Colmenares R.F., Naib Z.M. Significance of lymphocytic pools in the routine vaginal smear. *Obstet. Gynec.* 26:909, 1965.

42. Czygan P.J., Schulz K.D. Studies on the anti-oestrogenic and oestrogen like action of clomiphene citrate in women. *Gynec. Invest.* 3:126, 1972.

43. Danos M.L. Post-partum cytology: observations over a four year period. *Acta Cytol.* 12:308, 1968.

44. Diagnostic accuracy of vaginal smears. Letter to the Editor. *Lancet* 2:7894, 1974.

45. Dominguez A., Giron J.J. Toxoplasma cysts in vaginal smears. *Acta Cytol.* 20:269, 1976.

46. Esin G.S. Vaginal smear in a case of local infection with herpes simplex virus. *Zbl. Gynak.* 94:863, 1972.

47. Etherston W.C., Meyers A., Spekhard M.E. Adenocarcinoma of the vaginal in young women. The stilbestrol-adenosis-adenocarcinoma of the vaginal syndrome. *Wisconsin Med. J.* 71:87, 1972.

48. Eustratiades M.G., Batrinos M. The diagnostic significance of parabasal cells. II. Rate of their disappearance and reappearance under estrogen administration and withdrawl. *Acta Cytol.* 19:100, 1975.

49. Fentanes de Torres E., Benitez-Bribiesca, L. Cytologic detection of vaginal parasitosis. *Acta Cytol.* 17:252, 1973.

50. Forsberg J.G. Estrogen, vaginal cancer and vaginal development. *Amer. J. Obstet. Gynec.* 113:83, 1972.

51. Frable W.J., Smith J.H., Perkins J., Foley C. Vaginal cuff cytology. Some difficult diagnostic problems. *Acta Cytol.* 17:135, 1973.

52. Garcia-Valdecasas R., Rodriguez-Rico L., Linares J., Galera H., Salvatierra V. Malignant melanoma of the vagina. A case diagnosed cytologically. *Acta Cytol.* 18:535, 1974.

53. Garret R. Extrauterine tumor cells in vaginal and cervical smears. *Obstet. Gynec.* 14:21, 1959.

54. Gaudefroy M., Coliche D. Deux cas de vaginite dûe à un parasite encore inconnu en pathologie gynécologique. *J. Sciences Médicales de Lille* 89:1, 1971.

55. Gompel C., Horanyi Z., Simonet M.L. Ultrastructure of clear-cell carcinoma of the vagina and cervix. Report of a case with unusual structural findings. *Acta Cytol.* 20:262, 1976.

56. Graham R.M., Meigs J.V. The value of the vaginal smear. *Amer. J. Obstet. Gynec.* 58:843, 1949.

57. Graham R.M., Van Niekerk W.A. Vaginal cytology in cancer of the ovary. *Acta Cytol.* 6:496, 1962.

58. Gunning J.E., Ostergard D.R. Value of screening procedures for the detection of vaginal adenosis. *Obstet. Gynec.* 47:268, 1976.

59. Harrison V., Peat G. Fetal growth in relation to vaginal cytology. *Acta Cytol.* 18:210, 1974.

60. Heber K.R. Vaginal smear patterns in women taking ethynodiol diacetate as an oral contraceptive. *Med. J. Aust.* 1:379, 1970.

61. Heber, K.R. The effect of progestogens on vaginal cytology. *Acta Cytol.* 19:103, 1975.

62. Herbst A.L., Robboy S.J., Scully R.E., Poskanzer D.C. Clear-cell adeno-carcinoma of the vagina and the cervix in girls: Analysis of 170 Registry cases. *Amer. J. Obstet. Gynec.* 119:713, 1974.

63. Herbst A.L., Ulfelder H., Poskanzer D.C. Adenocarcinoma of the vagina. Association of maternal stilbestrol therapy with tumor appearance in young women. *New Engl. J. Med.* 284:878, 1971.

64. Hugoson A., Winberg E., Angström T. Cytologic findings in vaginal and oral smears from pregnant women. *Acta Cytol.* 16:111, 1972.

65. Huss K.S. Maternal diethylstilbestrol a time bomb for child? *JAMA* 218:1564, 1971.

66. Jenkins D.M., Goulden R. Psammoma bodies in cervical cytology smears. *Acta Cytol.* 21:112, 1977.

67. Johnston W.W. The cytopathology of mycotic infections. *Lab. Med.* 2:34, 1971.

68. Kaufman R.H., et al. Cervical and vaginal cytology in the child and adolescent. *Pediatr. Clin. North Amer.* 19:547, 1972.

69. Kauraniemi, T.V. The effect of oral contraceptive on hormonal cytology. *Annales Chir. et Gynaecol. Fenniae* 63:6, 1974.

70. Keebler C.M., Wied G.L. The estrogen test: an aid in differential cytodiagnosis. *Acta Cytol.* 18:482, 1974.

71. Krause W., Böhm W., Müller W. Vergleichende zytologische und histologische Studien nach Behandlung mit einer ovulationsunterdrückenden Ostrogen gestagen Kombination (ovosiston) *Gynaecologia* 166:432, 1968.

72. Lewis B.V., Chapman P.A. Cytological diagnosis of a primary malignant melanoma of the vagina. *Brit. J. Obst. Gynec.* 82:74, 1975.

73. Lloyd H.E.D., Feinberg R. Lymphoid follicular cells of uterine cervix in vaginal smears. *Acta Cytol.* 10:467, 1966.

74. Luksch F. Leukemia cells in vaginal smears. *Acta Cytol.* 8:95, 1964.

75. Malvi S.G., Sirsat S.M. A correlative study of the cytomorphologic and cytochemical changes accompanying maturation and growth in the human vaginal epithelium. *Indian J. Med. Res.* 62:220, 1975.

76. Masubuchi S. Jr., Nagai I., Hirata M., Kubo H., Masubuchi K. Cytologic studies of malignant melanoma of the vagina. *Acta Cytol.* 19:527, 1975.

77. McDonald R.R., et al. Cervical mucus, vaginal cytology and steroid excretion in recurrent abortion. *Obstet. Gynec.* 40:394, 1972.

78. McLennan M.T., McLennan C.E. Hormonal patterns in vaginal smears from puerperal women. *Acta Cytol.* 19:431, 1975.

79. Meisels A. Le diagnostic cytohormonal durant la grossesse. *Laval Med.* 34:551, 1963.

80. Meisels A. Computed cytohormonal findings in 3307 healthy women. *Acta Cytol.* 9:328, 1965.

81. Meisels A. The menopause: A cytohormonal study. *Acta Cytol.* 10:49, 1966.

82. Meisels A. The maturation value. *Acta Cytol.* 11:249, 1967.

83. Merchant S., Murad T.M., Dowling E.A., Durant J. Diagnosis of vaginal carcinoma from cytologic material. *Acta Cytol.* 18:494, 1974.

84. Metha P.V. Study correlating endometrial biopsies and vaginal cytology in one hundred tubectomized women. *Acta Cytol.* 19:330, 1975.

85. Meyer A.A., Okagaki T. Microspectrophotometric study of post-irradiation dysplasia in vaginal smears. *Cancer* 30:964, 1972.

86. Moracci E., Berlingieri D. Hormonal evaluation of vaginal smears from artificial vagina. *Acta Cytol.* 17:131, 1973.

87. Nasiell M. Histiocytes and histiocytic reaction in vaginal cytology. *Cancer* 14:1223, 1961.

88. Navab A., Koss L.G., La Due J.S. Estrogen-like activity of digitalis. Its effect on the squamous epithelium of the female genital tract. *JAMA*, 194:30, 1965.

89. Ng A.B.P. Reagan J.W., Hawliczek S., Wentz W.B. Cellular detection of vaginal adenosis. *Obstet. Gynec.* 46:323, 1975.

90. Nissen E.D., Godstein A.I. Stilbestrol therapy in pregnancy. Relationship to vaginal neoplasia in offspring. *Int. J. Gynaec. Obstet.* 11:138, 1973.

91. Nix H.G., Wright H.L. Mesonephric adenocarcinoma of the vagina. *Amer. J. Obstet. Gynec.* 99:893, 1967.

92. Nyklicek A. Vaginal cytology and amnioscopy in prolonged pregnancies. *Acta Cytol.* 16:48, 1972.

93. Palumbo L. Jr., Shingleton H.M., Fishburne J.I. Jr., Pepper F.D. Jr., Koch G.G. Primary carcinoma of the vagina. *Southern Med. J.* 62:1048, 1969.

94. Papanicolaou G.N., Observations on the origin and specific functions of the histiocytes in the female genital tract. *Fertil. Steril.* 4:472, 1953.

95. Papanicolaou G.N., Shorr E. The action of ovarian follicular hormones in the menopause as indicated by vaginal smears. *Amer. J. Obstet. Gynec.* 31:806, 1936.

96. Papanicolaou G.N., Traut H.F. Diagnostic value of vaginal smears in carcinoma of uterus. *Amer. J. Obstet. Gynec.* 42:193, 1941.

97. Pundel J.P. *Précis de Colpocytologie hormonale.* Masson, Paris, 1966.

98. Pundel J.P., Van Meensel F. *Gestation et cytologie Vaginale.* Masson, Paris, 1951.

99. Rakoff A. The vaginal cytology of gynecologic endocrinopathies. *Acta Cytol.* 5:153, 1961.

100. Reagan J.W., Lin, F. An evaluation of the vaginal irrigation technique in the detection of uterine cancer. *Acta Cytol.* 11:374, 1967.

101. Reyniak J.V., Sedlis A., Stone D., Connell E. Cytohormonal findings in patients using various forms of contraception. *Acta Cytol.* 13:315, 1969.

102. Rodgerson E.B. Vulvovaginal papillomas and *Trichomonas Vaginalis. Obstet. Gynec.* 40:327, 1972.

103. Rosenblatt, R. La cytologie vaginale au cours des deuxième et troisième trimestres des grossesses et son rôle dans le dépistage précoce de la souffrance foetale. *Gynaecologia* 168:393, 1969.

104. Rubio C.A. Estrogenic effect in vaginal smears in cases of carcinoma in situ and microinvasive carcinoma of the uterine cervix. *Acta Cytol.* 17:361, 1973.

105. Rubio C.A. Sigurdson A., Zajicek J. Viability tests in exfoliated cells from vaginal and oral epithelia. *Acta Cytol.* 17:32, 1973.

106. Ruffolo E.H., Foxworthy D., Fletcher J.C. Vaginal adenocarcinoma arising in vaginal adenosis. *Amer. J. Obstet. Gynec.* 111:167, 1971.

107. Sandberg E.C. The incidence and distribution of occult vaginal adenosis. *Amer. J. Obstet. Gynec.* 101:322, 1968.

108. Schneider V., Friedrich E., Schindler A.E. Hormonal cytology: a correlation with plasma estradiol, measured by radioimmunoassay. *Acta Cytol.* 21:37, 1977.

109. Schnell J.D., Voigt W.H. Are yeasts in vaginal smears intracellular or extracellular? *Acta Cytol.* 20:343, 1976.

110. Sen D.K., Langley F.A. Vaginal cytology as a monitor of fetal well-being in early pregnancy. *Acta Cytol.* 16:116, 1972.

111. Seshadri R., et al. Changes in vaginal cytology in women with androgenic lesions of the adreno-cortical-gonadal axis. *Indian J. Pathol. Bacteriol.* 14:142, 1971.

112. Silverberg S.G., De Giorgi L.S. Clear cell carcinoma of the vagina. *Cancer* 29:1680, 1972.

113. Smolka H. Ein bemerkenswerter vaginaler Zellbefund bei latenter Toxoplasmose. *Zbl f. gyn.* 75:730, 1953.

114. Song Y.S. The significance of positive vaginal smears in extrauterine carcinomas. *Amer. J. Obstet. Gynec.* 71:341, 1957.

115. Stafl A., Mattingly R.F. Vaginal adenosis: a precancerous lesion? Obstet. Gynec. 120:666, 1974.

116. Stern E., Longo L.L. Identification of herpes simplex virus in a case showing cytological features of viral vaginitis. *Acta Cytol.* 7:295, 1963.

117. Symposium on cytologic terminology. *Acta Cytol.* 2:26, 1958.

118. Symposium on hormonal cytology. *Acta Cytol.* 12:112, 1968.

119. Symposium B. Hormonal cytology during pregnancy and the postpartum period. *Acta Cytol.* 3:269, 1959.

120. Symposium Effects of endogenous estrogens on the vaginal epithelium. *Acta Cytol.* 4:13, 1960.

121. Taft P.O., Robboy S.J., Herbst A.L., Scully R.E. Cytology of clear-cell adenocarcinoma of genital tract in young females: Review of 95 cases from the Registry. *Acta Cytol.* 18:279, 1974.

122. Ten Gerge B.S. Value of vaginal cytologic study during use of oral contraceptive for timely detection of cancer of uterus. *Geneesk. Gids* 1:560, 1970.

123. Teter J. The use of selected cytologic indices for evaluation of estrogenicity of synthetic compounds. *Acta Cytol.* 16:366, 1972.

124. Tsukada Y., Hewett W.J., Barlow J.J., Pickren J.W. Clear-cell adenocarcinoma "mesonephroma" of the vagina. Three cases associated with maternal synthetic nonsteroid estrogen therapy. *Cancer* 29:1208, 1972.

125. Vooijs P.G., Ng A.B.P., Wentz W.B. The detection of vaginal adenosis and clear cell carcinoma. *Acta Cytol.* 17:59, 1973.

126. Wachtel E.G. The cytology of amenorrhea. *Acta Cytol.* 10:56, 1966.

127. Wachtel E.G. Vaginal cytology after the menopause. In Van Keep P.A., Lauritzen, *Estrogens in the Post-Menopause. Front. Hormone Res.*, 3:63, 1975.

128. Wied G.L. Evaluation of endocrinologic condition by means of exfoliative cytology. In Gold J.J. (Ed.), *Gynecologic Endocrinology.* Harper & Row, New York, 1968, Ch. 8, p. 133.

129. Wied G.L. The effect of physiological sex hormones on the vaginal epithelium of patients with inactive ovaries. *Acta Cytol.* 1:75, 1957.

130. Wied G.L. Importance of the site from which vaginal smears are taken. *Amer. J. Clin. Path.* 25:742, 1955.

131. Wood C., Osmond-Clarke F., Murray M. Vaginal cytology in pregnancy. *J. Obstet. Gynaec. Brit. Commw.* 68:778, 1961.

Cervix

132. Abell M.R., Ramirez J.A. Sarcomas and carcinosarcomas of the uterine cervix. *Cancer* 31:1176, 1973.

133. Abitbol M.M., Benjamin F., Gastillo N. Management of the abnormal cervical smear and carcinoma in situ of the cervix during pregnancy. *Amer. J. Obstet. Gynec.* 117:904, 1973.

134. Acosta A., Kaplan A.L., Kaufman R.H. Gynecologic cancer in children. *Amer. J. Obstet. Gynec.* 112:944, 1972.

135. Adam E., Levy A.H., Rawls W.E., Melnick J.L. Seroepidemiologic studies of herpesvirus type 2 and carcinoma of the cervix. I. Case control matching. *J. Nat. Cancer Inst.* 47:941, 1971.

136. Adam E., Sharma S.D., Zeigler O., Iwamoto K., Melnick J.L., Levy A.H., Rawls W.E. Seroepidemiologic studies of herpesvirus type 2 and carcinoma of the cervix. II Uganda. *J. Nat. Cancer Inst.* 48:65, 1972.

137. Albrechtsen R., Koch F. Prognostic accuracy of cervical cone biopsy. Prediction of complete or incomplete removal of cervical carcinoma in situ. *Danish Med. Bull.* 19:119, 1972.

138. Alexander E.R. Possible etiologies of cancer of the cervix other than herpesvirus. *Cancer Res.* 33:1485, 1973.

139. Alling Moller K. Organisation of population screening for cervical carcinoma in the county of Maribo, 1967–1969. *Danish Med. Bull.* 19:117, 1972.

140. Alousi M.A., Ballard L.A., Reilly J.V., Alousi S.S. Microinvasive carcinoma and inflammatory lesions of the cervix uteri: Histologic and cytologic differentiation. *Acta Cytol.* 11:132, 1967.

141. Amstey M.S., Patten S.F. Carcinoma of the cervix and genital herpes virus infection. *Obstet. Gynec.* 39:202, 1972.

142. Anderson W.A., Gunn S.A. Cancer of the cervix. Further studies of the patient-obtained vaginal irrigation smear. *Cancer* 17:102, 1967.

143. Arffmann E., Jacobsen J.C.N. Specific cervical cytodiagnosis in gynaecological malignancies. *Danish Med. Bull.* 22:24, 1975.

144. Arneson A.N., Schellhas H.F. Multiple primary cancers in patients treated for carcinoma of the cervix. *Amer. J. Obstet. Gynec.* 106:1155, 1970.

145. Aurelian L., Royston I., Davis H.J. Antibody to genital herpes simplex virus: association with cervical atypia and carcinoma in situ. *J. Nat. Cancer Inst.* 45:455, 1970.

146. Barron B.A., Richart R.M. Statistical model of the natural history of cervical carcinoma. II. Estimates of the transition time from dysplasia to carcinoma in situ. *J. Nat. Cancer Inst.* 45:1025, 1970.

147. Bartels P.H., Bibbo M., Bahr G.F., Taylor J., Wied G.L. Cervical cytology: descriptive statistics for nuclei of normal and atypical cell types. *Acta Cytol.* 17:449, 1973.

148. Bashir-Farahmand J., Taft P.D., McArthur J.W. Cyclic exfoliation of the endocrinal epithelium in the bonnet macaque. *Acta Cytol.* 20:167, 1976.

149. Benson P.A. Psammoma bodies found in cervico-vaginal smears. *Acta Cytol.* 17:64, 1973.

150. Berggren O. Association of carcinoma of the uterine cervix and *Trichomonas vaginalis* infestations. *Amer. J. Obstet. Gynec.* 105:166, 1969.

151. Berry A. Multispecies schistosomal infections of the female genital tract detected in cytology smears. *Acta Cytol.* 20:361, 1976.

152. Bertalanffy F.D. Cell renewal as the basis of diagnosis exfoliative cytology. *Amer. J. Obstet. Gynec.* 85:383, 1963.

153. Bhattacharya P.K., Bartels P.H. Bibbo M., Taylor J., Wied G.L. Estimation procedure for the cellular composition of cervical smears. *Acta Cytol.* 19:366, 1975.

154. Bibbo M., et al. The numerical composition of cellular samples from the female reproductive tract. I. Carcinoma in situ. *Acta Cytol.* 19:438, 1975.

155. Bibbo M., et al. The numerical composition of cellular samples from the female reproductive tract. II. Cases with invasive squamous carcinoma of uterine tract. *Acta Cytol.* 20:249, 1976.

156. Bibbo M., et al. The numerical composition of cellular samples from the female reproductive tract. III. Cases with mild and moderate dysplasia of uterine cervix. *Acta Cytol.* 20:565, 1976.

157. Bibbo M., Keebler C.M., Wied G.L. The cytologic diagnosis of tissue repair in the female genital tract. *Acta Cytol.* 15:133, 1971.

158. Boyes D.A. Age for routine cervical Papanicolaou screening tests. *JAMA* 232:961, 1975.

159. Bradburn G.B., Webb C.F. Cyclic variations in the endocervix. *Amer. J. Obstet. Gynec.* 62:997, 1951.

160. Broders A.C. Carcinoma: grading and practical application. *Arch. Path. Lab. Med.* 2:376, 1926.

161. Bryans F.E., Boyes D.A., Fidler H.K. The influence of a cytological screening program upon the incidence of invasive squamous cell carcinoma of the cervix in British Columbia. *Amer. J. Obstet. Gynec.* 88:898, 1964.

162. Butler B.E., Stanbridge C., Wells R. Personal communication, 1977.

163. Calanog A., Soll S., Gordon M., et al. Comprehensive cytologic screening in patients undergoing voluntary abortions. *Amer. J. Obstet. Gynec.* 118:102, 1974.

164. Carcinoma of the cervix and herpesvirus. Letter to the Editor. *Brit. Med. J.* 2:548, 1972.

165. Catalano L.W., Johnson L.D. Herpesvirus antibody and carcinoma in situ of the cervix. *J. Amer. Ass. Cancer Res.* 217:447, 1971.

166. Chatfield W.R., Bremner A.D. Intrauterine sponge biopsy. *Gynec. Obstet.* 39:323, 1972.

167. Child A.P., Spriggs A.I., Berry R.J. DNA-synthesizing cells in cervical scrapings. A potential but presently unusable approach to screening. *Acta Cytol.* 16:37, 1972.

168. Christopherson W.M. The changing patterns of cervix cancer. *Cancer* 21:283, 1971.

169. Christopherson W.M., Parker J.E. Cervix cancer death rates and mass cytologic screening. *Cancer* 26:808, 1970.

170. Christopherson W.M., Scott M.A. Trends in mortality from uterine cancer in relation to mass screening. *Acta Cytol.* 21:5, 1977.

171. Clark A.H., McKee E.E., Dixon D.C. Identification of trophozoites from *Entamoeba histolytica* by cytology techniques. *Acta Cytol.* 16:429, 1972.

172. Cobb C.M., Zuckerman J.E., Annis M. In vitro localization of 67 Ga in exfoliated epithelial cells from the uterine cervix. *Cancer Res.* 33:1578, 1973.

173. Coleman D.V. A case of tuberculosis of the cervix. *Acta Cytol.* 13:104, 1969.

174. Coleman S.A., Rube I.F., Kashgarian M., Erickson C.C. An appraisal of the irrigation cytology method for uterine cancer detection. *Acta Cytol.* 14:502, 1970.

175. Coppleson L.W., Brown B. Estimation of the screening error data from the observed detection rates in repeated cervical cytology. *Amer. J. Obstet. Gynec.* 119:953, 1974.

176. Cramer D.W. The role of cervical cytology in the declining morbidity and mortality of cervical cancer. *Cancer* 34:2018, 1974.

177. Crapanzano J.T. Office diagnosis in patients with abnormal cervicovaginal cytosmears: correlation of colposcopic, biopsy and cytologic findings. *Amer. J. Obstet. Gynec.* 113:967, 1972.

178. Creasman W.T., Rutledge F. Carcinoma in situ of the cervix. *Obstet. Gynec.* 39:373, 1972.

179. Cuyler W.K.L., Kaufman L.A., Carter B., Ross R.A., Thomas W.L., Palumbo L. Genital cytology in obstetric and gynecologic patients. A four-year study. *Amer. J. Obstet. Gynec.* 62:262, 1951.

180. Dabancens A., Prado R., Larraguibel R., Zanartu J. Intraepithelial cervical neoplasia in women using intrauterine devices and long-acting injectable progestogens as contraceptives. *Amer. J. Obstet. Gynec.* 119:1052, 1974.

181. Dance E.F., Fullmer C.D. Extrauterine carcinoma cells observed in cervico-vaginal smears. *Acta Cytol.* 14:187, 1970.

182. Davies S.W., Kelly R.M. Intraepithelial carcinoma of the cervix uteri in women aged under 35 years. *Brit. Med. J.*4:525, 1971.

183. Daw E. Extragenital adenocarcinoma metastatic to cervix uteri. *Amer. J. Obstet. Gynec.* 114:1104, 1972.

184. de Brux J., Dupre-Froment J. Essai d'histogénèse et de pronostic des dysplasies et du prétendu carcinome intra-epithélial.*Presse Méd.* 68:1753, 1960.

185. Dickinson L.E. Control of cancer of the uterine cervix by cytologic screening. *Gynecologic Oncology* 3:1, 1975.

186. Dickinson L. Evaluation of the effectiveness of cytologic screening for cervical cancer. III. Cost-benefit analysis. *Mayo Clin. Proc.* 47:550, 1972.

187. Dickinson L., Mussey M.E., Kurland L.T. Evaluation of the effectiveness of cytologic screening for cervical cancer. II. Survival parameters before and after inception of screening. *Mayo Clin. Proc.* 47:545, 1972.

188. Dickinson L., Mussey M.E., Soule E.H., Kurland L.T. Evaluation of the effectiveness of cytologic screening for cervical cancer. I. Incidence and mortality trends in relation to screening. *Mayo Clin. Proc.* 47:534, 1972.

189. Do cervical smears save lives? *Brit. Med. J.* 2:585, 1969.

190. Donohue L.R., Meriwether W. Colposcopy as a diagnostic tool in the investigation of cervical neoplasias. *Amer. J. Obstet. Gynec.* 113:107, 1972.

191. Dowdy A.H., Lagasse L.D., Sperling L., Barker W.F. A combined screening program for the detection of carcinoma of the cervix and carcinoma of the breast. *Surg. Gynec. Obstet.* 131:93, 1970.

192. Dunn J.E. Jr. Preliminary findings of the Memphis-Shelby county uterine cancer study and their interpretation. *Amer. J. Public Health* 48:861, 1958.

193. Edwards D. Gynaecological abnormalities found at a cytology clinic. *Brit. Med. J.* 4:218, 1974.

194. Eisenstein R., Battifora H. Lymph follicles in cervical smears. *Acta Cytol.* 9:344, 1965.

195. Epstein N. The significance of cellular atypia in the diagnosis of malignancy of ulcers of the female genital tract. *Acta Cytol.* 16:483, 1972.

196. False-positive cytology in ectopic pregnancy. *New Engl. J. Med.* 291:1142, 1974.

197. Farr G.H., Hajdu S.I. Exfoliative cytology of metastatic neuroblastoma. *Acta Cytol.* 16:203, 1972.

198. Feldman M., Poulsen R., Shepherd L., Marshall K.G. The occurrence of isolated dysplastic and carcinoma in situ type cells in cervical smears from patients with dysplasia and carcinoma in situ: significance to prescreening using image processing technique. *Acta Cytol.* 17:395, 1973.

199. Ferenczy A., Richart R.M. Scanning electron microscopy of the cervical transformation zone. *Amer. J. Obstet. Gynec.* 115:151, 1973.

200. Fidler H.K., Boyes D.A., Loeb D.R. Intra-epithelial carcinoma of the cervix, 214 cases, with emphasis on investigation by cytology and cone biopsy. *Canad. Med. Ass.* 77:79, 1957.

201. Fidler H.K., Boyes D.A., Worth A.J. Cervical cancer detection in British Columbia, a progress report. *J. Obstet. Gynaecol. Br. Commonw.* 75:392, 1968.

202. Figge D.C., Bennington J.L., Schweid A.I. Cervical cancer after initial negative and atypical vaginal cytology. *Amer. J. Obstet. Gynec.* 108, 422, 1970.

203. Findler H.K., Boyes D.A. Patterns of early invasion from intraepithelial carcinoma of the cervix. *Cancer* 12:673, 1959.

204. Forsberg J.G. Cervico-vaginal epithelium: Its origin and development. *Amer. J. Obstet. Gynec.* 115:1025, 1973.

205. Fox C.H. Time necessary for conversion of normal to dysplastic cervical epithelium. *Obstet. Gynec.* 31:749, 1968.

206. Frable W.J., McDonald B., Smith J. "The mini-coner." Evaluation as a biopsy technique in patients with abnormal cervical cytology. *Acta Cytol.* 17:410, 1973.

207. Frable W.J., Smith J.H., Perkins J., Foley C. Vaginal cuff cytology. Some difficult diagnostic problems. *Acta Cytol.* 17:135, 1973.

208. Friedell G.H., Hertig A.T., Younge P.A. *Carcinoma in Situ of the Uterine Cervix.* Thomas Springfield, Ill., 1960.

209. Frost J.K. Diagnostic accuracy of the cervical smears. *Obstet. Gynec. Survey* 24:893, 1969.

210. Gallager H.S., Simpson C.B., Ayala A.G. Adenoid cystic carcinoma of the uterine cervix. Report of 4 cases. *Cancer* 27:1398, 1971.

211. Gompel, C. Is there a physiological cell type which may be defined as "androgenic cell type"? *Acta Cytol.* 1:83, 1957.

212. Gompel C. Possibilités d'appréciation de l'évolution d'un cancer génital par la cytologie exfoliatrice après radiothérapie. *Bull. Soc. Royale Belge Gynec. Obstet.* 28:71, 1958.

213. Gompel, C. Le cancer in situ du col utérin. *Acta Union Internationalis Contra Cancrum:* 14:447, 1958.

214. Gondos B., Marschall D., Ostergard D.R. Endocervical cells in cervical smears. *Amer. J. Obstet. Gynec.* 114:833, 1972.

215. Gondos B., Towsend D.E., Ostergard D.R. Cytologic diagnosis of squamous dysplasia and carcinoma of the cervix. *Amer. J. Obstet. Gynec.* 110:107, 1971.

216. Graham R.M. Definition of radiation response on normal squamous cells (RR cells). *Acta Cytol.* 3:347, 1959.

217. Graham R.M. The method of prognosis after irradiation by means of exfoliative cytology. *Acta Cytol.* 3:400, 1959.

218. Graham R.M., Meigs J.V. The value of the vaginal smear. *Amer. J. Obstet. Gynec.* 58:843, 1949.

219. Graham S., Priore R.L., Schueller E.F., Burnett W. Epidemiology of survival from cancer of the cervix. *J. Nat. Cancer Inst.* 49:639, 1972.

220. Green T.H. Further trial of a cytologic method for selecting either radiation or radical operation in the primary treatment of cervical cancer. Experience with 264 consecutively treated patients. *Amer. J. Obstet. Gynec.* 112:544, 1972.

221. Gupta P.K., Hollander D.H., Frost J.K. Actinomycetes in cervico-vaginal smears: an association with I.U.D. usage. *Acta Cytol.* 20:295, 1976.

222. Haour P., Conti C. Cytochemistry of irradiated cells. *Acta Cytol.* 3:357, 1959.

223. Heller C.J. *Neisseria gonorrhoeae* in Papanicolaou smears. *Acta Cytol.* 18:338, 1974.

224. Heller C.J., et al. Squamous cell changes associated with the presence of Candida sp. in cervical-vaginal Papanicolaou smears. *Acta Cytol.* 15:379, 1971.

225. Hermann G., Deininger J.T. Correspondence to the Editor. Vorticella, an unusual protozoa found on endocervical smear. *Acta Cytol.* 7:129, 1963.

226. Herrera I., Valenciano L., Sanchez-Garrido F., Botella-Llusia J. On findings of virus-like structures in uterine cervical carcinoma. *Acta Cytol.* 18:45, 1974.

227. Highman W.J. Cervical smears in tuberculous endometritis. *Acta Cytol.* 16:16, 1972.

228. Hill E.C. Clear cell carcinoma of the cervix and vagina in young women. *Amer. J. Obstet. Gynec.* 116:470, 1973.

229. Hinselmann H. Beitrag zur Ordnung und Ableitung der Leukoplakien des weiblichen genitaltraktes. *Z. Geburtsch. Gynäk* 101:142, 1932.

230. Hinselmann H. Die klinische und mikroskopische Frühdiagnose des Portiokarzinoms. *Arch. f. Gynaekol.* 156:239, 1934.

231. Holley M.R. Management of the patient with an abnormal Papanicolaou smear. *Amer. J. Obstet. Gynec.* 110:979, 1971.

232. Hulka B.S. et al. Predictors of atypical smear status among women attending a cervical cancer screening program. *Amer. J. Epidem.* 94:564, 1971.

233. Hulka B.S., Redmond C.K. Factors related to progression of cervical atypias. *Amer. J. Epidem.* 93:23, 1971.

234. Husain O.A.N. The irrigation smear. A comparative trial of vaginal irrigation pipette and spatula smears in the detection of cervical cancer. *Amer. J. Obstet. Gynec.* 106:138, 1970.

235. Husain O.A.N. Quality control in cervical cytology. *J. Clin. Path.* 27:935, 1974.

236. Jones H.W., Droegemueller W., Makowski E.L. A primary melanocarcinoma of the cervix. *Amer. J. Obstet. Gynec.* 111:959, 1971.

237. Jordan S.W., Evangel E., Smith N.L. Ethnic distribution of cytologically diagnosed herpex simplex genital infections in a cervical cancer screening program. *Acta Cytol.* 16:363, 1972.

238. Kanbour A., Doshi N., Romadella R. Significance of lymphoid follicular cells in cervico-vaginal smears (abstract). *Acta Cytol.* 20:583, 1976.

239. Kaufman R.J., et al. Cervical cytology in the teen-age patient. *Amer. J. Obstet. Gynec.* 108:515, 1970.

240. Kay S., Frable W.J., Hume D.M. Cervical dysplasia and cancer developing in women on immunopression therapy for renal homotransplantation. *Cancer* 26:1048, 1970.

241. Kessler I.I., Kulcar Z., Zimolo A., Grgurevic M., Strnad M., Goodwin B.J. Cervical cancer in Yugoslavia. II. Epidemiologic factors of possible etiologic significance. *J. Nat. Cancer Inst.* 53:51, 1974.

242. Kleger B., Prier J.E., Rosato D.J., McGinnis A.E. Herpes simplex infection of the female genital tract. I. Incidence of Infection. *Amer. J. Obstet. Gynec.* 102:745, 1968.

243. Klinken L., Koch F., Albrechtsen R. Comparison of pipette and smear methods in population screening for carcinoma of the uterine cervix. *Danish Med. Bull.* 19:138, 1972.

244. Koch F., Albrechtsen R. Punch biopsy or conization. *Danish Med. Bull.* 19:127, 1972.

245. Koss L.G. Significance of dysplasia. *Clin. Obstet. Gynec.* 13:873, 1970.

246. Koss L.G. Detection of carcinoma of the uterine cervix. *JAMA* 222:699, 1972.

247. Koss L.G., Durfee G.R. Unusual patterns of squamous epithelium of uterine cervix; cytologic and pathologic study of koilocytotic atypia. *Ann. N.Y. Acad. Sci.* 63:1245, 1956.

248. Koss L.G., Phillips A.J. Summary and recommendations of the workshop on uterine-cervical cancer. *Cancer* 33:753, 1974.

249. Koss L.G., Stewart F., Foote F.W. Some histological aspects of behaviour of epidermoid carcinoma in-situ and related lesions of the uterine cervix. A long-term prospective study. *Cancer* 16:1160, 1963.

250. Kurman R.J., Prabha A.C. Thyroid and parathyroid glands in the vaginal wall: report of a case. *Amer. J. Clin. Path.* 59:503, 1973.

251. Lewis C.E. Consumer control of carcinoma of the cervix. *Amer. J. Obstet. Gynec.* 119:669, 1974.

252. Llanes A.T., Farre C.B., Ferenczy A., Richart R.M. Scanning electron microscopy of normal exfoliated squamous cervical cells. *Acta Cytol.* 17:507, 1973.

253. Lucas W.E., Benirschke K., Lebherz T.B. Verrucous carcinoma of the female genital tract. *Amer. J. Obstet. Gynec.* 119:435, 1974.

254. Mac Gregor J.E. Cervical carcinoma. The beginning of the end? *Lancet* 2:1296, 1967.

255. MacGregor, J.E., Fraser M.E., Mann E.M. Improved prognosis for cervical cancers due to comprehensive screening. *Acta Cytol.* 16:14, 1972.

256. Malhotra S.L. A study of carcinoma of uterine cervix with special reference to its causation and prevention. *Brit. J. Cancer* 25:62, 1971.

257. Marsan C. Possibilités et limites du cyto-diagnostic des cancers du col utérin irradies. *Bull. Fed. Soc. Gynec. Fr.* 23:238, 1971.

258. Martin P.L. How preventable is invasive cervical cancer. A community study of preventable factors. *Amer. J. Obstet. Gynec.* 113:541, 1972.

259. Martzloff K.H. Carcinoma of the cervix uteri: a pathological and clinical study with particular reference to the relative malignancy of the neoplastic process as indicated by the predominant type of cancer cell. *Bull. Johns Hopk. Hosp.* 34:141, 1923.

260. Mason P.R., Super H., Fripp P.J. Comparison of four techniques for the routine diagnosis of *Trichomonas vaginalis infection. J. Clin. Path.* 29:154, 1976.

261. Masubuchi K., Significance of the role of cytology in population screening of cancers. *Acta Cytol.* 19:334, 1975.

262. Masubuchi K., Nemoto H. Epidemiologic studies on uterine cancer at Cancer Institute Hospital, Tokyo, Japan. *Cancer* 30:268, 1972.

263. May D. Error rates in cervical cytological screening tests. *Brit. J. Cancer* 29:106, 1974.

264. McLennan M.T., McLennan C.E. Significance of cervicovaginal cytology after radiation therapy for cervical carcinoma. *Amer. J. Obstet. Gynec.* 121:96, 1975.

265. Meisels A., Ayotte D. Cells from the seminal vesicles: contaminants of the V-C-E smear. *Acta Cytol.* 20:211, 1976.

266. Meisels A., Fortin R. Condylomatous lesions of the cervix and vagina. I. Cytologic patterns. *Acta Cytol.* 20:505, 1976.

267. Merchant S., Murad T.M., Dowling E.A., Durant J. Diagnosis of vaginal carcinoma from cytologic material. *Acta Cytol.* 18:494, 1974.

268. Miller D.F. The impact of hormonal contraceptive therapy on a community and effects on cytopathology of the cervix. *Amer. J. Obstet. Gynec.* 115:978, 1973.

269. Moghissi K.S., Mack H.C., Porzak J.P. Epidemiology of cervical cancer study of a population. *Amer. J. Obstet. Gynec.* 100:607, 1968.

270. Nahmias A.J., Josey W.E., Naib Z.M., Luce G.M., Guest B.A. Antibodies to herpesvirus hominis type I and II in humans. Women with cervical cancer. *Amer. J. Epidem.* 91:547, 1970.

271. Naib Z.M. Exfoliative cytology of viral cervico-vaginitis. *Acta Cytol.* 10:126, 1966.

272. Naib Z.M., Mazukawa N. Identification of condyloma acuminata cells in routine vaginal smears. *Obstet. Gynec.* 13:735, 1961.

273. Naib Z.M., Nahmias A.J., Josey W.E. Cytology and histopathology of cervical herpes simplex infection. *Cancer* 19:1026, 1966.

274. Nasiell K., Dudkiewicz J., Nasiell M., Hjerpe A., Silfversward C. The occurrence of *Bacillus vaginalis Döderlein and cytolysis in dysplasia, carcinoma in situ, and invasive carcinoma of the uterine cervix. Acta Cytol.* 16:21, 1972.

275. Nasiell K., Nasiell M., Vaclavinkova V., Roger V., Hjerpe A. Follow-up studies of cytologically detected precancerous lesions (dysplasia) of the uterine cervix. In *Health Control in Detection of Cancer.* Almquist & Wiksell International, Stockholm, 1976, p. 244.

276. Nasiell M. Hodgkin's disease limited to the uterine cervix. *Acta Cytol.* 8:16, 1964.

277. Nelson J.H., Hall J.H. Detection diagnostic evaluation and treatment of dysplasia and early carcinoma of the cervix. *Cancer* 20:150, 1970.

278. Ng A.B.P., Atkin N.B. Histological cell type and DNA value in the prognosis of squamous cell cancer of uterine cervix. *Brit. J. Cancer* 28:322, 1973.

279. Ng A.B.P., Reagan J.W. Microinvasive carcinoma of the uterine cervix. *Amer. J. Clin. Path.* 52:511, 1969.

280. Ng A.B.P., Reagan J.W. Microinvasive carcinoma of the uterine cervix. In Wied G.L., Koss C.G., Reagen J.W. *Compendium on Diagnostic Cytology.* 4th Ed. Editors' Tutorials of Cytology, Chicago, 1976.

281. Ng A.B.P., Reagan J.W., Lindner E.A. The cellular manifestations of microinvasive squamous cell carcinoma of the uterine cervix. *Acta Cytol.* 16:5, 1972.

282. Noda K., et al. Histopathologic criterion of dysplasia of the uterine cervix and its biological nature. *Acta Cytol.* 20:224, 1976.

283. Nyirjesy I. Atypical or suspicious cervical smears. An aggressive diagnostic approach. *JAMA* 222:691, 1972.

284. Okagaki T., Meyer A.A., Sciarra J.J. Prognosis of irratiated carcinoma of cervix uteri and nuclear DNA in cytologic postirradiation dysplasia. *Cancer* 33:647, 1974.

285. Opperman A., et al. Cervico-vaginal smears in screening for uterine cancer. *Rev. Fr. Gynec. Obstet.* 66:643, 1971.

286. Ortiz R., Newton M. Colposcopy in the management of abnormal cervical smears in pregnancy. *Amer. J. Obstet. Gynec.* 109:46, 1971.

287. Papanicolaou G.N. *Atlas of Exfoliative Cytology.* Harvard University Press, Cambridge, Mass., 1963.

288. Papanicolaou G.N., Traut H.F. Diagnosis of uterine cervix by the vaginal smear. Cambridge, Mass., Harvard University Press, 1943.

289. Patten S.F. Jr. Diagnostic Cytology of the Uterine cervix. Monograph. Karger, Basel, 1969.

289a. Poulsen H.E., Taylor C.W., Sobin L.H. *Types histologiques des tumeurs du tractus génital féminin.* Organisation Mondiale de la Santé, 1975.

290. Pundel J.P. *Precis de colpocytologie hormonale.* Masson, Paris, 1966.

291. Purola E., Savia E. Cytology of gynecologic condyloma acuminatum. *Acta Cytol.* 21:26, 1977.

292. Qizilbash A.H. Papillary squamous tumors of the uterine cervix. A clinical and pathological study of 21 cases. *Amer. J. Clin. Path.* 61:508, 1974.

293. Qizilbash A.H. Chronic plasma cell cervicitis. A rare pitfall in gynecologic cytology. *Acta Cytol.* 18:198, 1974.

294. Rawls W.E., Adam E., Melnick J.L. An analysis of seroepidemiological studies of herpres virus type 2 and carcinoma of the cervix. *Cancer Res.* 33:13477, 1973.

295. Rawls W.E., Tompkins W.A., Melnick J.L. The association of herpes virus type II and carcinoma of the uterine cervix. *Amer. J. Epidem.* 89:547, 1969.

296. Reagan J.W. Cellular pathology and uterine cancer. *Amer. J. Clin. Path.* 62:150, 1974.

297. Reagan J.W., Bell B.A., Neuman J.L., Scott R.B., Patten S.F. Dysplasia in the uterine cervix during pregnancy: An analytical study of the cells. *Acta Cytol.* 5:17, 1961.

298. Reagan J.W., Hamonic M.J., Wentz W.B. Analytical study of the cells in cervical squamous cell cancer. *Lab. Invest.* 6:241, 1957.

299. Reagan J.W., Ng A.B.P. The cellular manifestations of uterine carcinogenesis. In Norris, J.J. Hertig, A.T., Abell M.P. (Eds.), *Symposium on the Uterus.* International Academy of Pathology Monograph Series. 1973, p. 320.

300. Reagan J.W., Scott R.B. The detection of cancer of the uterine cervix by cytological study. *Amer. J. Obstet. Gynec.* 62:1347, 1951.

301. Reagan J.W., Seidemann I.L., Saracusa Y. Cellular morphology of carcinoma in-situ and dysplasia or atypical hyperplasia of uterine cervix. *Cancer* 6:224, 1953.

302. Rhine S.A., Cain J.L., Clearly R.E., Plamer C.G., Thompson J.F. Prenatal sex detection with en-

docervical smears: successful results utilizing Y-body fluorescence. *Amer. J. Obstet. Gynec.* 122:155, 1975.

303. Richarts R.M., Baron, B.A. A follow-up study of patients with cervical dysplasia *Amer. J. Obstet. Gynec.* 105:386, 1969.

304. Richarts R.M., Vaillant H. Influence of cell collection techniques upon cytological diagnosis. *Cancer* 18:1474, 1965.

305. Roberts T.H., Ng A.B.P. Chronic lymphocytic cervicitis: cytologic and histopathologic manifestations. *Acta Cytol.* 19:235, 1975.

306. Robinson S.C. Cervical cancer incidence doubles (abstract). *Proceedings of the 6th International Congress of Cytology.* Tokyo, 1977. p. 86.

307. Rothbard M.J., Markam E.H. Leiomyosarcoma of the cervix: report of a case. Amer. J. Obstet. Gynec. 120:853, 1974.

308. Rubio C.A. The false positive smear. *Acta Cytol.* 19:212, 1975.

309. Rubio C.A. Who is responsible for the false negative smear? *Acta Cytol.* 19:319, 1975.

310. Rubio C.A. The exfoliating cervical epithelial surface in dysplasia, carcinoma in situ and invasive squamous carcinoma. I. Scanning electron microscopic study. *Acta Cytol.* 20:144, 1976.

311. Rubio C.A. The cervical epithelial surface. III. Scanning electron microscopic study in atypias and invasive carcinoma in mice. *Acta Cytol.* 20:375, 1976.

312. Rubio C.A., Lagerlof B. Proliferating and non-proliferating compartments in cervical dysplasia and carcinoma in situ. *Acta Path. Microbiol. Scand.* 83A:189, 1975.

313. Rummel H.H., Frick R., Heberling D. Long-term cytological observations in cervical dysplasia (abstract). *The 6th International Congress of Cytology.* Tokyo, 1977, p. 124.

314. Ryden S.E., Silverman E.M., Goldman R.T. Adenoid cystic carcinoma of the cervix presenting as a primary bronchial neoplasm. *Amer. J. Obstet. Gynec.* 120:846, 1974.

315. Sahiar B.E., Malvi S.G., Affandi Z.M., Gullar S.U. The value of exfoliative cytology in the detection of uterine malignancy in 3028 women. *Indian J. Cancer* 4:116, 1967.

316. Salm R. Superficial intra-uterine spread of intra-epithelial cervical carcinoma. *J. Path.* 97:719, 1969.

317. San Cristobal A., Roset S., Blay C. Finding of ciliated protozoa genus Vorticella on cervical and endocervical smears. *Acta Cytol.* 20:387, 1976.

318. Sassy-Dobray G., Keszler P., Kompolthy K. Experiences with respect to intraoperative cytodiagnosis. *Acta Cytol.* 16:478, 1972.

319. Savage E.W. Microinvasive carcinoma of the cervix. *Amer. J. Obstet. Gynec.* 113:708, 1972.

320. Scaife B. Survey of cervical cytology in general practice. *Brit. Med. J.* 3:200, 1972.

321. Sedlis A., Cohen A., Sall S. The fate of cervical dysplasia. *Amer. J. Obstet. Gynec.* 107:1065, 1970.

322. Sedlis A., Walter A.T., Balin H., Hontz A., Lo Sciuto L. Evaluation of two simultaneously obtained cervical cytological smears. A comparison study. *Acta Cytol.* 18:291, 1974.

323. Shanta V., Krishnamurthi S. The aetiology of carcinoma of the uterine cervix in south India: a preliminary report. *Brit. J. Cancer* 23:693, 1969.

324. Shennan D.W., Gelfand M. Bilharzia ova in cervical smears. A possible additional route for the passage of ova into water. *Trans R. Soc. Trop. Med. Hyg.* 65:95, 1971.

325. Shilkin K.B. Adenoid basal carcinoma of the cervix uteri. *J. Clin. Path.* 13:301, 1972.

326. Shingleton H.M., Gore H., Straughin J.M., Austin M., Littleton H.J. The contribution of endocervical smears to cervical cancer detection. *Acta Cytol.* 19:261, 1975.

327. Shuster M., et al. Irrigation cervico-vaginal cytology in industrial health screening examinations. *Ind. Med. Surg.* 40:30, 1971.

328. Siebert S., Jobert M.E., Besancon D. Cytologic features of cervicovaginal smears from patients treated with oral contraceptives. *Rev. Franç. Gynec.* 67:489, 1972.

329. Simmons P.D., Vosmik F. Cervical cytology in non-specific genital infection. *Br. J. Ven. Dis.* 50:313, 1974.

330. Smolka H. Cytological changes of the endocervical epithelium as a results of the physiological presence, deficiency or absence of endogenous estrogen stimulation. *Acta Cytol.* 4:46, 1960.

331. Song H.A. Herpes simplex virus infection detected on routine gynecologic cell specimens. *Acta Cytol.* 13:354, 1969.

332. Spriggs A.I. Population screening by the cervical smear. *Nature* 238:135, 1972.

333. Stafl A., Friedrich E.G., Mattingly R.F. Detection of cervical neoplasia: reducing the risk of error. *Cancer* 24:23, 1974.

334. Stern E., Forsythe A.M., Youkeles L., Dixon W.J. A cytological scale for cervical carcinogenesis. *Cancer Res.* 34:2358, 1974.

334a. Symposium on dyskaryosis. *Acta Cytol.* 1:19, 1957.

335. Takahashi M., Kamatsu T., Uei Y. Correlation between the prognosis and the cytology during irradiation for cervical cancer (abstract). *Proceedings of the 6th International Congress of Cytology.* Tokyo, 1977, p. 113.

336. Takeuchi A., McKay D.G. The area of the cervix involved by carcinoma in situ and anaplasia (atypical hyperplasia). *Obstet. Gynec.* 15:134, 1960.

337. Talerman A. Sarcoma botryoides presenting as a polyp on the labium majus. *Cancer* 2:994, 1973.

338. Tasker J.T., Collins J.A. Adenocarcinoma of the uterine cervix. *Amer. J. Obstet. Gynec.* 118:344 1974.

339. Taylor J., Bahr G.F., Bartels P.H., Bibbo M., Richards D.L., Wied G.L. Development and evaluation of automatic nucleus finding routines: thresholding of cervical cytology images. *Acta Cytol.* 19:289, 1975.

340. Te Linde R.W. Demonstration of the relationship of carcinoma in situ to invasive carcinoma of the cervix. *Amer. J. Obstet. Gynec.* 115:1022, 1973.

341. Teplitz R.L., Valco Z., Rundall T. Comparative sequential cytologic changes following in vitro infection with herpes virus types I and II. *Acta Cytol.* 15:455, 1971.

342. Terris M., Wilson F., Nelson J.H. Relation of circumcision to cancer of the cervix. *Amer. J. Obstet. Gynec.* 117:1056, 1973.

343. The value of cervical cytology. *Lancet* 2:1236, 1972.

344. Thompson B.H., Woodruff J.D., Davis H.H., Juli C.G., Silva II.F.G. Cytopathology, histopathology and colposcopy in the management of cervical neoplasia. *Amer. J. Obstet. Gynec.* 114:329, 1972.

345. Townsend E.D., Ostergard D.R., Mishell D.R., Hirose F.M. Abnormal Papanicolaou smears. Evaluation by colposcopy biopsies and endocervical curettage. *Amer. J. Obstet. Gynec.* 108:429, 1970.

346. Tweeddale D.N., Langenbach S.R., Roddick J.W., Holt M.L. The cytopathology of microinvasive squamous cancer of the cervix uteri. *Acta Cytol.* 13:447, 1969.

347. van Niekerk W.A., Cervical cytological abnormalities caused by folic acid deficiency. *Acta Cytol.* 10:67, 1966.

348. Vesterinen E., Saksela E., Vaheri A. Adenovirus induced pathological alterations in human cervical epithelial cells in vitro (abstract). *Proceedings of the 6th International Congress of Cytology.* Tokyo, 1977, p. 124.

349. Von Bertalanfly, L., Masin M, Masin F. A new rapid method for diagnosis of vaginal and cervical cancer by fluorescence microscopy. *Cancer* 11:873, 1958.

350. Von Haam E. A comparative study of accuracy of cancer cell detections by cytological methods. *Acta Cytol.* 6:508, 1962.

351. Von Haam E., Scarpelli D.G. Experimental carcinoma of the cervix: A comparative cytologic and histologic study. *Cancer Res.* 15:449, 1955.

352. Wakefield J., Yule R., Smith A., Adelstein A. Relation of abnormal cytological smears and carcinoma of cervix uteri to husband's occupation. *Brit. Med. J.* 2:142, 1973.

353. Walters W.D., Reagan J.W. Epithelial dysplasia of the uterine cervix in pregnancy. *Amer. J. Clin. Path.* 26:1314, 1956.

354. Warren S. The grading of carcinoma of the cervix uteri as checked by autopsy. *Arch. Path.* 12:783, 1931.

355. Weisbrot I.M., Stabinsky C., Davis A.M. Adenocarcinoma in situ of the uterine cervix. *Cancer* 29:1179, 1972.

356. Wentz W.B. Survival in cervical cancer with respect to cell type. *Cancer* 12:384, 1959.

357. Whitaker D. The role of cytology in the detection of malignant lymphoma of the uterine cervix. *Acta Cytol.* 20:510, 1976.

358. Who should be screened for cervical cancer? *N. Engl. J. Med.* 294:223, 1976.

359. Wied G.L. Importance of the site from which cytologic smears are taken. *Amer. J. Clin. Path.* 25:742, 1955.

360. Wied G.L., Bahr G.F. Vaginal, cervical and endocervical smears on a single slide. *Obstet. Gynec.* 14:362, 1959.

361. Wied G.L., Legoretta G., Mohr D., Rauzy Y. Cytology of invasive cervical carcinoma and carcinoma in situ. *Ann. N.Y. Acad. Sci.* 97:759, 1962.

362. Wilbanks G.D., Ikomi E., Prado B., Richart R.M. An evaluation of a one-slide cervical cytology method for the detection of cervical intraepithelial neoplasia. *Acta Cytol.* 12:157, 1968.

363. Williams A.E., Jordan J.A., Allen J.M., Murphy J.F. The surface ultrastructure of normal and metaplastic cervical epithelia and of carcinoma in situ. *Cancer Res.* 33:504, 1973.

364. Wookey B.E.P. Exfoliative cytology in general practice. *Brit. Med. J.* 3:31, 1971.

365. Wynder E.L. Will husband's circumcision reverse cervical cancer class. III. Papanicolaou test results. *JAMA* 232:961, 1975.

366. Yajima A., et al. Cellular changes of dysplasia accompanied with its progression to malignancy (abstract). *Proceedings of the 6th International Congress of Cytology.* Tokyo, 1977, p. 114.

367. Yokoyama Y., Kishigami T., Noda S., Takahashi Y., Hongo J., Ishii S., Hayakawa K., Kitada A., Hashimoto H., Taketani S. The sponge-cytocylinder smear. A cytologic method from mass screening. *Amer. J. Obstet. Gynec.* 109:119, 1971.

368. Youkeles L., Forsythe A.B., Stern E. Evaluation of Papanicolaou smear and effect of sample biopsy in follow-up of cervical dysplasia. *Cancer Res.* 36:2080, 1976.

369. Ziabkowski T.A., Naylor B. Cyanophilic bodies in cervico-vaginal smears. *Acta Cytol.* 20:340, 1976.

Corpus Uteri

370. Abate S.D., Edwards C.L., Vellios F. A comparative study of endometrial jet-washing technic and endometrial biopsy. *Amer. J. Clin. Path.* 58:118, 1972.

371. Abrams R.Y., Spritzer T. Endometrial cytology in patients using IUCD. *Acta Cytol.* 10:240, 1966.

372. Anderson D.G., Eaton C.J., Galinkin L.J., Newton C.W., Haines J.P., Miller N.F. The cytologic diagnosis of endometrial adenocarcinoma. *Amer. J. Obstet. Gynec.* 125:378, 1976.

373. Becker S.N. Keratin bodies and pseudokeratin bodies. Endometrial adeno-acanthoma versus ligneous vaginitis. *Acta Cytol.* 20:486, 1976.

374. Berg J.W., Durfee G.R. The cytological presentation of endometrial carcinoma *Cancer* 11:158, 1958.

375. Berry A. Evidence of gynecologic bilharziasis in cytologic material. *Acta Cytol.* 15:482, 1971.

376. Berry A.V., Livni N.M., Epstein N. Some observations on cell morphology in the cytodiagnosis of endometrial carcinoma. *Acta Cytol.* 13:530, 1969.

377. Bibbo M., Rice A.M., Wied G.L., et al. Comparative specificity and sensitivity of routine cytologic examinations and the Gravlee jet wash technic for diagnosis of endometrial changes. *Obstet. Gynec.* 43:253, 1974.

378. Bibbo M., Shanklin D.R., Wied G.L. Endometrial cytology on jet wash material. *J. Reprod. Med.* 8:90, 1972.

379. Boram L.H., Erlandson R.A., Hajou S.I. Mesodermal mixed tumor of the uterus. A cytologic, histologic and electron microscopic correlation. *Cancer* 30, 1295, 1972.

380. Boschann H.W. Advantages and disadvantages of intrauterine brush technique for endometrial cytology. *Acta Cytol.* 2:572, 1958.

381. Buschmann C., Hergenrader M., Porter D. Keratin bodies. A clue in the cytological detection of endometrial adenoacanthoma. Report of two cases. *Acta Cytol.* 18:297, 1974.

382. Clark A.H., McKee E.E., Dixon D.C. Identification of trophozoite form of *Ontamoeba histolytica* by cytologic techniques. *Acta Cytol.* 16:429, 1972.

383. De Borges R. Findings of microfilarial larval stages in gynecologic smears. *Acta Cytol.* 15:476, 1971.

384. Dibona D.D., Knab D.R. Use of the Gravlee jet washer in routine screening of asymptomatic postmenopausal women. *Amer. J. Obstet. Gynec.* 119:681, 1974.

385. Dowling E.A., Gravlee L.C., Hutchins K.E. A new technique for the detection of adenocarcinoma of the endometrium. *Acta Cytol.* 13:496, 1969.

386. Dupre-Froment J., de Brux J. Valeur du frottis endométrial dans l'évaluation des déséquilibres ovariens. *Rev. Cytol. Clin.* 2:7, 1969.

387. Factor S.M. Papillary adenocarcinoma of the endometrium with psammoma bodies. *Arch. Path.* 98:201, 1974.

388. Fornari M.L. Cellular changes in the glandular epithelium of patients using IUCD. A source of cytologic error. *Acta Cytol.* 18:341, 1974.

389. Gauthe P. Adénocarcinome utérin. Confrontation cytologique et histologique systématique (à partir de 2850 diagnostics) *Rev. Cytol. Clin.* 3:16, 1970.

390. Gladwell P., Duncan P., Barham K., Kenny J. Amnioscopy of late pregnancy with fetal membrane and decidual cytology. *Acta Cytol.* 18:333, 1974.

391. Gompel C. The ultrastructure of the human endometrial cell studied by electron microscopy. *Amer. J. Obstet. Gynec.* 84:1000, 1962.

392. Gompel C. Ultrastructure of endometrial carcinoma. *Cancer* 28:745, 1971.

393. Gore H., Hertig A.T. Premalignant lesions of the endometrium. *Clin. Obstet. Gynec.* 5:1148, 1962.

394. Gravlee L.C. Jet-irrigation method for a diagnosis of endometrial adenocarcinoma: its principle and accuracy. *Obstet. Gynec.* 34:168, 1969.

395. Gusberg S.B., Kaplan A.L. Precursors of corpus cancers. IV. Adenomatous hyperplasia as stage O of carcinoma of the endometrium. *Amer. J. Obstet. Gynec.* 87:662, 1963.

396. Hagenfeld K., Johannisson E. The effect of I U copper on the DNA content in isolated human endometrial cells. *Acta Cytol.* 16:472, 1972.

397. Hall H.H., Stone M.L., Sedlis A., Chaban I. The intrauterine ring for contraception control. *Fert. Steril.* 15:618, 1964.

398. Hammond D.O., Seckinger D.L., Ledrew D.I. Endometrial cellular study: A new method using membrane filtration. *Acta Cytol.* 11:181, 1967.

399. Harris M.J., Bibbo M., Rao C., Wied G.L. Cytopreparatory techniques for the endometrial jet wash specimens. *Acta Cytol.* 16:508, 1972.

400. Hart W.R., et al. Cytologic findings in stilbestrol exposed females with emphasis on detection of vaginal adenosis. *Acta Cytol.* 20:7, 1976.

401. Hertig A.T., Sommers S.C., Bengloff H. Genesis of endometrial carcinoma. III. Carcinoma in situ. *Cancer* 2:964, 1949.

402. Hibbard L.T., Schwinn C.E. Diagnosis by endometrial jet washings. *Amer. J. Obstet. Gynec.* 111:1039, 1971.

403. Highman W.J. Calcified bodies and intrauterine device. *Acta Cytol.* 15:473, 1971.

404. Hofmeister F.J. Endometrial sampling. *J. Reprod. Med.* 4:33, 1970.

405. Hustin J. *Etude du contexte endocrinien du cancer endométrial en post-ménopause. Evaluation de la thérapeutique progestative dans cette affection.* Thesis, Liège, 1973.

406. Hustin J. Morphology and DNA content of endometrial cancer nuclei under progestogen treatment. *Acta Cytol.* 20:556, 1976.

407. Isaacs J.H., Wilhoite R.W. Aspiration cytology of the endometrium: office and hospital sampling procedures. *Amer. J. Obstet. Gynec.* 118:679, 1974.

408. Ishihama A., Kagabu T., Imai T., Shima M. Cytologic studies after insertion of intrauterine contraceptive device. *Acta Cytol.* 14:35, 1970.

409. Jimenez-Ayala M., Vilaplana E., Becerro De Bengoa C., Zomeno M., Moreno S., Granados M. Endometrial and endocervical brushing techniques with a menhosa cannula. *Acta Cytol.* 19:557, 1975.

410. Johannisson E., Engstrom L. Cytological diagnosis of endometrial disorders with a brush technique. *Acta Obstet. Gynec. Scand.* 50:141, 1971.

411. Johnsson J.E., Stormby N.G. Cytological brush technique in malignant disease of the endometrium. *Acta Obstet. Gynec. Scand.* 47:38, 1968.

412. Kanbour A., Klionsky B., Cooper R. Cytohistologic diagnosis of uterine jet wash preparations. *Acta Cytol.* 18:51, 1974.

413. Koss L.G., Durfee G.R. Cytologic diagnosis of endometrial carcinoma. *Acta Cytol.* 6:519, 1962.

414. Lanier A.P., Noller K.L., Decker D.G., Elveback L.R., Kurland N. T. Cancer and stilbestrol. A follow-up of 1719 persons exposed to estrogens in utero and born 1943–1959. *Mayo Clin. Proc.* 48:793, 1973.

415. Liu W. Cytodiagnosis of endometrial carcinoma. *Canad. J. Med. Technol.* 33:26, 1971.

416. Lukeman J.M. An evaluation of the negative pressure "jet washing" of the endometrium in menopausal and postmenopausal patients. *Acta Cytol.* 18:462, 1974.

417. McGowan L. The current efficacy for the cervico-vaginal cytosmear to detect endometrial cancer. *Acta Cytol.* 8:434, 1964.

418. Marsan C., Sicard A. Le cytodiagnostic du cancer du corps utérin par aspiration endométriale. *Presse Méd.* 68:799, 1960.

419. Masukawa T., Wada Y., Mattingly R.F., Kuzma J.F. Cytologic detection of minute ovarian endometrial and breast carcinomas, with emphasis on clinical pathological approaches. *Acta Cytol.* 17:316, 1973.

420. Milan A.R., Markley R.L. Endometrial cytology by a new technic. *Obstet. Gynec.* 42:469, 1973.

421. Moyer D.L., Mishell D.R., Bell J. Reaction of human endometrium to the intrauterine device. *Amer. J. Obstet. Gynecl.* 106:799, 1970.

422. Ng A.B.P., Reagan J.W., Cechner R.L. The precursors of endometrial cancer: a study of their cellular manifestations. *Acta Cytol,* 17:439, 1973.

423. Ng A.B.P., Reagan J.W., Hawliczek S., Wentz B.W. Significance of endometrial cells in the detection of endometrial carcinoma and its precursors. *Acta Cytol.* 18:356, 1974.

424. Ober W. Morphological changes in the uterus associated with steroid contraceptions and intra-uterine contraceptive device. *Amer. J. Obstet. Gynec.* 106:799, 1970.

425. Parsi B., Le Treut A., Dilhuydy M.H., Meuge C., Trojani M. Technique simple de prélèvement endo-uterin. A propos de 100 examens. *Rev. Clin. Cytol.* 8:29, 1975.

426. Picoff R.C., Meeker C.I. Psammoma bodies in the cervicovaginal smear in association with benign papillary structures of the ovary. *Acta Cytol.* 14:45, 1970.

427. Reagan J.W., Ng A.B.P. The Cells of Uterine Adenocarcinoma. Vol. 1. 2nd Ed. (revised). Karger, Basel, 1973.

428. Reyniak J.V., Sedlis A., Stone D., Connell E. Cytohormonal findings in patients using various forms of contraception. *Acta Cytol.* 13:315, 1969.

429. Sagiroglu N., Sagiroglu E. Cytology of intrauterine contraceptive device. *Acta Cytol.* 14:58, 1970.

430. Schiffer M.D., Elguesebal A., Sultana M., Allen A.C. Actinomycosis infections associated with intra-uterine devices. *Obstet. Gynec.* 45:67, 1975.

431. So-Bosita J.L., Lebherz T.B., Blair O.M. Endometrial jet washer. *Obstet. Gynec.* 36:287, 1970.

432. Vassilakos P., Wyss R., Wenger D., Riotton G. Endometrial cytohistology by aspiration technic and by Gravlee jet washer. *Obstet. Gynec.* 45:320, 1975.

433. Vellios F. Endometrial hyperplasia, precursors of endometrial carcinoma. In *Pathology Annual.* Appleton-Century-Crofts, New York, 1972, pp. 201–227.

434. White A.J., Buchsbaum H.J., Rodman N.F. Accuracy of the Gravlee jet washer in detecting endometrial adenocarcinoma. *Amer. J. Obstet. Gynec.* 116:1169, 1973.

Uterine Tube and Ovary

435. Ångström T., Kjellgren O., Bergman F. The cytologic diagnosis of ovarian tumors by means of aspiration biopsy. *Acta Cytol.* 16:336, 1972.

436. Benson P.A. Cytologic diagnosis in primary carcinoma of fallopian tube. *Acta Cytol.* 18:429, 1974.

437. Beyer-Boon M.E. Psammoma bodies in cervico-vaginal smears: an indicator of the presence of ovarian carcinoma. *Acta Cytol.* 18:41, 1974.

438. Differding J.T. Psammoma bodies in a vaginal smear. *Acta Cytol.* 11:199, 1967.

439. Dudkiewicz J. Quantitative and qualitative changes of epithelial cells of fallopian tubes in women according to the phase of menstrual cycle. A cytologic study. *Acta Cytol.* 14:531, 1970.

440. Funkhouser J.W., Hunter K.K., Thompson N.J. The diagnostic value of cul-de-sac aspiration in the detection of ovarian carcinoma. *Acta Cytol.* 19:538, 1975.

441. Graham R.M., van Niekerk W.A. Vaginal cytology in cancer in the ovary. *Acta Cytol.* 6:496, 1962.

442. Lehto L. Cytology of the human fallopian tube. *Acta Obstet. Gynec. Scand.* 42 (Suppl. 4):3, 1963.

443. Luecke A., Klebs E. Beitrag zur Ovariotomie und zur Kentnis des Abdominal Geschwuelste. *Virch. Arch. Path. Anat.* 41:1, 1867..

444. Picoff R.C., Meeker C.I. Psammoma bodies in the cervico-vaginal smear in association with benign papillary structures of the ovary. *Acta Cytol.* 14:45, 1970.

445. Wachtel E. The cytology of tumors of the ovary and fallopian tube. *Clin. Obstet. Gynec.* 4:1159, 1961.

446. Wachtel E. La cytologie des tumeurs de l'ovaire. *Rev. Cytol. Clin.* 2:7, 1969.

447. Wagman H., Brown C.L. Ovarian cytology: an application of cytology in an attempt at the early detection of ovarian carcinoma. *Brit J Cancer* 25:81, 1971.

448. Woodruff J.D., Pauerstein C.J. *The Fallopian Tube. Structure, Function, Pathology and Management.* Williams & Wilkins, Baltimore, 1969.

6
Mammary Cytology

INTRODUCTION

The frequency of the dysplastic and tumoral lesions of the breast justifies the length of this chapter. Breast cancer is indeed the most frequent of all cancers in women, representing 25 to 30 percent of all primary tumors. Annually, 35 women out of 10,000 develop breast cancer; it can occur in all age groups but is more frequent after 45.[15] Its preferential localization is the superior external quadrant (approximately 50 percent of the cases) (Figure 6.1). The exact etiology of breast cancer is not understood, but it is known that hormonal factors (hormonal imbalance) affect breast tissue. The favorable long-term influence of endocrine factors during pregnancy is illustrated by the higher frequency of cancers in women who have not had children. The concept of hormone dependency of certain breast tumors is based on various experimental and clinical studies. This concept has a practical value in the hormonal treatment in certain types of mammary carcinoma.

The recognition of the lesions, particularly at their beginning stages, may be accomplished by different types of examinations: clinical, radiologic (e.g., mammography and thermography), cytologic, and histological. However, definite diagnoses are made by comparison and integration of results obtained by these various methods.

For a cytological examination, cellular elements may be obtained in three ways:

- by collection of nipple discharge
- by needle puncture
- by direct imprint of biopsied tissue or surgical specimens

STRUCTURE OF THE MAMMARY GLAND

Histology

The mammary gland consists of a glandular system embedded in a connective and adipous stroma. The galactophoric canalicular tree originates at the nipple in the lactiferous sinus and branches into the lobules, the functional unit of the mammary gland (Figure 6.2). The lobule is formed by the terminal portions of a group of lactiferous ducts and by the intersitial stroma (Plate 6.1).

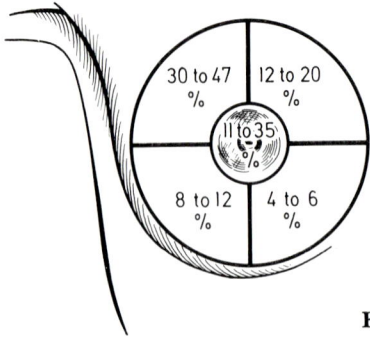

Figure 6.1. Data concerning the site of the primary tumor.

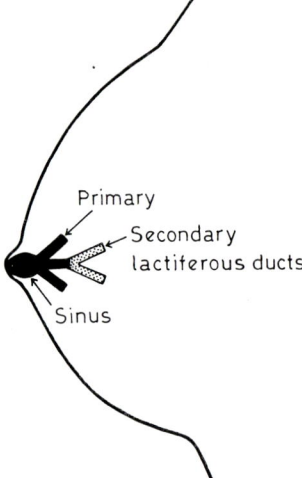

Figure 6.2. Schematic representation of the glandular system.

The lactiferous duct is bordered by columnar epithelium and is covered by an external layer of myloepithelial cells. Surrounding the canalicular epithelial structures, there is a loose intralobular connective tissue stroma, poor in collagen fibers and sensitive to hormonal stimulations. The perilobular stroma contains more collagen fibers and more adipose tissue.

During lactation, the acinar cells of the terminal portion of the lactiferous ducts acquire secretory properties. Proteins are produced by the endoplasmic reticulum and are eliminated through the cellular membrane via reverse pinocytosis, while lipids are eliminated in the form of membrane-bound vesicles.

Cytology

Different cellular elements are encountered on smears obtained from the breast, they are:

- epithelial cells of the canalicular lactiferous duct system
- myoepithelial cells
- connective tissue cells of the stroma and blood cells

The epithelial cells have a different appearance depending on whether they desquamate into a cystic canal or are aspirated from the ductal epithelium.

When the galactophoric cells are desquamated into the lumen of a canal or a cyst, they appear as very large, microvacuolized elements with an abundant, foamy, pale green cytoplasm (Plates 6.2 and 6.3). The cells reach an upper limit of 30μ in diameter. The nuclei are small and regular and show a thinly distributed chromatin as well as a prominent nucleolus. They measure 6 to 8 μ in diameter. If cellular degeneration is at an advanced stage, nuclear lysis and cytoplasmic necrosis are observed. The quantity of these desquamated foamy elements is of diagnostic significance: they are numerous in benign cystic formations.

When the cells are detached from the ductal epithelium by the aspiration, they appear as regular columnar or cuboidal elements that have preserved the structure they normally have in the ducts (Plate 6.4). A round, elongated nucleus is surrounded by a cyanophilic cytoplasm and measures approximately 20 μ in height and 5 to 8 μ in diameter. Apocrine metaplasia shows large foamy cells with a finely vacuolated, eosinophilic cytoplasm (Plates 6.5 and 6.6). By electron microscopy, the cytoplasm is seen to contain numerous cytoplasmic organelles including mitochondria.

Adipocytes appear as clusters of large cells (30 to 40 μ in diameter) containing a clear vacuolated cytoplasm, emptied of its lipids by fixation. The nuclei are small, rounded, and generally found at the cell periphery; their structure is not easily visible, but they seem to exhibit a dense and homogeneous chromatin (Plate 6.7).

The stromal cells are elongated fibroblastic elements that desquamate individually. They have a fusiform nucleus and a pale cytoplasm which is often poorly delineated (Plate 6.7).

Certain authors recognize the presence of myoepithelial cells, especially when these are hyperplastic.[25] They appear as branching sheets of cells, but their identification is often difficult because they may be confused with the stromal cells.

Blood cells (leukocytes and erythrocytes) and histiocytes vary in number depending on clinical circumstances.

Finally, it is not rare to find some squamous cells from the cutaneous epithelium whose presence is due to the trauma of the puncture.

Pathologic cells in needle punctures or nipple discharges will be encountered in four main types of lesions: inflammatory lesions, dystrophic and dysplastic lesions, benign tumors, and malignant tumors.

INFLAMMATORY LESIONS

The most frequent inflammatory lesions of the mammary gland are abcesses, lipophagic granuloma, and ruptured lactiferous cysts with secondary granulomatous inflammatory phenomena. A case of mammary tuberculosis diagnosed by cytology has been reported.[47] Smears taken from these lesions are rich in leukocytes, histiocytes, and red blood cells which indicate the inflammatory nature of the lesion. The modified lactiferous cells show anomalies in size and form, lysis, and a marked cytoplasmic vacuolation (Plate 6.8). They are dispersed in the inflammatory exudate. Their identification is not always easy due to necrosis, and they are better preserved when they originate from a cystic fluid. Moderate anisonucleosis, anisokaryosis, and hyperplasia may be present.

FIBROCYSTIC DYSPLASIA

Fibrocystic dysplasia is the term under which we group all the dystrophic and dysplasic alterations of the mammary gland. It is characterized by different histologic elementary lesions present in various quantities. These features are ductal cysts, proliferation of duct epithelium including papillary patterns, stromal fibrosis, adenosis (which represents a benign proliferation of ductules and acini), and finally apocrine metaplasia (Plate 6.9). Aspirated cystic fluid contains many cellular elements, which may be isolated or arranged in clumps, and sometimes leukocytes. The fluid may contain fresh blood if the needle has hit a blood vessel or old partially hemolyzed blood if intracystic bleeding has occurred.

The cells originating from cyst walls frequently exhibit signs of cellular degeneration, such as nucleolysis and cytolysis, cellular fragmentation, and loss of tinctorial affinities[1, 38] (Plate 6.10). The nuclei are rounded or elongated and measure 6 to 8 μ in diameter. The chromatin is quite dense but the nucleolus is distinguishable. The normal nucleocytoplasmic ratio and the homogeneous chromatin distinguish these cells from neoplastic ones (Plate 6.11). Certain of the desquamated cells take an histiocytic appearance and acquire macrophagic properties. The nucleus maintains its size, but the cytoplasm increases in volume and forms vacuoles. Needle aspirations of massive epithelial proliferations produce irregularly arranged aggregates. Multinucleated histiocytes may be observed (Plate 6.12).

Although the stroma is quantitatively much more abundant in these lesions, paradoxically the epithelial structures constitute the majority of the aspirated elements (Plate 6.13).

Certain authors have reported significant qualitative and quantitative differences in the desquamation typical of fibrocystic dysplasia and fibroadenoma. A high proportion of epithelial cells and the presence of stromal cells would characterize the fibroadenoma.[61, 62]

In cases of fibrocystic dysplasias where hyperplasia of the epithelial cords dominates the histologic picture (Plate 6.14), the cytologic aspiration is rich in cells of various sizes and forms of nuclei and cytoplasm (Plate 6.15). In atypical hyperplasia, these anomalies constitute a genuine difficulty in the differential diagnosis of malignant tumors. Regular disposition of cells in clusters and relatively moderate cellular alterations are in favor of benignity (Plate 6.16).

BENIGN TUMORS

The most common benign tumors of the breast are fibroadenoma, lipoma, giant fibroadenoma, and the intraductal papilloma.[9] Needle aspiration of these lesions produces a variable quantity of cells, depending on the nature of the lesion.

In the *fibroadenoma,* the smear is rich in cells and exhibits some clumps of ductal elements, as well as some stroma fibroblasts and adipocytes (Plate 6.17). The presence of an abundant desquamation of benign cells arranged in dense clumps and of naked nuclei come from the ductal epithelium and to a lesser degree from myoepithelial cells and the stroma. The clumps are composed of regular, round cells that may have a honeycomb appearance (Plates 6.18 and 6.19).

The *giant form of fibroadenoma* (cystosarcoma phyllodes)[4] is characterized by its richness in connective tissue cells and their structural anomalies (Plate 6.20). The connective tissue cells exhibit elongated nuclei 10 to 15 μ in length, a thinly dispersed chromatin, one or two prominent nucleoli, a well-delineated membrane, and a pale cytoplasm whose borders are sometimes well defined. The epithelial elements also show some alterations of hyperplasic nature (Plate 6.21). Malignant forms of the giant fibroadenoma are rare and exhibit cells that are clearly neoplastic.

Cells of *intraductal papilloma* (Plates 6.22 and 6.23) desquamate in nipple discharge as rounded aggregates whose centers are thickened by the piling up of the cells; the structure is visible only at the periphery of such aggregates where there is a single layer of cells.[29,45] They have a rounded nucleus with a thinly distributed chromatin, a visible nucleolus, and an abundant cytoplasm which is pale, and vacuolated and often has a hazy periphery (Plates 6.24–6.26). Sometimes a large size cytoplasmic vacuole displaces the nucleus to the periphery. The chromatin structure may be altered in densely packed papillary cell formations and exhibits a homogeneous ground-glass aspect.

The regular form of the nucleus, the aspect of the chromatin, and the preservation of a normal nucleocytoplasmic ratio are suggestive of the benign nature of the cells. Some degree of nuclear atypias may be present in highly cellular and proliferating papillomatous lesions. Microcalcifications have been observed rarely and originate from small foci of cellular necrosis.

In the presence of apocrine metaplasia, the cells are remarkable because of their large size and the foamy appearance of the cytoplasm; their nucleus is small and regular. Desquamation in aggregates is the rule.

If the cellular anomalies become very apparent, one should suspect the existence of ductal papillary carcinoma. A biopsy always should be taken in the presence of cellular and papillary nipple secretions.

Lipomas are composed of clumps of large, easily recognized adipocytes. The small nucleus is flattened by a voluminous, clear cytoplasm, drained of its lipids by fixation.

Some other rare benign lesions are the *granular rhabdomyoma,* the *histiocytoma,* and the *sebaceous cyst.* When dealing with a sebaceous cyst or an epidermoid inclusion cyst, one observes marked inflammatory infiltrates, cellular debris, and masses of eosinophilic granular or amorphous substance. This latter is either sebaceous or keratinous material.

MALIGNANT TUMORS

Histology

Malignant tumors of the breast reveal a great variety of histologic types. Table 6.1 summarizes the classification proposed by WHO. Some types have been classified individually because they represent a very typical histologic entity or because they have a specific clinical behavior. The great majority of the epithelial tumors (infiltrating carcinomas) originate in the galactophoric cells (Plate 6.29).

The noninfiltrating forms (lobular in situ carcinoma and intraductal carcinoma) are rare (Plate 6.27), and with a few exceptions, their exact diagnosis cannot be made cytologically.[26] They are detected by chance in routine biopsies of fibrocystic dysplasia or of any clinical or mammographic anomaly requiring histologic evaluation.[11] The cellular characteristics of malignancy are localized in the ducts or lobules and are

Table 6.1. WHO Classification of Malignant Breast Tumors

Carcinoma

    intracanalicular and intralobular non infiltrating carcinoma

    infiltrating carcinoma

    special forms of carcinoma

        medullary carcinoma

        papillary carcinoma

        cribriform carcinoma

        mucoid carcinoma

        lobular carcinoma

        squamous carcinoma

        Paget's disease

        carcinoma originating in cellular intracanalicular fibroadenoma (cystosarcoma phyllodes)

Sarcoma

    sarcoma originating in a intracanalicular cellular fibroadenoma (cystosarcoma phyllodes)

    other types of sarcoma

Epitheliosarcoma

Unclassified tumors

---

discrete (Plate 6.28). The risk of developing an infiltrative carcinoma is not negligible, and the tumor is often diffuse.

Certain histologic forms of adenocarcinoma reflect their particularity in the cytology of breast aspiration, for example, mucinous (Plate 6.34) carcinoma, adenoid cystic carcinoma, medullary carcinoma, and adenocarcinoma with squamous metaplasia. Sarcomas are rare.

Cytology

Needle aspiration smears are generally rich in cells, and the accumulation and overlapping of cells in thick aggregates are characteristic features of malignancy. The cells are arranged in a disorderly manner in the cohesive clusters.

The cells exhibit the usual cytologic criteria of malignancy: anisocytosis, anisonucleosis, hyperchromatism, nucleolar hypertrophy, increase in the nucleo-cytoplasmic ratio, anomalies of cellular aggregates (Plates 6.30–6.33). Sometimes the cells are closely packed, and the nuclear shapes are modified by the pressure of adjacent elements (nuclear molding). Cellular degeneration and necrosis manifest themselves by the homogeneous structure and "ballooning" of nuclei, by karyorrhexis, and by disappearance of cytoplasm. Mitotic figures are not a common finding.

Cellular necrosis and the presence of histiocytes and blood elements are frequent. An inflammatory diathesis is indicative of neoplasia but may also be found in secondary infection of benign cystic lesions; in the latter eventuality the presence of large, foamy columnar cells points to a benign origin.

Certain tumoral histologic forms may be recognized in the cytology of needle aspiration.

- Mucoid carcinoma is characterized by the abundant mucous medium in which the cellular clumps bathe and by the presence of cytoplasmic vacuoles which are sometimes voluminous and displace the nucleus (Plates 6.34 and 6.35)
- Medullary carcinoma may reveal the simultaneous presence of epithelial malignant cells and lymphocytic elements (Plate 6.36)
- Adenocarcinoma with squamous metaplasia exhibits characteristic benign looking epidermoid elements.
- Nipple imprints in Paget's disease may sometimes reveal the presence of large columnar type neoplastic cells with a clear cytoplasm, a large hyperchromatic usually round nucleus, and a well-defined nucleolus.
- Malignant lymphomas reveal aggregates of small cells whose characteristics are those of myelocytic, histiocytic, or lymphocytic cells (Plate 6.37).
- Fibrosarcomas are composed of cells having hyperchromatic and fusiform nuclei and pale and elongated cytoplasm. Smears are rich in randomly scattered isolated cells.
- Rhabdomyosarcomas and liposarcomas are extremely rare and exhibit anaplastic cells (Plates 6.38 and 6.39).
- Stromal sarcomas exhibit various cellular patterns: small regular elements or anaplastic epithelial and mesodermal elements including striated muscle and cartilage. The polymorphism and the indeniable malignant aspect of the exfoliated cells may suggest a sarcomatous origin.

Certain metastatic tumors can be detected cytologically. This is the case for the malignant melanoma which is easily recognizable if pigmented.

POSTRADIATION TREATMENT LESIONS

Cytology is valuable in the surveillance of irradiated cases since it is an innocuous procedure that permits repeated sampling.

Radiation favors the appearance of the following anomalies: gigantism, multinucleation, hyperchromatism and irregularities of nuclear size, cytoplasmic vacuolation, cells of bizarre shape, and abundant leukohistiocytic infiltration.

A differential diagnosis therefore must distinguish between irradiated neoplastic cells that are becoming necrotic and active neoplastic cells. The preservation of hyperchromatic nuclei with large nucleoli is indicative of persistent tumor cells. Normal epithelial cells disappear rapidly after irradiation and thus present no problems in this respect. Extensive acellular fibrosis resulting from radiation is often responsible for a negative needle aspiration.

ADVANTAGES AND LIMITATIONS OF THE METHOD

The cytologic method is one of the principal means of diagnosing mammary disorders. Its effective use, however, depends on the observation of certain principles:

- It requires the presence of a mass that can be punctured.
- The cytologic, histologic, radiologic, and clinical data must all be compared when

diagnostic difficulties arise; especially when some contradictory results are obtained, for example, positive mammography with a negative cytology.

• A negative examination is of little value.

• Major therapy should not be instituted without histologic verification.

Among the advantages of the method are its rapidity of execution (an indicative result can be obtained in a few hours), its painless nature, the simplicity of fixating and staining, and the possibility of its use on a large scale.[44] The technique permits, in many cases, an evaluation of the degree of differentiation of tumor cells. It is useful in the diagnosis of inoperable cases submitted to radiotherapy.

A disadvantage of the cytologic method is the impossibility of evaluating the topographic criteria of the lesion. This is very true in localized or lobular lesions with very moderate cellular anomalies. Even in expert hands, the percentage of false negatives can reach 3 to 5 percent. These false negatives can be due to any of the following reasons:

• discrete cytological alterations in differentiated tumors

• rarity of tumoral elements in the highly fibrous forms

• a puncture performed in a nontumoral zone or in a neighboring benign lesion

• difficulty of puncturing a mass of small volume

A negative cytologic examination, particularly in cases of suspicious clinical or radiologic reports, should be considered with reservation and not be accepted as a definitive answer.

The frequency of false-positive examinations depends on the experience of the cytopathologist and on the difficulties of differential diagnosis presented by certain benign lesions. Some benign lesions show a marked cellularity accompanied by discrete cellular alterations. The greater intensity of these cellular atypias distinguishes malignant tumor from intraductal papilloma, fibroadenoma, and epithelial hyperplasia of fibrocystic disease.

False-positive percentage should not exceed 1 percent in a competent laboratory. The danger of false-positive diagnosis is eliminated if all radical therapy is sanctioned by a biopsy.

Fears were raised that the biopsy procedure would entail a risk of transplanting cancerous elements along the path of the needle.[37] This fear seems unfounded and in any case, if cancer cells are indeed found, treatment should ensue immediately.

DIFFERENTIAL DIAGNOSIS

The major difficulties of differential diagnosis are the following:

• Well-differentiated epitheliomas vs fibroadenoma and papilloma: desquamation is abundant in each of these lesions; the existence of cellular anomalies and the disorderly arrangement within cell clusters point to malignancy.

• Giant fibroadenoma vs fibrosarcoma: the cells are quite similar. Nuclear anomalies are predominant in the fibrosarcoma.

• Lipophagic granuloma vs carcinoma: an inflammatory and necrotic smear is sometimes found in both types of lesions. The absence of marked cellular atypias and the presence of foreign giant cells and histiocytes point to a diagnosis of granuloma.

SUMMARY

Indications for the method:

- detection of cytological elements in the aspiration of clinically evident mass or nipple discharge

What should be noted:

- presence and morphology of ductal cells
- presence and morphology of the stromal cells
- possible presence of blood cells and inflammatory cells

What the lesions are that may be diagnosed by cytology:

- inflammatory and degenerative lesions: fibrocystic dysplasia
- benign tumors: intracanalicular papilloma, fibroadenoma, nipple adenoma, tumors of the soft tissues: lipoma, fibroadenoma, etc.
- malignant tumors: adenocarcinomas, sarcomas, metastases

What should be reported:

- description of the elements present
- specification of the benignity or malignancy of the sample
- observation of those cellular structures and modes of desquamation that permit a more precise diagnosis

Value of the cytologic method:

- In clinically obvious malignant tumors, cytology offers a rapid confirmation that many consider sufficient to establish the definitive diagnosis.
- In lesions that are not clinically evident, particularly in small or poorly localized tumors, cytology is less exact and the margin of error can be as high as 10 percent

If the cytopathologist remembers the indications for and the limitations of the cytologic method, and if he bases diagnosis on clinic, radiologic, histologic, and cytologic grounds, he will not commit any flagrant mistakes.[20, 31, 46, 48] The latter arise when one bases too many decisions on too few criteria observed on smears too poor in cells.

TECHNIQUES

Collection techniques vary with the type of cytology in question. In discharge cytology, slides are applied directly to the nipple, fluid is expelled and then smeared. Fixation (alcohol-ether or commercial aerosol) must be rapidly performed. The stains used are Shorr's, hematoxylin, or Papanicolaou's and possibly more specific stains (i.e., PAS, sudan, toluidine blue). Some laboratories use May Grunwald-Giemsa, applied to air-dried smears.

For needle puncture, one uses a thin needle (22 gauge) with a stylet adapted to the syringe. The nodule is punctured and aspirated by moving the needle in and out while maintaining negative pressure. Lastly, the needle is carefully withdrawn and the negative pressure released. This maneuver avoids the dispersion of the biopsy fluid on the walls of the syringe, which is especially undesirable if the quantity of product collected is small. In such cases, it is best to wash the syringe and the needle with several milliliters of physiologic saline, which is then centrifuged. If the tumor is highly fibrous, a needle of larger diameter may be required. The next step is to spread the sample on a slide uniformly. The fixative agents and stains are the same as those used in drainage cytology.

Imprint smears of surgical specimens is a rapid procedure for examination of suspicious lesions and is adjuvant to frozen sections.[13, 35]

BIBLIOGRAPHY

1. Abramson D. J. A clinical evaluation of aspiration of cysts of the breast. *Surg. Gynec. Obstet.* 139:531, 1974.

2. Ashikari R., et al. A clinicopathologic study of atypical lesions of the breast. *Cancer* 33:310, 1974.

3. Ashton P.R., Hollinsworth A.S., Johnston W.W. The cytopathology of metastatic breast cancer. *Acta Cytol.* 19:1, 1975.

4. Blichert-Toft M., Hansen J.P.H., Hansen H., Schiodt T. Clinical course of cytosarcoma phyllodes related to histologic appearance. *Surg. Gynec. Obstet.* 140:929, 1975.

5. Boquoi E., Krebs S., Kreuzer G. Feulgen-DNA-cytophotometry on mammary tumor cells from aspiration biopsy smears. *Acta Cytol.* 19:326, 1975.

6. Chu E.W., Hoye R.C. The clinician and the cytopathologist evaluate fine needle aspiration cytology. *Acta Cytol.* 17:413, 1973.

7. Cornillot M., Cappelaere P., Clay A., Verhaeghe M. Le diagnostic cytologique des tumeurs solides du sein par la ponction-aspiration à l'aiguille fine. Confrontations anatomo-pathologiques. *Rev. Cytol. Clin.* 1:9, 1968.

8. De Brux J., Dupre-Froment J. Cytologie Mammaire. *Vie Med.* 48:1087, 1967.

9. Demay C. Cytologie des affections mammaires bénignes. *Bull. Mem. Soc. Clin. Paris* 62:238, 1972.

10. Fechner R.E. Oral contraceptive effects on the breast. *JAMA* 224, 1973.

11. Frank H.A., Hall F.M., Steer M.L. Preoperative localizations of nonpalpable breast lesions demonstrated by mammography. *New Engl. J. Med.* 295:259, 1976.

12. Franzen S., Zajicek J. Aspiration biopsy in diagnosis of palpable lesions of the breast. *Acta Radiol.* 7:241, 1968.

12a. Geier G.R., Korner B.H., Schuhmann R. Differential cytology of breast cancer. *Expl. Cell Biol.* 45:167, 1977.

13. Godwin J.T. Rapid cytologic diagnosis of surgical specimens. *Acta Cytol.* 20:111, 1976.

14. Groll M., Takeda M., Rakoff A. Breast and vaginal hormonal cytology in patients with breast secretions. *Acta Cytol.* 19:429, 1975.

15. Haagensen C.D. *Diseases of the breast.* 2nd Ed. Saunders, Philadelphia, 1971.

16. Hajdu S.I., Melamed M.R. The diagnostic value of aspiration smears. *Amer. J. Clin. Path.* 59:350, 1973.

17. Hengen H.A. Persistent bilateral galactorrhea in patient with atypical (class II) breast cancer cytologic findings. *JAMA* 232:962, 1975.

18. Hutter R.V.P. The pathologist's role in minimal breast cancer. *Cancer* 28:1527, 1971.

19. Kern W.H., Dermer G.B. The cytopathology of hyperplastic and neoplastic mammary duct epithelium. Cytologic and ultrastructural studies. *Acta Cytol.* 16:120, 1972.

20. Kreuzer G., Boquoi E. Aspiration biopsy cytology, mammography and clinical exploration: A modern set up in diagnosis of tumors of the breast. *Acta Cytol.* 20:319, 1976.

21. Kreuzer G., Boquoi E., Meyer R.D., Zajicek J. Studies on aspiration biopsy cytologic method in diagnosis of mammary carcinomas. *Abstracts 2nd European Congress on Cytology Societies.* Akademiai Kiado, Budapest, 1972, p. 92.

22. Kreuzer G., Zajicek J. Cytologic diagnosis of mammary tumors from aspiration biopsy smears. III. Studies on 200 carcinomas with false negative or doubtful cytologic reports. *Acta Cytol.* 16:249, 1972.

23. Lapey J.D. Lipid-rich mammary carcinoma. Diagnosis by cytology. *Acta Cytol.* 21:120, 1977.

24. Laumonier R., Hemet J. La cytologie des tumeurs mammaires. *Rev. Cytol. Clin.* 1:19, 1968.

25. Linsk J., Kreuzer G., Zajicek J. Cytologic diagnosis of mammary tumors from aspiration biopsy smears. *Acta Cytol.* 16:130, 1972.

26. Ludwig A.S., Okagaki T., Richart R.M., Lattes R. Nuclear DNA content of lobular carcinoma in situ of the breast. *Cancer* 31:1553, 1973.

27. Masukawa T. Discovery of psammoma bodies and fungus organisms in the nipple secretion with improved breast cytology technique. *Acta Cytol.* 16:408, 1972.

28. Masukawa T. Breast cytology. *Amer. J. Med. Tech.* 39:397, 1973.

29. Masukawa T., Lewinson E.F., Frost J.K. The cytologic examination of breast secretions. *Acta Cytol.* 10:261, 1966.

30. Masukawa T., Wada Y., Mattingly R.F., Kuzma J.F. Cytologic detection of minute ovarian, endometrial and breast carcinomas with emphasis on clinical pathological approaches. *Acta Cytol.* 17:316, 1973.

31. Moskowitz M., Milbrath J., Gartside P., Zermeno A., Mandel D. Lack of efficacy of thermography as a screening tool for minimal and stage I breast cancer. *N. Engl. J. Med.* 295:249, 1976.

32. Mouriquand J. L'empreinte mammaire. *Rev. Cytol. Clin.* 1:23, 1968.

33. Murad M.T., Snyder M.E. The diagnosis of breast lesions from cytologic material. *Acta Cytol.* 17:418, 1973.

34. Nordenskjold B., Lowhagen T., Westerberg H., Zajicek J. H-thymidine incorporation into mammary carcinoma cells obtained by needle aspiration before and during endocrine therapy. *Acta Cytol.* 20:137, 1976.

35. Pickren J.W., Burke E.M. Adjuvant cytology to frozen section *Acta Cytol.* 7:164, 1963.

36. Rajcic V. Cytologic studies of aspiration biopsy of the breaet. *Minerva Ginec.* 23:417, 1971.

37. Robbins G.F., Brothers J.H., Eberhart W.F., Quan S. Is aspiration biopsy of breast cancer dangerous to the patient? *Cancer* 7:774, 1954.

38. Rosemond G., Maier W., Brobyn T. Needle aspiration of breast cysts. *Surg. Gynec. Obstet.* 128:351, 1969.

39. Rosen P., et al. Diagnosis of carcinoma of the breast by aspiration biopsy. *Surg. Gynec. Obstet.* 134:837, 1972.

40. Scarff R.W., Torloni H. *Histologic Typing of Breast Tumors.* WHO, Geneva 1968.

41. Smith G.V., Shirley R.L. Management of benign lesions related to the diagnosis of early breast cancer. In Reid D.E., Christian C.D. (Eds.), *Controversy in Obstetrics and Gynecology.* Saunders, Philadelphia, 1974, II.

42. Spriggs A.I., Jerrome D.W. Intracellular mucous inclusions. A feature of malignant cells in effusions in the serous cavities, particularly due to carcinoma of the breast. *J. Clin. Path.* 28:929, 1975.

43. Stavric G.D., Tevcev D.T., Kaftandjiev D.R., Novak J.J. Aspiration biopsy cytologic method in diagnosis of breast lesions. A critical review of 250 cases. *Acta Cytol.* 17:188, 1973.

44. Tribe C. A comparison of rapid methods including imprint cytodiagnosis for the diagnosis of breast tumors. *J. Clin. Path.* 26:273, 1973.

45. Troisier S. Les écoulements anormaux du mamelon. *Vie. Med.* 48:1033, 1967.

46. Van Bogaert L.J., Mazy G. Reliability of the cyto- radio-clinical triplet in breast pathology diagnosis. *Acta Cytol.* 21:60, 1977.

47. Vassilakos P. Tuberculosis of the breast: cytologic findings with fine-needle aspiration. *Acta Cytol.* 17:160, 1973.

48. Verhaeghe M., Cornillot M., Herbeau J., Wurtz A., Verhaeghe G. Le triple diagnostic cyto-radio-clinique dans les tumeurs du sein (a propos de 2460 cas). *Mém. Acad. Chir.* 95:48, 1969.

49. Vessey M.P., Doll R., Jones K. Oral contraceptives and breast cancer. Progress report of an epidemiological study. *Lancet* 1:941, 1975.

50. Vilaplana E., Jimenez-Ayala M. The cytologic diagnosis of breast lesions. *Acta Cytol.* 19:519, 1975.

51. Wallgren A., Silfversward C., Zajicek J. Evaluation of needle aspirates and tissue sections as prognostic factors in mammary carcinoma. *Acta Cytol.* 20:313, 1976.

52. Wallgren A., Zajicek J. Cytologic presentation of mammary carcinoma on aspiration,biopsy smears. *Acta Cytol.* 20:469, 1976.

53. Wallgren A., Zajicek J. The prognostic value of the aspiration biopsy smear in mammary carcinoma. *Acta Cytol.* 20:479, 1976.

54. Webb A.J. The diagnostic cytology of breast carcinoma. *Brit. J. Surg.* 57:259, 1970.

55. Wellings S.R., Jensen H.M. On the origin and progression of ductal carcinoma in the human breast. *J. Cancer Invest.* 50:1111, 1973.

56. Winship T. Aspiration biopsy of breast cancers by the pathologist. *Amer. J. Clin. Path.* 52:438, 1969.

57. Zajdela A. Valeur et intérêt du diagnostic cytologique dans les tumeurs du sein par ponction. Etude de 600 cas confrontés cytologiquement et histologiquement. *Arch. Anat. Path.* 11:85, 1963.

58. Zajdela A., Durand J.C., Veith F. Aspect cytologique de quelques variétés particulières d'épithéliomas mammaires. *Bull. Cancer* 62:227, 1975.

59. Zajdela A., Ghossein N.A., Pilleron J.P., Ennuyer A. The value of aspiration cytology in the diagnosis of breast cancer: experience at the Foundation Curie. *Cancer* 35:49, 1975.

60. Zajdela A., Pilleron J.P., Ennuyer A., Maublanc M.A. Cytodiagnostic des lésions rares du sein par ponction à l'aiguille fine. *Rev. Cytol. Clin.* 4:185, 1972.

61. Zajicek J., Bartels P.H., Bahr G., Bibbo M., Jakobsson P.A., Wied G.L. Computer analysis of needle aspirates from breast carcinomas during radiotherapy. *Acta Cytol.* 17:179, 1973.

62. Zajicek J., Caspersson T., Jakobsson P., Kudinowski J., Linsk j., Us-Krasovec M. Cytologic diagnosis of mammary tumor from aspiration biopsy smears. Comparison of cytologic and histologic findings in 2111 lesions and diagnostic use of cytophotometry. *Acta Cytol.* 14:370, 1970.

7
Effusions
Into Body Cavities

The studies of cells of effusional fluids was one of the first domains of cytology to draw the interest of clinicians seriously.

The first microscopic studies of the cytology of effusions are now more than a century old. Among the pioneer works published on this matter, were those of Lucke and Klebs (1867), Quincke (1875), Ehrlich (1882), Chuquet (1879), Widal and Ravaut (1900), and Dopfer and Touton (1910). The reader may find the more recent references in the bibliography.

EMBRYOLOGY

The mesothelial cavities arise from the intraembryonic coelom, which is mesodermal in origin. The mesenchyme or primitive connective tissue, gives rise to the mesothelium. Anatomically speaking, the mesothelial cavities include the pericardium, the pleura, the peritoneum, and the vaginal tunic of the testicle, which arises from the peritoneum.

HISTOLOGY

The pericardial, pleural, and peritoneal cavities are lined with mesothelium that lies on loose connective tissue. The mesothelium is a simple epithelium arising from the mesenchyme which in turn arises from the mesoblast. The mesothelial lining was formerly considered to be a pseudoepithelium; this concept is no longer valid. It is formed by elongated cells running parallel to its surface; their nuclei are spindle-shaped or rounded; they have a visible nucleolus and a finely dispersed chromatin. The cytoplasm is pale and flattened (Plate 7.1). The underlying connective tissue contains collagen fibers, reticulin fibers, fibroblasts, macrophages, and mastocytes.

Electron microscopy has not revealed any structural peculiarities in this type of cell.[17,36,55] Let us note, however, that the cytoplasm contains glycogen droplets, and that numerous microvilli and pinocytotic vesicles are present. The microvilli appear less numerous in desquamated cells.

The Pathologic Anatomy of Mesothelial Lesions

Normally, mesothelial cavities contain a few detached cells and a small quantity of serous secretion, whose role is to reduce friction between the parietal and visceral layers.

Histologic lesions of mesothelial coverings must be grouped under differing clinical headings; among these are desequilibrium of fluid dynamics and effusions due to trauma, inflammation, and tumors.

Fluid accumulation in mesothelial cavities can arise from multiple causes: interference with the blood circulation, neoplastic invasion of lymph vessels, and invasion and destruction of the mesothelial lining of tumor cells. The presence of fluid in the cavities provokes the following histologic modifications: vascular congestion, edema, proliferation and hyperplasia of mesothelial cells, and formation of fibrinous, purulent, or hemorrhagic exudates (Plates 7.2 and 7.10).

In certain cases the cellular composition of a liquid effusion allows the determination of the nature of the disorder. Transudates are sterile, nearly free of cellular elements, and contain less than 3g of protein per 100 ml. Exudates, however, are not sterile, contain many cells, and have an elevated protein concentration.

Infections of mesothelial coverings (pericarditis, pleurisy, peritonitis) can be acute or chronic, and specific or nonspecific. They provoke congestion and edema of the mesothelial wall with infiltration of leukocytes, plasmocytes, and histiocytes. The excess of fluid in the cavities modifies the form of the cells; their flattened appearance, a product of the continual sliding contact of the two mesothelial layers, is transformed into a cubic or columnar one.

A biopsy allows precise determination of the nature of an inflammation and exposes specific lesions such as giant-cell tubercular granulomas. Complications of inflammations are fibrosis as well as thickening and adhesion of the visceral and parietal sheets.

Primitive tumoral lesions are rare. However, we distinguish benign tumors (mesotheliomas, lipomas, angiomas) and malignant ones (malignant mesotheliomas). Metastatic tumoral lesions are more frequent. Pulmonary, mammary, digestive, and ovarian tumors have the greatest propensity to invade mesothelial tissue.

Histologically, malignant mesotheliomas appear as fibrosarcomas, either accompanied by massive cellular proliferation or arranged in tubular glandular papillary cords. It is the pseudoepithelial aspect that characterizes these tumors. The cellular elements are scattered in an abundant mesenchymal stroma. The coexistence of these epithelial type structures and the mesenchymal stroma characterizes the malignant mesothelioma.

The etiology of this highly malignant tumor is not known. However, the correlation between both pleural and peritoneal mesotheliomas and contact with asbestos fibers has been established by the elevated frequency of occurrence of such lesions among workers in the asbestos industry.[59]

Metastatic tumors may appear as discrete lymphatic emboli with preservation of the overlying mesothelium (Plate 7.13). In this eventuality, the irritation of the mesothelium provokes the accumulation of irritated benign cells and fluids without malignant cells. This is why effusions can be negative even in the presence of clinical obvious tumors. When the metastatic formations become larger they invade and destroy the mesothelial epithelium allowing cancer cells to pass into the fluids.

Special methods such as fluorescence, tissue culture, and histochemistry have been studied thoroughly and instituted to solve the diagnostic problems, but none of them is conclusive. For example, a strong nonspecific esterase activity can be demonstrated not only in malignant cells but also in mesothelial cells from patients with pulmonary embolism.[2]

CYTOLOGY

The cytology of effusions is one of the most difficult to evaluate and, as such, must be approached with caution. The most competent cytopathologists will sometimes fail to determine the exact nature of cellular elements. This is reflected in the false- positive and false-negative results reported in the literature.

Nontumorous Cellular Elements Found in Effusional Fluids

The Mesothelial Cell

The diameter of a mesothelial cell may vary from 15 to more 80μ in the multinucleated varieties. The cell has a polyhedral shape that is responsible for the mosaiclike appearance of clusters (Plates 7.3, 7.6, and 7.19). When isolated, the mesothelial cell appears rounded and often shows small pseudopods (Plate 7.20). The round or oval nucleus may measure between 9 and 25μ in diameter and has finely dispersed chromatin and a visible nucleolus measuring roughly 2μ in diameter (Plate 7.8). Mitoses may be observed (Plate 7.4), and the cytoplasm is basophilic. Alcian blue/PAS-positive granules may be present in the cytoplasm. When a mesothelial cell is dying, its cytoplasm becomes pitted with vacuoles which may displace the nucleus and modify the cellular shape: the borders of the cell sometimes may be indistinguishable (Plate 7.5).

As mentioned, previously multinucleated forms also may be found. These are relatively large cells with 10 or more nuclei and a dense, intensively stained cytoplasm. These cells sometimes form aggregates which prefigure tubular structures or cords. The existence of sheets of various sizes and of all the intermediate forms between the isolated cell and the sheet itself are reliable criteria for the identification of mesothelial cells.

"Reactive" mesothelial cells exhibiting structural anomalies appear in different clinical conditions such as chronic inflammatory processes, liver cirrhosis, and collagen diseases (Plates 7.7 and 7.21). The intensity of these alterations increases with repeated aspirations; it is particularly true in the evolution of a cirrhotic effusion. False positives have been reported in such cases. Benign signet ring cells are observed in cirrhosis.

When clusters of tumoral cells are scattered among groups of mesothelial cells, the malignancy may be determined by the size of the cells, modifications in the chromatin structure, the density of the sheets, and the overlapping of the cells. The quantity of desquamated mesothelial cells varies with pathologic circumstances.

Macrophages

The term "macrophage" refers to functional characteristics rather than to specific morphologic ones. The macrophage absorbs and stores vital stains, as well as various particles, and cellular debris (Plates 7.3, 7.19, and 7.21). It may be of either mesothelial or hemopoietic (monocyte) origin.

The diameter of a macrophage varies from 15 to 60μ with a nuclear diameter of between 7 and 10μ. The nucleus is usually irregular but may be rounded, and its chromatin is dense and finely dispersed. The pale basophilic cytoplasm contains vacuoles and, often, cellular debris, such as red blood cells, carbon, iron, hemosiderin, and melanin. It is indeed the highly characteristic cytoplasm and the type of pigment present that permit identification of the macrophage.

Lymphocytes

Lymphocytes are found in variable quantities in inflammatory conditions — principally chronic ones — and in effusions of neoplastic origin. They are also found in great numbers in tubercular pleurisy. In maligant lymphomas, lymphosarcomas, and lymphoid luekemias, the density of lymphocytic elements constitutes a means of diagnosis.

It has been shown that effusions of tumoral origin contain both T-type and B-type lymphocytes.

Some authors[72] have suggested that an increase in the percentage of lymphocytes in pleural effusions may correspond to a favorable clinical cause. More data should be available to determine the significance in tumor immunology of lymphocytes present in effusions.

Leukocytes

Neutrophilic granular leukocytes are found in large quantities in acute inflammatory effusions, such as empyema abscess, rheumatoid disorders, and cancer. Eosinophilic leukocytes are found in certain clinical manifestations such as epithelial tumors, Hodgkin's disease, allergic conditions, pneumothorax, pulmonary infarcts, and viral and parasitic infections (Plate 7.9).

Plasmocytes

Plasma cells are found in chronic infections and, particularly, in tuberculosis, malignant tumors, cardiac effusions, and rheumatoid arthritis.

Mastocytes

Mast cells are larger than mesothelial cells. Their cytoplasm contains numerous metachromic granules that stain well, notably with toluidine blue.

Erythrocytes

Erythrocytes can be found in practically all effusions.

Squamous Cells

Superficial and keratinized squamous cells of cutaneous origin result from a contamination of the fluid during incision and drainage.

Tumoral Cellular Elements

The criteria inherent in the proper diagnosis of effusional cancerous cells reside solely in the distinction between the mesothelial and the cancerous cell. Because of its cytologic nature and its exfoliation into a liquid, the cellular anarchy of mesothelial elements tends to overshadow the classic diagnostic criteria of cytologic malignancy. Hyperchromatism, increased nuclear and nucleolar volume, changes in cell size and shape, and multinucleation are found in irritated or hyperplastic mesothelial cells as well as in cancerous cells. The changes involved are only minimal quantitative ones, and the evaluation of such changes is therefore partially subjective. This difficulty is aggravated when tumoral elements are undifferentiated cells.

The most valuable criteria for identification of cytologic malignancies are the following: an increase in the nucleocytoplasmic ratio, an abnormal and irregular chromatin distribution, the presence of atypical mitoses, multinucleation, nucleolar hypertrophy, and abnormalities of the form and size of the cytoplasm (Plates 7.15, 7.16, and 7.18).

The clustering of cells, with their nuclei tightly packed, the presence of glandular, tubular, or papillary structures, and the phagocytic properties of cancerous cells also represent important diagnostic criteria (Plate 7.22). The value of a diagnosis of malignancy is considerably enhanced when examination reveals such primitive epithelial structures. Usually mesothelial elements do not form cellular aggregates. The presence of specific cellular products such as melanin, mucus, and psammoma bodies also points to a cancerous origin.

A frequent source of confusion to the cytopathologist is the existence of highly atypical mesothelial cells in effusions of a patient whose cancer is clinically undeniable (Plates 7.17 and 7.20). Although all other diagnostic considerations leave no room for doubt, firm cytologic proof is lacking. Postmortem biopsies have explained this phenomenon: histology reveals numerous lymphatic emboli beneath a mesothelial covering whose integrity has been respected. Mesothelial cellular exfoliation has been considerably increased by the underlying tumoral proliferation in the stroma (Plate 7.13).

Furthermore, effusions often reveal the simultaneous presence of mesothelial and neoplastic cells, which in no way simplifies the task of diagnosis. To surmount this perplexing obstacle, one must, first of all, work only with slides that have been carefully fixed and stained.

In this field of cytology a high quality technique is most important to improve the interpretation of smears. In this vein, let us mention that Dekker and Bupp[14] have shown that collecting cells from a centrifuged sediment with a cotton swab produces greater cell concentration and better cell morphology than slides prepared by the pipette technique. The choice of fixatives and stains naturally varies from one cytopathologist to another, but it is essential that one remains consistent in one's techniques to assure consistently reproducible and acceptable results. This means that specimens must be rapidly delivered to the laboratory, fixed, concentrated, and stained. We do not recommend a drying technique, although this procedure is advocated by different authors.[33,63]

Primary Tumors

Primary tumors of the mesothelium (mesotheliomas) can be divided into two varieties: a slow-growing benign localized tumor consisting of elongated fibroblastlike forms exist. Epithelial-like cells form clusters of moderate size with various nuclear atypias. Papillary structures with the occasional presence of a vascular stroma are very characteristic, but cytologically they can also represent papillary components of ovarian or thyroid carcinomas.[4] If the cytoplasmic vacuoles are mucicarmine-positive, one can rule out malignant mesothelioma. In malignant fibrous mesothelioma, sarcoma-like cells are bizarre, elongated, and stellate with round or fusiform hyperchromatic nuclei. The cytoplasm is often poorly preserved. These cells represent malignant fibroblasts (Plates 7.11, 7.12, and 7.24). Some authors have stressed the diagnostic significance of the presence of a collagen core in the center of cell aggregates.[70]

A diagnosis of mesothelioma is not always suggested by cytology; clinical and radiological data may help to confirm the diagnosis.

Metastatic Tumors

In the cytology of effusions, it is not always possible to identify the organ of origin of metastatic cells.

Adenocarcinomas are the most easily recognized because they tend to form tubular, papillary, or glandular formations that recall the structure of the original tumor. Examples are tubular or glandular formations in adenocarcinomas of bronchial, mammary, digestive, or ovarian origin. Ovarian and gastric tumors result in effusions that are generally rich in cells with voluminous nucleoli (Plate 7.15).

Mammary tumors may result in rounded and spherical aggregates that suggest the structure of the embryonic blastula (Plate 7.22). The nuclei are moderately hyperchromatic and the nucleoli, hypertrophied. In paraffin-embedded centrifugation bands, glandlike structures are more common because spheroidal formations are well preserved by the embedding procedure (Plate 7.14). Cells from lobular carcinoma are identified as small, round cells with a large hyperchromatic nucleus.

Squamous undifferentiated carcinomas of bronchial or esophageal origin are more difficult to identify because the signs of squamous maturation are not obvious as in differentiated forms.

Malignant lymphomas give characteristic cellular images, particularly when lymphosarcomas and lymphoid or myeloid leukemias are concerned.[5] Lymphocytes, lymphoblasts, and cells of the myelocyte lineage are present in great numbers (Plate 7.23). One error to avoid is mistaking a chronic inflammatory fluid rich in lymphocytes for a malignant lymphocytic lymphoma. Compared with the small cells of lobular cancer of the breast, the cytoplasm is more abundant, the nucleolus is less prominent, and the cells tend to form aggregates. The small cells of malignant lymphomas also should not be confused with oat cells of the lung. A distinct oat-cell organization has been described which suggests the histologic type of primary tumor: it consists of rows of malignant cells similar to the arrangement of the vertebrae in the spinal column.

The cytologic diagnosis of Hodgkin's disease is more difficult because Reed-Sternberg cells are rarely found in effusional fluids.

In medullary carcinoma of the breast, one finds voluminous, isolated cells with highly characteristic lobular nuclei and sometimes a significant number of leukocytes.

A differential diagnosis among cells of mammary, pancreatic, ovarian, or thyroid origin presents a delicate problem even for the most competent cytologist. As mentioned before, the cytologic differential diagnosis between epithelial papillary carcinomas and malignant epithelial-like mesothelioma is a hazardous initiative.

Some rare tumors may be suggested by examination of a direct smear or after centrifugation. Among these are sarcomas, malignant melanomas, and neuroblastomas. Primitive cancers that give the highest percentage of positive effusions come from the mammary glands, the lungs, the ovary, the gastrointestinal tract, and lymphatic tissue.

Efficiency of the Method

Correctly performed, cytologic evaluations give correct diagnoses in approximately 60 to 90 percent[10,15,33,53,60] of the cases of cancer, with slightly better results with ascitic fluid than with pleural effusion. The 10 to 40 percent margin of error is constituted by false negatives. (The false-positive rate should not exceed 5 percent). Chromosome analysis combined with the standard cytologic method gives significantly better results than either technique alone.[16]

Differential Diagnosis

The difficulties often encountered in making a differential diagnosis between mesothelial and tumoral cells have inspired numerous attempts at clarification. Neoplastic elements tend to form aggregates and show cellular piling; their cytologic anomalies are usually more obvious. The abundance of cytoplasm is characteristic of mesothelial cells. An increase in the nucleocytoplasmic ratio points to malignancy.

The difficulty is greater when atypical mesothelial cells are compared with differentiated tumoral elements that do not exhibit nuclear anomalies and desquamate as single cells. Some adenocarcinomas of the breast and of the ovary are in this category. When neoplastic cells are of relatively small size, one must consider oat-cell carcinoma of the lung, lymphoma, neuroblastoma, Wilms' tumor, Ewing's sarcoma, and embryonic rhabdomyosarcoma. Neuroblastomas classically exhibit rosette formations of small cells with scanty cytoplasm. However, this rosette arrangement has been reported among mesothelial as well as neoplasic elements.

Confusion between small undifferentiated carcinoma cells and active lymphoid cells and lymphoblasts can be avoided by a careful analysis of cellular structures. Small epithelial cells tend to desquamate in clumps. Lymphoid cells exhibit a more profuse cyanophilic cytoplasm and prominent nucleoli. Mitoses may be frequent in the active proliferation of mesothelial cells caused by inflammation.

The benign conditions that may cause problems of differential diagnosis with tumorous conditions are: chronic heart feature, pulmonary infarcts and emboli, pericarditis, lung abscess, cirrhosis, renal failure with uremia, and postradiation effusions.

The presence of alcian blue/PAS-positive granules in the cytoplasm of mesothelial cells appears as a nonspecific morphologic characteristic (Plate 7.7). Pfitzer and Huth[56] have not observed any corresponding ultrastructural change, and they assume that the material (mucopolysaccharides and glycoproteides) has been dissolved during fixation and inclusion.

Various histochemical methods have been proposed to improve the accuracy of the method but none has definitely settled this problem of differential diagnosis.[3,40]

Particular centrifugal methods[68] may represent a significant usefulness in cytologic interpretation especially when dealing with a low concentration of cellular material.

The advantages of the air-dried smear method have been reported by Spriggs and Boddington[64] and consist of better preservation of cytoplasm, no nuclear retraction, and a pale purple color of the nucleus with Giemsa stain (Plates 7.16 and 7.20). Traditionally Giemsa stain is a tool of hematologists. Both methods (air-dried and alchol-fixed) have their qualities: if one is accustomed to either one, he should stick to it.

Comparison of the results of direct cytologic examination of a centrifugation band and paraffin embedding technique leads us to conclude that both methods are valuable and should be used simultaneously to increase the sensitivity of the method.

One rule should never be overlooked: always obtain the clinical data. For example, the morphologic effects of radiotherapy and chemotherapy, and the presence of cellular anomalies in cirrhosis perfectly mimic tumor cell alterations.

Summary

Indications for the method:

- recognition of mesothelial cells and metastatic or primary malignant elements
- recognition of alterations of mesothelial cells

What should be noted:

- distinction between acellular transudates and exudates
- mesothelial cells; their presence is a pathologic condition even if they do not exhibit morphologic changes
- macrophages, histiocytes, leukocytes, and red blood cells and their relative quantities
- cancer cells: metastatic or primary tumors

What should be reported:

- correlation of the abundance of different benign cells to certain specific causal infections
- qualitative and quantitative description of the cellular elements present in the effusion; evaluation of the quantity and the type of pigments (hemosiderin, bile pigments, melanin, etc.)
- description of the cellular type and the mode of exfoliation of cancer cells
- correlation of the cytologic observations with the clinical status
- evaluation of the effectiveness of various therapeutic agents (chemotherapy, radiotherapy): disappearance of neoplastic elements and or appearance of cellular anomalies in benign cells

Without specific cellular criterial and clinical data, one should not try to guess the primary site of the malignant cells.

BIBLIOGRAPHY

1. Bakalos D., Constantakis, N., Tsicricas T. Distinction of mononuclear macrophages from mesothelial cells in pleural and peritoneal effusions. *Acta Cytol.* 18:20, 1974.

2. Bakalos D., Constantakis N., Tsicricas T. Recognition of malignant cells in pleural and peritoneal effusions. *Acta Cytol.* 18:118, 1974.

3. Bauer Z., Milic N., Handl S., Koprčina M. The results of some cytochemical reactions in metastatic malignant tumor cells in pleural and peritoneal effusions. *Acta Cytol.* 21:141, 1977.

4. Becker S.N., Pepin D.W., Rosenthal D.L. Mesothelial papilloma: a case of mistaken identity in a pericardial effusion. *Acta Cytol.* 20:266, 1976.

5. Bellingham M.E., Rawlinson D.G., Berry P.F., Kempson R.L. The cytodiagnosis of malignant lymphomas and Hodgkin's disease in cerebrospinal, pleural and ascitic fluids. *Acta Ctyol.* 19:547, 1975.

6. Berge T., Gröntoft O. Cytologic diagnosis of malignant pleural mesothelioma. *Acta Cytol.* 9:207, 1965.

7. Berge T., Nelsten S. Cytological diagnosis of cancer in pleural and ascitic fluid. *Acta Cytol.* 10:138, 1966.

8. Black L.F. The pleural space and pleural fluid. *Mayo Clin. Proc.* 47:493, 1972.

9. Boddington M.M., Spriggs A.I., Morton J.A., Mowat A.G. Cytodiagnosis of rheumatoid pleural effusions. *J. Clin. Path.* 24:95, 1971.

10. Ceelen G.H. The cytologic diagnosis of ascitic fluid. *Acta Cytol.* 8:175, 1964

11. Creasman W.T., Rutledge F. The prognostic value of peritoneal cytology in gynecologic malignant disease. *Amer. J. Obstet. Gynec.* 110:773, 1971

12. Danner D.E., Gmelich J.T. A comparative study of tumor cells from metastatic carcinoma of the breast in effusions. *Acta Cytol.* 19:509, 1975.

13. De Brux J.A., Dupre-Froment J., Mintz N. Cytology of the peritoneal fluids sampled by coelioscopy or by cul-de-sac puncture. *Acta Cytol.* 12:395, 1968.

14. Dekker A., Bupp P.A. Cytology of serous effusions: a comparative study of two slightly different preparative methods. *Acta Cytol.* 20:394, 1976.

15. Dekker A., Graham T., Bupp P.A. The occurrence of sickle cells in pleural fluid: report of a patient with sickle cell disease. *Acta Cytol.* 19:251, 1975.

16. Dewald G., Dines D.E., Weiland L.H., Gordon H. Usefulness of chromosome examination in the diagnosis of malignant pleural effusions. *N. Engl. J. Med.* 295:1494, 1976.

17. Domagala W., Wdyke S. Transmission and scanning electron microscopic studies of cells in effusions. *Acta Cytol.* 19:214, 1975.

18. Domagala W.M., Koss L.G., Emeson E.E. Immunological aspects of cytology of serous cavity fluids (asbtract). *Proceedings of the 6th International Congress of Cytology.* Tokyo, 1977, p.153.

19. El-Mahdi A., Levene A., Lott S. Observations and management of malignant pleural effusions in breast. *The Johns Hopkins Medical Journal* 132:44, 1973.

20. Figuera J.M. Presence of microfilariae of *Mansonella ozzardi* in ascitic fluid. *Acta Cytol.* 17:73, 1973

21. Foot N.C. Identification of types and primary sites of metastatic tumors from exfoliated cells in serous fluids. *Amer. J. Path.* 30:661, 1954.

22. Foot N.C. The identification of neoplastic cells in serous effusions. *Amer. J. Path.* 32:961, 1956.

23. Foot N.C., Holmquist N.D. Supravital staining of sediments of serous effusions; simple technique for rapid cytological diagnosis. *Cancer* 11:151, 1958.

24. Foot N.C. The identification of mesothelial cells in sediments of serous effusions. *Cancer* 12:429, 1959.

25. Grunze H. The comparative diagnostic accuracy, efficiency and specificity of cytologic techniques used in the diagnosis of malignant neoplasm in serous effusions of the pleural and pericardial cavities. *Acta Cytol.* 8:150, 1964.

26. Hansen H.H., Bender R.A., Shelton B.J. The cyto-centrifuge and cerebrospinal fluid cytology. *Acta Cytol.* 18:259, 1974.

27. Jacobson E.S. A case of secondary echinococcosis diagnosed by cytologic examination of pleural fluid and needle biopsy of pleura. *Acta Cytol.* 17:76, 1973.

28. Järvi O.H., Kunnas R.J., Laitio M.T., Tyrkko J.E.S. The accuracy and significance of cytologic cancer diagnosis of pleural effusions. (A follow up study of 338 patients). *Acta Cytol.* 16:152, 1972.

29. Johada E., Bartels P.H., Bibbo M., Bahr G., Holzner J.H., Wied G.L. Computer discrimination of cells in serous effusions. I. Pleural fluids. *Acta Cytol.* 17:94, 1973.

30. Johnson W.D. The cytological diagnosis of cancer in serous effusions. *Acta Cytol.* 10:161, 1966.

31. Kcettel W.C., Pixley E.E., Buchsbaum II.J. Experience with peritoneal cytology in the management of gynecologic malignancies. *Amer. J. Obstet. Gynec.* 120:174, 1974.

32. Kern W.H. Benign papillary structures with psammoma bodies in culdocentesis fluid. *Acta Cytol.* 13:178, 1969.

33. Krivinkova H., Ponten J., Blöndal T. The diagnosis of cancer from body fluids. *Acta Path. Microbiol. Scand.* (Series A) 84:455, 1976.

34. Lopez-Cardozo P. A critical evaluation of 3000 cytology analyses of pleural ascitis and pericardial fluid. *Acta Cytol.* 10:455, 1966.

35. Luse S.A., Reagan J.W. A histological study of effusions. *Cancer* 7:1155, 1954

36. Luse S.A., Reagan J.W. A histological study of effusions. II. Effusions associated with malignant tumors. *Cancer* 7:1167, 1954.

37. Luse S.A., Reagan J.W. A histologic and electron microscopic study of effusions associated with malignant disease. *Ann. N.Y. Acad. Sci.* 63:1331, 1956.

38. Luttwak E.M. Quantitative studies of ascitic fluid. *Surg. Gynec. Obstet.* 136 (2) :269, 1973

39. Mandelbaum F.S. The diagnosis of malignant tumors by paraffin sections of centrifuged exsudates. *J. Lab. Clin. Med.* 2:580, 1917.

40. Marsan C., Cayphas J. The aid of some histochemical stains in the identification of mesothelial cells. Preliminary results. *Acta Cytol.* 18:252, 1974.

41. Marsan C., Roujeau J. Les épanchements à cellules mésotheliales. *Rev. Cytol. Clin.* 4:25, 1971.

42. Masin F., Masin M. Cytodiagnosis of effusions by the desorption technic. *Acta Cytol.* 9:380, 1965.

43. McGowan L., Bunnag B. A morphologic classification of peritoneal fluid cytology in women. *Intern. J. Gynaec. Obstet.* 11:173, 1973.

44. McGowan L., Bunnag B. Morphology of mesothelial cells in peritoneal fluid from normal women. *Acta Cytol.* 18:205, 1974.

45. McGowan L., Bunnag B., Arias L.B. Peritoneal fluid cytology associated with benign neoplastic ovarian tumors in women. *Amer. J. Obstet. Gynec.* 113:961, 1972.

46. NcGowan L., Davis R.H. Cytology of serous fluid in the pelvic peritoneal cavity of regnant and post partum women. *Amer. J. Clin. Path.* 51:150, 1969.

47. McGowan L., Davis R.H. Peritoneal fluid cellular pattern in obstetrics and gynecology. *Amer. J. Obstet. Gynec.* 106:979, 1970.

48. McGrew E.A., Nanos S. The cytology of serous effusions. In Wied G.L., Koss L.G., Reagan J.W., Editors'. *Compendium on diagnostic cytology.* 4th Ed., Tutorials of cytology, Chicago, 1976, p. 370.

49. Melamed M.R. They cytologic presentation of malignant lymphomas and related diseases in effusions. *Cancer* 16:413, 1963.

50. Morris H.H.B., Bennett M.J. The classification and origin of amniotic fluid cells. *Acta Cytol.* 18:149, 1974.

51. Murat M. Electron microscopic studies of cells in pleural and peritoneal effusions. *Acta Cytol.* 17:401, 1973.

52. Murphy W.M., Ng A.B.P. Determination of primary site by examination of cancer cells in body fluids. *Amer. J. Clin. Path.* 58:479, 1972.

53. Naylor B. The exfoliative cytology of diffuse malignant mesothelioma. *J. Path. Bact.* 86:293, 1963.

54. Naylor B., Schmidt R.W. The case for exfoliative cytology of serous effusions. *Lancet* 1:711, 1964.

55. Nosanchuk J.S., Naylor B. A unique cytologic picture in pleural fluid from patients with rheumatoid arthritis. *Amer. J. Clin. Path.* 50:330, 1968.

56. Pfitzer P., Huth F. Alcian PAS. Positive granules in mesothelioma and mesothelial cells. *Acta Cytol.* 10:205, 1966.

57. Quincke H. Uber die Geformten Bestandteile von Transsudaten. *Dtschr. Arch. f. klin. Med.* 30:580, 1882.

58. Ramsey S.J., Tweedale D.N., Bryant L.R., Braunstein H. Cytologic features of pericardial mesothelium. *Acta Cytol.* 14:283, 1970.

59. Salhadin A., Nasiell M., et al. The unique cytologic picture of oat cell carcinoma in effusions. *Acta Cytol.* 20:298, 1976.

60. Selikoff I.J., Churg J., Hammond E.C. Relation between exposure to asbestos and mesothelioma. *New Engl. J. Med.* 272:560, 1965.

61. Spriggs A.I. *The Cytology of Effusions in the Pleural, Pericardial and Peritoneal Cavities.* Heinemann, London 1957.

62. Spriggs A.I. A simple density gradient method for removing red cells from haemorrhagic serous fluids. *Acta Cytol.* 19:470, 1975.

63. Spriggs A.I., Boddinton M.M. *The Cytology of Effusions in the Pleural, Pericardial and Peritoneal Cavities and of Cerebrospinal Fluid.* 2nd Ed. Heinemann, London, 1968.

64. Spriggs A.I., Boddington M.M. Oat-cell bronchial carcinoma. Identification of cells in pleural fluid. *Acta Cytol.* 20:525, 1976.

65. Spriggs A.J., Jerrome D.W. Aspects particuliers de surface des cellules dans les épanchements. *Rev. Cytol. Clin.* 2:7, 1969.

66. Spriggs A.I., Jerrome D.W. Intracellular mucous inclusions. *J. Clin. Path.* 28:929, 1975.

67. Spriggs A.J. Meek G.A. Surface specializations of free tumor cells in effusions. *J. Path. Bact.* 82:151, 1961.

68. Teplitz R.L., Minami R., Yokota S. Separation of malignant from normal cells in effusions (abstract). *Proceedings of the 6th International Congress of Cytology.* Tokyo, 1977, p.86.

69. Vellios F., Griffin J. Examination of body fluids for tumor cells. *Amer. J. Clin. Path.* 24:676, 1954.

70. Whitaker D., Shilkin K.B. The cytology of malignant mesothelioma in Western Australia (abstract). *Proceedings of the 6th International Congress of Cytology.* Tokyo, 1977, p.86.

71. Woyke S., Domagala W., Olszewski W. Ultrastructure of hepatoma cells detected in peritoneal fluid. *Acta Cytol.* 18:130, 1974.

72. Yamagishi K., Tajima M., Suzuki A., Kimura K. Relation between cell compositum of pleural effusions in patients with pulmonary carcinomas and their clinical courses. *Acta Cytol.* 20:537, 1976.

8
The Respiratory System

EMBRYOLOGY

The primitive respiratory system arises from a fold in the anterior portion of the embryonic pharynx. This fold constitutes the entodermal laryngotracheal tube and gives rise to the larynx, the trachea, and the lungs (Figure 8.1). Thus, the embryonic entodermal pulmonary apparatus, along with the splanchnic mesoderm that surrounds it, comprise the various tissue of the respiratory system: the endoderm gives rise to the epithelial linings, and the mesenchyme is responsible for the pulmonary cartilages, muscles, connective tissue, and vessels. The progressive branching of the primary bronchi results in the bronchioles and the alveoli of the pulmonary tree. The intraembryonic coelom forms the pleura, which is comprised of visceral and parietal portions, both of mesodermal origin.

For practical purposes we have included buccal cytology in this chapter. The mouth or stomodeum is a depression on the surface ectoderm, facing the anterior part of the foregut (entoderm). These two layers form the buccopharyngeal membrane which ruptures during the third week.

HISTOLOGY

There are two types of epithelium in the respiratory system. The mouth, pharynx, larynx, vocal cords, and the epiglottis are covered with the stratified squamous nonkeratinizing type, which lies on a vascularized connective tissue stroma. Columnar stratified epithelium lines the nasal cavities and sinuses, the nasopharynx, the trachea, and the bronchi. In the upper respiratory tract, this epithelium is comprised of four cell layers (Plate 8.1) but gradually diminishes to a single layer in the alveoli (Plate 8.2). This columnar epithelium includes ciliated cells, mucous or goblet cells, and deeply, basal-type cells. These epithelial cells form a continuous lining perforated with pores ranging between 10 and 15μ (Colin's pores).

Beneath the epithelium is a vascularized connective tissue stroma containing smooth muscle cells and reticular, elastic, and collagen fibers. Lymphatic capillaries are abundant.

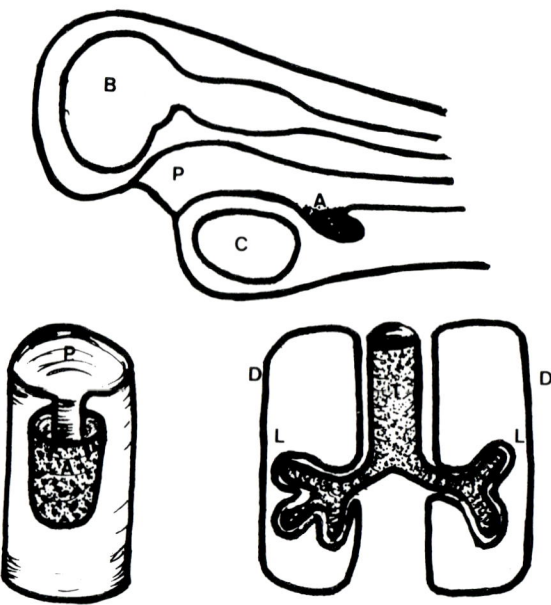

Figure 8.1. Embryology of the primitive respiratory system. *A,* Laryngotracheal tube; *B,* brain; *C,* pericardial cavity; *D,* pleura; *L,* lung buds; *P,* pharynx; *T,* trachea

The alveolar walls, where gaseous exchange occurs, are comprised of capillary endothelium, the basal lamina membrane and the alveolar cell itself.

CYTOLOGY OF THE RESPIRATORY TRACT

The nature and quantity of respiratory cells appearing on smears depend on the anatomic region examined and on the mode of sampling. Oral, pharyngeal, and laryngeal samples yield predominantly squamous cells, whereas expectorations and tracheobronchial aspirations result in the columnar type: the presence of the latter is necessary to confirm the pulmonary nature of the sample.

The squamous elements are comprised mostly of the superficial, nonkeratinized type, the intermediate and parabasal cells appearing only when erosion has occurred due to mechanical, inflammatory, or tumoral aggressions. These cells may arise from any portion of the respiratory tree.

The ciliated columnar cells measure from 30 to 40μ in height and are recognized by the more or less well-preserved cilia on their apical borders (Plate 8.3). The nuclei are small (10μ) and may be rounded or oval. The chromatin is dense and the nucleolus prominent. The cells which are found either in clusters or as isolated elements are more numerous and better preserved in bronchial aspirations than in sputum. Lipofuscin pigment may be present in the cytoplasm.

The mucosecretory or goblet cells are characterized by their clear cytoplasm, which elaborates a dilated, apical vacuole whose volume varies with the quantity of mucus present. The nuclei are small (5 to 10μ) and dense; they are round or, if crowded by the cytoplasmic mucus, crescent-shaped (Plate 8.4).

The basal cells of the deep germinal layer appear in pathologic circumstances leading to erosion of more superficial cells. However, they may be found in sputum even in the absence of such epithelial modifications, if local trauma has occurred (e.g., bronchoscopy).

Macrophages (or pneumocytes) are round cells, measuring between 25 and 35μ in diameter (Plate 8.4). On a smear, they appear as isolated elements whose nuclei may be round or kidney-shaped and either centrally or peripherally located. Their chromatin is arranged in small clumps and their nucleolus is prominent. Their abundant, pale cytoplasm is finely vacuolated and may contain particles of carbon, hemosiderin, or lipids. These cells probably arise from two different sources: they may be medullary cells migrating through the mucosa or simply modified alveolar cells. They are always present in sputum and authenticate the pulmonary origin of the specimen.

The nature of the engulfed particles observed in macrophages may be of diagnostic significance: hemosiderin is observed in cardiac cells; carbon and silica, in silicoanthracosis; lipids, in cases of necrosis and lipoid pneumonia. Those macrophages that differentiated from alveolar cells acquire only transitory phagocytic ability. The true macrophage, which is essential to the body's immune system, has its own characteristic enzymatic and lysosomial apparatus. It has been shown that true macrophages exert cytostatic effects on tumoral cells.

The presence of megakaryocytes has at times been described: these cells enter the bloodstream and may become trapped in the pulmonary capillaries.

A few white blood cells, particularly polymorphonuclear leukocytes, are virtually always observed on smears. However, an increased concentration of a particular type of leukocyte is often in itself of diagnostic value. Polymorphonuclear leukocytes are numerous in acute bronchiopneumonic inflammatory lesions and in ulcerative vegetating tumors. Lymphocytes are characteristic of chronic infections, such as tuberculosis. Eosinophilic leukocytes are characteristic of asthma.

Histiocytes are present in all chronic infections; there are multinuclear forms which, here again, may have a diagnostic significance (e.g., tuberculosis, foreign body, giant-cell granuloma, etc.).

Other Identifiable Elements

Viral inclusions (herpes, cytomegalic infection) may be observed but, in practice, are not frequently discovered.[131,185]

Curschmann's spirals are a mold of bronchiolar cavities formed by the precipitation of mucus and proteins (Plate 8.5). As the name implies, they have a highly characteristic spiral appearance. Their presence is indicative of inflammatory phenomena of the bronchial and bronchiolar walls and is frequent in asthmatics.

Charcot-Leyden crystals are shaped like two long pyramids placed base to base. They derive from nuclei of eosinophilic leukocytes[194] (Plate 8.6) and are associated with asthma.

Amylaceous bodies are intraalveolar protein condensations.

Presence of acantholytic cells (Tzanck's cells) has been described in scrapings of pemphigus vesicles. These cells exhibit nuclear pleomorphism and a pathologic chromatin distribution which makes them resemble tumor cells.

Gingival and jugal modifications brought on notably by poor dentition provoke foci of leukoplakia, chronic modifications, and pseudopapillary proliferations whose clinical manifestations simulate those of malignant lesions.

Mycoses, chiefly monialiasis are easily identified by their spores or hyphae which are identical to those described in genital cytology.

The increase of the nuclear and cytoplasmic volumes described in certain cases of pernicious anemia constitutes, in our opinion, a rather vague and subjective diagnostic criterion of limited clinical value. Also such nuclear modifications may be observed even in normal epithelium as well as in cases of malnutrition other than folic acid deficiency.

Smears of the oral mucosa may be used to determine the presence of the sex heterochromatin (Barr body), which appears in the resting cell as a nodule adhering to the inner membrane of the nuclear envelope (see Chapter 4).

Smears of the nasal mucosa taken from patients suffering from allergic rhinitis are extremely rich in cells and show a relative increase in the number of mucosecretory elements[104] (Plate 8.8).

TUMORS OF THE ORAL CAVITY, NASOPHARYNX, PARANASAL SINUSES, AND LARYNX

Oral Cavity

Not only are lesions of the oral cavity easily accessible to direct examination but any oral fungating or ulcerating lesion lends itself readily to cytologic sampling procedure. It is in the dentist's office that the cytologic method is the most useful: it is simple, rapid, and painless and provides an efficient diagnostic orientation. We strongly encourage adaptation of the method as it allows for a ready diagnostic procedure for noninvasive lesions, dysplasia, and intraepithelial carcinoma, all of which naturally have more optimistic prognoses than the invasive forms.

The great majority of tumors are of the squamous keratinized type. All the various lesions proper to this type of epithelium may be observed from simple dysplasia and intraepithelial carcinoma to squamous invasive carcinoma (Plate 8.9). Spinocellular carcinoma represents about 90 percent of the malignant tumors of the mouth. It is localized in order of diminishing frequency on the tongue, the lips, the oral floor, the oral mucosa, the gums, and the palate.[137] Among the predisposing factors, are the use of tobacco and alcohol, the existence of leukoplakia, and inadequate oral hygiene.[18] In Asia, the chewing of the betel nut constitutes another factor. Certain populations of India have the curious habit of smoking cigars with the lit end held inside the mouth (reverse smoking[57, 136]) and this predisposes to carcinoma of the palate.

Leukoplakia is a clinically defined condition characterized by a white spot which constitutes a hyperkeratosis. Applied to leukoplakia, cytologic procedure reveals superficial squamous eosinophilic elements which are sometimes anuclear. If there are cellular anomalies, and if they are numerous and severe, they may constitute an intraepithelial carcinoma.

Squamous oral carcinoma exhibits the same cellular anomalies as those described for squamous genital carcinoma: anomalies of size and shape of both the nucleus and cytoplasm, abnormal chromatin distribution, nucleolar hypertrophy, multinucleation, abnormal nucleocytoplasmic ratio, and cytoplasmic maturation modifications (precocious or irregular keratinization).

Certain highly differentiated and surface-keratinized lesions may yield false-negative results upon scraping: needle puncture of such lesions eliminates this problem by obtaining cells from the underlying tumor.

Certain infrequently occurring tumors which together represent about 10 percent of all oral lesions must nonetheless be mentioned: glandular carcinoma, adenoid and cystic carcinomas, labial basocellular carcinoma, mixed salivary gland tumors, and even more rarely malignant lymphomas, sarcomas, fibrosarcomas, and reticulosarcomas. The detection of all these lesions may be realized cytologically. The presence of cells exhibiting suspicious malignant characteristics necessitates a follow-up histologic examination to determine the tumor's nature.

The cytologic surveillance of irradiated tumors is a simple procedure permitting an early detection of possible recurrence.[174] Postradiation cellular anomalies are identical to those described in Chapter 5.

The Nasopharyngeal Region

The standard indication for the cytologic method in this domain is the detection of suspected tumoral cells in light of exaggerated mucous secretion or hemorrhage or in cases of clinical suspicion. These secretions may be collected directly by using cotton swabs or by pipette aspiration. Sinus needle puncture also may provide cytologic material.

We describe spinocellular carcinomas, Schneiderian or basal carcinomas, mucinous carcinomas, papillary carcinomas, and finally, columnar cell carcinomas.[7] Sarcomas are extremely rare. Because the nasopharyngeal region has both squamous and glandular structures, this multiple classification is justified. The cytologic anomalies observed vary with the tumor in question, but the general criteria of malignancy are the same. Histologic methods must be used to determine the exact nature of the tumor.

Nasopharyngeal carcinomas are most frequent among the Chinese. Certain extremely rare tumors such as the chordoma have been the object of cytologic studies.

The squamous epithelium of the tonsillar crypts may exhibit marked inflammatory and necrotic cellular modifications which should not be confused with malignant modifications.

Cytology may prove useful during laryngoscopy, either as a preliminary diagnostic procedure or because biopsy is impracticable for technical reasons. We have observed cases of carcinoma of the vocal cords where swab specimens demonstrated the lesion's malignancy.

Most malignant laryngeal tumors are squamous carcinomas. Here, as with other squamous epithelia, dysplastic lesions and intraepithelial carcinomas may be encountered.

For salivary gland tumors, see Chapter 11.

INFLAMMATORY PROCESSES
OF THE BRONCHIO-ALVEOLAR TREE

Histology

The bronchio-alveolar mucosa can undergo structural modifications which may be either specific to a certain pathogenic agent or quite nonspecific. The integrity of the mucosa is necessary to ensure the protection normally afforded by mucous secretion, the ciliary apparatus, and the surface lipoproteins (superfactant whose role is to reduce the surface tension). If the integrity of this protective ciliated mucous layer is modified, lesions appear. The intensity of such lesions depends on the nature of the pathogenic agent and on the duration of its action.

An increase in the number of mucosecretory cells occurs in all infections and more particularly so when bronchiectasis and mucous hypersecretion occur (Plate 8.10).

Hyperplasia of the basal cells is a nonspecific reaction comparable to that found in other epithelia, such as that of the cervix. The excessive proliferation of these cells causes a thickening of the mucosa with persistence of the columnar lining (Plate 8.11). This superficial layer may remain or undergo desquamation at foci of erosion. Such hyperplasia constitutes a reversible phenomenon whose intensity depends on the nature of the aggression. It is common in allergic conditions such as hay fever and asthma.[29]

Squamous metaplasia is the consequence of hyperplasia followed by epidermoid transformation of the basal cells. There are two possible mechanisms that may explain the appearance of metaplasia: hyperplasia of the basal cells followed by their direct and progressive transformation into squamous cells with elimination of the superficial columnar elements or a desquamation of the mucociliated layer with concomitant proliferation of the deeper layers. The first explanation is the more plausible. Metaplasia is frequent.[152] It is found in about 40 percent of all inflammatory pulmonary lesions including tuberculosis.[9] Papanicolaou[123] described a metaplastic cell type in his own sputum during an acute phase of a chronic respiratory inflammaroty condition: this cell is small with a dark ovoid nucleus and eosinophilic cytoplasm (Plate 8.13). This epidermoid metaplastic cell may be encountered not only in all chronic inflammatory infections but also in dysplasia and in malignant lesions. Later we shall discuss its relations with tumoral lesions.[153] Cellular structural modifications are multiple and consist of multinucleation, hyperchromatism, irregular disposition of heterochromatin, nuclear necrosis and fragmentation, and irregularity of nuclear and cytoplasmic size and shape[56] (Plate 8.12). These modifications are nonspecific and complicate the cytopathologist's task of differential diagnosis.

Ciliocytophtoria as described by Papanicolaou[125] is a particular form of ciliated cell degeneration, manifested by cellular debris, which may or may not contain a degenerating nucleus but which still exhibits a ciliated border[130, 131] (Plate 8.14). Ciliocytophtoria is found in infections, particularly viral ones. To our experience this degeneration is not frequently encountered.

Intranuclear eosinophilic viral inclusions are specific modifications to certain pathologic conditions. The following may be diagnosed cytologically: cytomegalic disease, herpes, adenovirus.[113] Ciliocytophtoria may be present.

Papillomatous proliferation of the bronchial mucosa or micropapillomatosis is the formation of small papillae consisting of a connective tissue axis with columnar epithelium possibly exhibiting squamous metaplasia. It is a hyperplastic reaction to diverse inflammatory aggressions.

The Cytology of Inflammatory Processes

Bacterial infections do not cause the appearance of any specific cytologic lesions, and one observes all the various modifications we have just described. With some luck, examinations of smears may reveal multinucleated cells of the Langhans type or epitheloid cells which suggest tuberculosis[56, 95, 142] as their cause. Bacteria, however, are rarely apparent on smears or histologic sections, and only a culture can provide a valuable diagnosis. Nonspecific modifications of chronic inflammatory processes are the appearance of rather numerous densely cellular clumps of hyperplastic cells.

Mycotic infections provoke nonspecific cellular alterations which may be quite marked. In certain cases, the causative agent can be observed, thus indicating the nature of the alterations. Candida and Aspergillus are the most frequent genera observed. Branching of the hyphae is characteristic of *Aspergillus Septal Lymphae*. Some cases of phycomycosis have been reported. Culture is necessary to identify the fungus definitely.

Opportunistic fungi are very common and usually represent saprophytes or contaminants, but in certain conditions they may produce infections. For example, patients who received chemotherapy or immunodepressive drugs frequently present such infections.[111]

Coccidioides immitis is characterized by the presence of spherules and endospores. Spherules may attain 100μ in diameter and contain the spores which range from 3 to 5μ in diameter.[78]

Cryptococcus neoformans is a yeast consisting of an ovoid or spherical budding organism which has a typical teardrop shape. It measures from 10 to 20μ in diameter.

Blastomyces dermatitidis is a round yeast measuring 10 to 15μ in diameter with characteristic refractile walls. Budding is frequent and the two resulting parts are closely connected.

Histoplasma capsulatum is a round fungus measuring 3 to 5μ in diameter and often engulfed in macrophages. Johnston and Frable advise use of the methenamine stain for its detection.[78]

Pneumocystis carinii organisms are better seen with methenamine silver stain. Numerous elements are present in the alveolar mucus. They are round or oval organisms measuring 5 to 10μ in diameter. Sometimes they exhibit a ringlike structure applied to the external membrane. I is a fungus or a protozoan.[66, 139]

The presence of large vacuolated macrophages (Plate 8.18) containing oil droplets is suggestive of lipid pneumonia. It is due to aspiration of oil contained in nose drops, in laxatives, or more rarely, in foreign bodies.

MALIGNANT TUMORS OF THE LUNG

The high frequency and clinical gravity of malignant tumors of the lung make them an extremely important chapter in cancerology. Indeed, in many geographic regions, pulmonary tumors are the most frequent of all cancers. According to the epidemiologic and vital statistics report of the WHO[202] the death rate from primary malignant tumors of the trachea, bronchi, and lung per 100,000 varies from 51 (Great Britain) to 6.5 (Japan). Intermediate figures are noted in the USA (27.3), Holland (28.7), France (17.9), and Germany (28.3).

During the last few decades, the frequency of lung cancer has soared dramatically: we are thus faced with the problem of its etiology and more specifically of determining what role tobacco and air pollution play in this rise. Asbestosis due to inhalation of asbestos fibers is another example of occupational disease. The incidence of bronchogenic carcinoma is significantly higher in the population processing this substance.

What is not completely understood is why certain heavy smokers and individuals living in the seriously polluted environments can sometimes do so with impunity. A possible explanation would be a protective, genetically determined enzymatic system whose role would be the degradation of cyclic hydrocarbons deposited in the lungs. These enzymes would be synthesized by the liver and the pulmonary alveolar cells and activated by the foreign hydrocarbons. Individual immunologic factors are suggested in these mechanisms.

Under the influence of such diverse irritants, the transformation of the respiratory tract epithelium occurs progressively, passing through the stages of epidermoid metaplasia[153,193] simple or aggravated dysplasia,[44,116] and intraepithelial carcinoma.[182,196,197] The relationship between these structural modifications and substances such as tobacco has been described by various authors;[4,150] Auerbach[5] confirmed the clinical observations by provoking these same lesions in beagles.

Pulmonary cytologic lesions such as dysplasia and in situ carcinoma, quite comparable to those described in the cervix present the same problems of differential diagnosis. These lesions were described later for the lung than for other sites doubtless because of their relative anatomic inaccesibility.

Today the various steps of the malignant transformation of the bronchial mucosa are well recognized (Figure 8.2). The separation of the superficial layers of the epithelium (slit formation) and the subsequent degeneration of these desquamated cells (ciliocytophtoria) would explain the development of epidermoid metaplasia.[117]

Histology

Classification systems for pulmonary tumors are numerous; we recommend use of the nomenclature adapted by the WHO[202] since its international distribution would provide different laboratories with an effective common language (Table 8.1).

Epidermoid carcinomas, which arise from epidermoid metaplastic lesions, comprise half of all pulmonary tumors. They may be distinguished as differentiated, slightly differentiated and undifferentiated, depending on the degree of squamous maturation (Plate 8.22).

Undifferentiated small-cell carcinomas are comprised of large cells exhibiting no glandular differentiation and keratinization. They represent approximately 10 percent of pulmonary tumors.

Undifferentiated small-cell carcinomas account for about 25 percent of all pulmonary tumors. They have a rather unfavorable clinical prognosis. They include oat-cell carcinomas and other types of small anaplastic cells. This last group of tumors may exhibit a paraneoplastic syndrome characterized by endocrine metabolic, and hematologic disorders.

Glandular carcinomas or adenocarcinomas comprise 10 to 15 percent of all pulmonary tumors and are characterized by tubular or pseudoglandular formations.

Finally, there are the rare forms (listed in the table) which together comprise less that 10 percent of all pulmonary lesions.

Cytology

Epidermoid Carcinoma

Squamous cells in in situ carcinoma have been identified[13,126,177,197] and can be compared in many respects to the homologous lesion of the cervix. Dysplastic lesions may precede the development of carcinoma.[44] The diagnosis can be suggested cytologically

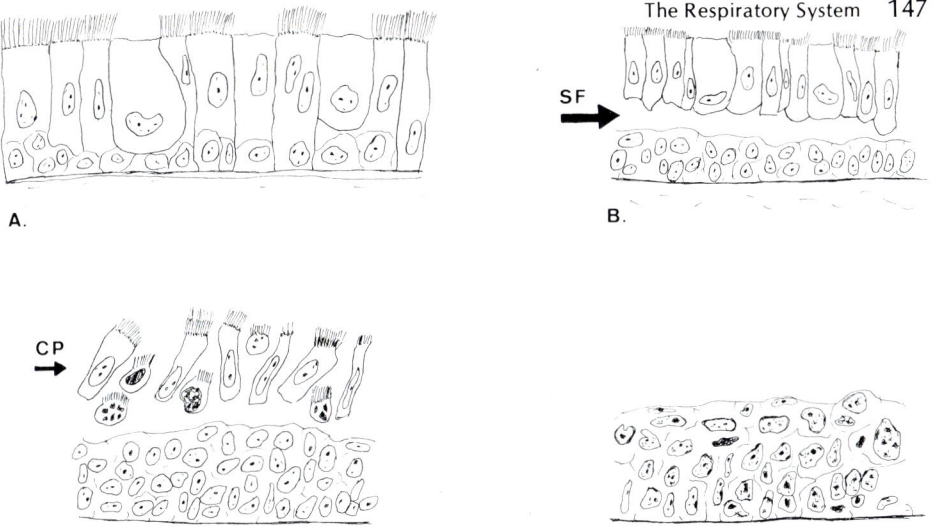

A.

B.

C.

D.

Figure 8.2. Development of epidermoid metaplasia. [117] *A*, normal bronchial mucosa, *B*, slit formation (SF) between the basal layers and the superficial layers of the epithelium; *C*, cellular degeneration of the superficial layers: Ciliocytophtoria (CP); *D*, atypical metaplasia or dysplasia.

Table 8.1. WHO Classification of Pulmonary Tumors

I.	Epidermoid carcinoma
II.	Small-cell anaplastic carcinoma
III.	Adenocarcinoma
IV.	Large-cell anaplastic carcinoma (giant cell carcinoma)
V.	Combined epidermoid and adenocarcinoma
VI.	Bronchial carcinoid
VII.	Bronchiolar alveolar carcinoma

in the presence of small malignant epidermoid cells with more discrete anomalies (Plate 8.21) than invasive cancers but with obvious keratinization. Large eosinophilic epidermoid cells from the surface of the lesion with nuclear atypias also may be recognized. The absence of an inflammatory diathesis also may be indicative of a noninvasive lesion.

Epidermoid malignant cells are the most frequently encountered and the most easily recognized of all pulmonary tumors. They desquamate in clumps or as large isolated elements. These have a well-preserved, keratinized, eosinophilic refractile cytoplasm with a dense hyperchromatic irregular nucleus (Plate 8.23). Nuclear pycnosis is frequent. Incomplete pearl keratin formation is observed. Nucleoli are not seen. The cytoplasmic keratin content of these cells explains why most of them are well preserved compared with other types of carcinoma (Plate 8.24). Epidermoid carcinomas in general are so typical that even the presence of a few cells is enough to make the diagnosis. For example, the presence of necrotic keratotic debris in inflammatory smears is highly suggestive of epidermoid carcinoma.

In poorly differentiated cases, the epidermoid character of the lesion is not as evident. The dense, large cyanophilic cytoplasm with well-defined cellular margins is very suggestive of an epidermoid origin. The chromatin structure is more apparent, and nucleoli can be seen more frequently compared with the differentiated form. The quantity of cells varies from one case to another. Cellular necrosis is frequent in centrally located elements of large cancerous cords.

The most common pathogenic agents are summarized in Table 8.2. The morphological modifications brought on by such agents are the following:

- an increased number of mucosecretory cells
- hyperplasia of the basal or the reserve cells
- squamous metaplasia
- papillomatous proliferation of the columnar epithelium

The cytological modifications include:

- multinucleation
- pycnosis, karyorrhexis, and chromatin modifications
- anomalies of nuclear and cytoplasmic size and shape
- cytoplasmic vacuolation
- intranuclear viral inclusions
- cellular necrosis
- ciliocytophtoria

Small Cell Anaplastic Carcinoma

Malignant desquamation is composed of isolated elements about 12μ in diameter, which may form clusters. Nuclei are hyperchromatic and are either naked or surrounded by a thin cyanophilic rim of cytoplasm. Cohesiveness and compression of cell groups create a typical image of nuclear molding. These elements are round, elongated, or polygonal. Nucleoli may be seen among the dense chromatin blocks. The small size of the elements is the most significant criterion of this type of tumor.[110] The mode of desquamation varies with the sampling technique used. In blended sputum most of the elements are isolated (Plate 8.25); on the contrary, in bronchial washings or fresh sputum, cells are better preserved, less compressed, and appear as larger elements arranged in small clumps. Inflammatory diathesis is a constant characteristic of these tumors.

Large Cell Anaplastic Carcinoma

Smears reveal the presence of large malignant elements with irregular hyperchromatic and often monstrous nuclei (Plate 8.26). Nucleoli are voluminous. The nucleocytoplasmic ratio is high. Contrary to the eosinophilic pattern of epidermoid carcinoma, the cytoplasm is cyanophilic, finely vacuolated with often indistinct

Table 8.2. Common Pathogenic Agents

Clinical agents: e.g., tobacco, mustard gas, silica, Carbon, asbestos
Microbial agents: viral, mycotic, and parasitic agents
Therapeutic agents: e.g., busulfan
Physical agents: e.g., ionizing radiation, abnormally rich O_2 atmosphere
Trauma

cellular borders. The marked structural modifications of nuclei and cytoplasm help distinguish between anaplastic large cancer cells and atypical alveolar pneumocytes.

Again in fresh sputum and bronchial washings, clumps are better preserved than in blended sputum. In the latter, suspicious or malignant elements are dispersed by the blending. Necrosis, cellular degeneration, and inflammatory cells are constantly present.

Adenocarcinoma

The adenocarcinoma type of tumor may be suggested when the cells exhibit a columnar or a cubic pattern with a "palisade" or glandular arrangement[162] (Plate 8.27). This characteristic configuration is more or less preserved according to the degree of cell degeneration and necrosis. In specimens where blending is not performed, there is better preservation of desquamated epithelial structures. Well-preserved cells have an eccentric, elongated nucleus with a finely granular chromatin; the cytoplasm is cyanophilic and sometimes vacuolated and plasmalemma borders are absent. Psammoma bodies may be associated with papillary adenocarcinoma.

This distinction between adenocarcinoma and the rare bronchiolar alveolar carcinoma is possible if the cells have been well preserved.[141] Bronchiolo-alveolar cells are smaller and more uniform in shape; they desquamate in tightly packed clusters with rare isolated cells (Plate 8.28) and provide an abundant desquamation in sputum or bronchial washings.[58] Cytoplasmic vacuoles are PAS positive for mucus.

Adenocarcinomas from bronchial epithelium and bronchial glands show larger cells with a more irregular chromatin structure and prominent nucleoli (Plate 8.29). Cellular density of smears is somewhat lower than in bronchiolar tumors. Mucous secretion may be evident in vacuolated cells.

In mixed carcinoma, the simultaneous presence of squamous and columnar atypical elements may orientate the diagnosis but this favorable condition is rare.

The malignant nature of some rare tumors such as adenomas of the carcinoid type, adenoid cystic carcinoma, sarcomas (Plate 8.30), lymphomas, and neoplasms consisting of both carcinomatous and sarcomatous elements may be diagnosed cytologically.[70, 98, 120] Histologic evaluation is required to define the type of tumor.

Carcinoid tumor shows very typical small regular elements.[93]

Metastatic Tumors

Metastatic tumors of the lungs are common and occur in about 15 percent of all malignancies. Most of the time, they are bilateral and multiple. The most common primary neoplasms are carcinoma of the breast, the digestive tract, the female genital tract (Plate 8.31), the kidney, the male genital tract, the bones (Plate 8.32), and any localization of malignant melanoma. Malignant melanoma is easily recognized when the large atypical tumor cells exhibit melanin pigment.

ACCURACY OF THE METHOD

Different results are obtained depending on the techniques of sampling, and on cellular concentrations and types. They vary from 40 to 95 percent according to the authors.

Accuracy can be improved if great care is applied in sampling. Every laboratory knows how difficult it is to obtain valuable sputum and not just saliva. The patient should be carefully taught how to provide good specimens, and if there is any doubt about the patient's cooperation, sampling should be performed in the presence of a technician or nurse familiar with those techniques. The use of aerosol inhalation is sometimes required. At least three specimens should be examined.

Bronchial washings and brushings are obtained during bronchoscopy and may help to obtain selected samples from the observed lesion.

Small peripheral lesions are rarely recognized cytologically; in these cases transthoracic needle aspiration may represent a more reliable technique.[69]

DIFFERENTIAL DIAGNOSIS

- Inflammatory alterations may be severe and create difficult diagnostic problems. Dense clusters of columnar cells with altered ciliated borders and with dark and condensed nuclei may arise from foci of bronchiectasis or bronchitis and simulate adenocarcinoma. Again the intensity of the nuclear and cytoplasmic alterations is more pronounced in malignant conditions. Clinical data provide valuable information that should be considered in the final diagnosis. Herpes simplex viral atypias (nuclear multiplication, homogenization of the nuclear content, and rarely, nuclear inclusions) are easily recognized if the altered cells are well preserved.
- Contaminates of the sputum: vegetable cells, large elements derived from food particles, have a characteristic dense, thick cellulose membrance, a dark nucleus, and brownish-yellow cytoplasm with a double-contoured membrane. Other contaminates are fungus spores, air pollen, and dusts. A palisade arrangement of the cells is frequent.
- Differentiation of epidermoid carcinoma from benign or atypical squamous metaplasia: epidermoid cells in large number or their constant presence in various specimens is an abnormal condition that requires further radiologic, clinical, and histologic control, particularly in patients from high-risk groups. Adenocarcinoma cells may be differentiated from epidermoid cancer cells; in the former nuclei are usually larger, the chromatin is dispersed in distinct blocks scattered over a clear nuclear matrix, and nucleoli are more prominent.
- Small cellular elements: in presence of small elements one should think of anaplastic small type carcinomatous cells. Size, shape, and structure of the nucleus, nucleolar size, the nuclear cytoplasmic ratio, and configuration of clumps are factors to be considered.
- Bronchio-alveolar cells: desquamation of bronchio-alveolar cells is evident in mucosal hyperplasia, pulmonary fibrosis (Hamman-Rich syndrome), pulmonary infarcts, and chronic inflammatory lesions. The benign appearance of the cells may be overlooked when cell preservation is poor. Nuclei of malignant cells are larger and chromatin structural modifications are more pronounced than in hyperplastic cells
- Certain very specific structures will help to make the diagnosis. Langhans' giant cells are indicative of a tuberculous lesion; they should not be confused with histiocytic foreign-body giant cells. Charcot-Leyden crystals and eosinophilic leukocytes are indicative of allergic conditions. Reed-Sternberg cells are indicative of Hodgkin's disease. Lipid-laden macrophages are indicative of lipid pneumonia. Hemosiderin-laden macrophages are indicative of chronic heart failure.

SUMMARY

Indications for the method:

- recognition of cellular alterations in inflammatory, allergic, and tumorous conditions
- evaluation of postradiation and chemotherapy modifications

What should be noted:

- presence of inflammatory cells and cellular inflammatory alterations
- eventually, the specific aspect of the inflammatory reaction (for example, giant cells)
- presence of pigments
- presence of hyperplastic, dysplastic, and neoplastic cells
- presence of an inflammatory diathesis accompanying tumors conditions
- presence and abundance of macrophages

What should be reported:

- nature of the inflammatory infiltrate
- presence of specific etiologic factors: parasite, viruses, etc.
- presence of contaminates
- presence of atypical and neoplastic cells possibly with a definition of the tumor type
- request for control in case of suspicious cells

BIBLIOGRAPHY

1. Allegra S., Broderick P.A., Corvese N. Oral cytology, seven year oral cytology screening program in the state of Rhode Island. Analysis of 6448 cases. *Acta Cytol.* 17:42, 1973.
2. Allen A.R., Fullmer C.D. Primary diagnosis of pulmonary echinococcosis by the cytologic technique. *Acta Cytol.* 16:212, 1972.
3. An S.H., Koprowska I. Primary cytologic diagnosis of asbestosis associated with bronchogenic carcinoma. Case report and review of literature. *Acta Cytol.* 6:391, 1962.
4. Auerbach O., Stout A.P., Garfinkel L. Changes in bronchial epithelium in relation to sex, age, residence, smoking and pneumonia. *New Eng. J. Med.* 267:11, 1962.
5. Auerbach O., Stout A.P., Hammond E.C., Garfinkel L. Changes in bronchial epithelium in relation to cigarette smoking and in relation to lung cancer. *New Engl. J. Med.* 265:253, 1961.
6. Bamforth J. The examination of the sputum and pleural fluid in the diagnosis of malignant disease of the lung. *Thorax* 1:118, 1946.
7. Beale L.S. Examination of sputum from a case of cancer of the pharynx and adjacent parts. *Arch. Med.* 2:44, 1860
8. Bell J.W. Positive sputum cytology and negative chest roentgenograms. A surgeon's dilemma. *Ann. Thorac. Surg.* 9:149, 1970.
9. Berkheiser S.W. Atypical bronchiolar proliferation and metaplasia associated with tuberculosis. *Dis. Chest* 45:522, 1964.
10. Berkson D.M., Snider G.L. Heated hypertonic aerosol in collecting sputum specimens for cytological diagnosis. *Jama* 173:135, 1960.
11. Bibbo M., Fennessy J.J., Lu C-T, Strauss F.H., Variakojis D., Wied G.L. Bronchial brushing technique for the cytologic diagnosis of peripheral lung lesions: a review of 693 cases. *Acta Cytol.* 17:245, 1973.

12. Bickerman H.A., Sproul E.E., Barach A.L. An aerosol method of producing bronchial secretions in human subjects: a clinical technic for the detection of lung cancer. *Dis. Chest* 33:347, 1958.

13. Black H., Ackerman Lauren V. The importance of epidermoid carcinoma in situ in the histogenesis of carcinoma of the lung. *Ann. Surg.* 126:44, 1952.

14. Blair O.M., Goldenberg D.M. A correlative study of bronchial cytology, bronchial washing carcino-embryonic antigen and plasma carcinoembryonic antigen in the diagnosis of bronchogenic cancer. *Acta Cytol.* 18:510, 1974.

15. Bonime R.G. Improved Procedure for the preparation of pulmonary cytology smears. *Acta Cytol.* 16:543, 1972.

16. Brenner S.A., Lambert R.L., Pablo G.E. Superheated aerosol induced sputum in the cytodiagnosis of lung cancer. *Acta Cytol.* 6:405, 1962.

17. Broderick P.A., Corvese N.L., Lachance T., Allard J. Giant cell carcinoma of lung: a cytologic evaluation. *Acta Cytol.* 19:225, 1975.

18. Brown A.M., Young A. The effects of age and smoking on the maturation of the oral mucosa. *Acta Cytol.* 14:566, 1970

19. Brun J., Perrin-Fayolle M., Kofman J., Jeanblanc F. Les résultats globaux de la cytologie exfoliative au cours de 200 cas de cancer broncho-pulmonaire primitif. *Poumon* XXY:101, 1969.

20. Bryan M.P., Bryan W.T.K. Cytologic diagnosis in allergic disorders. *The Otolaryngologic Clinics of North America,* October 1974, p.637.

21. Cardozo L.P., De Graaf S., De Boer M.J., Doesburg N., Kapsenberg P.D. The results of cytology in 1000 patients with pulmonary malignancy. *Acta Cytol.* 11:120, 1967.

22. Chang J.P., Anken M., Rusell W.O. Sputum cell concentration by membrane filtration for cancer diagnosis. A preliminary report. *Acta Cytol.* 5:168, 1961.

23. Chang S.C. Microscopic properties of whole mounts and sections of human bronchial epithelium of smokers and non-smokers. *Cancer* 10:1246, 1957.

24. Chayeb J., Gahan P.B., La Cour . L.F. Chemical changes in human bronchial epithelium and their relation to bronchial cancer. *Nature* 183:1743, 1959.

25. Clerf L.H., Herbut P.A. Diagnosis of bronchogenic carcinoma by examination of bronchial secretions. *Ann. Otol. Rhinol. Laryngol.* 55:646, 1946.

26. Clerf L.H., Herbut P.A. The value of cytological diagnosis of pulmonary malignancy. *Amer. Rev. Tuberc.* 61:60, 1950.

27. Coonez W., Dzuira B., Harper R., Nash G. The cytology of sputum from thermally injured patients. *Acta Cytol.* 16:433, 1972

28. Dahlgren S.E., Lind B. Comparison between diagnostic results obtained by transthoracic needle biopsy and by sputum cytology. *Acta Cytol.* 16:53, 1972.

29. Dudgean L.S., Wrigley C.H. On the demonstration of particles of malignant growth in the sputum by means of the wet-films method. *J. Laryngol. Otol.* 50:752, 1935.

30. Ellis H.D., Kernosky J.J. Efficiency of concentrating malignant cells in sputum. *Acta Cytol.* 7:372, 1963.

31. Eneroth C.M., Franzen S., Zajicek J. Aspiration biopsy of salivary gland tumors. A critical review of 910 biopsies. *Acta Cytol.* 11:355, 1965.

32. Eneroth C.M., Jakobsson P., Zajicek J. Aspiration biopsy of salivary gland tumors. V. Morphologic investigations on smears and histologic sections of acinic cell carcinoma. *Acta Radiol.* 310:85 (Suppl.), 1971.

33. Eneroth C.M., Zajicek J. Aspiration biopsy of salivary gland tumors. II. Morphologic studies on smears and histologic sections from oncocytic tumors (45 cases papillary cystadenolymphoma and 4 cases oncocytoma). *Acta Cytol.* 9:355, 1965.

34. Eneroth C.M., Zajicek J. Aspiration biopsy of salivary gland tumors. III. Morphologic studies on smears and histologic sections from 368 mixed tumors. *Acta Cytol.* 10:440, 1966.

35. Eneroth C.M., Zajicek J. Aspiration biopsy of salivary gland tumors. IV. Morphologic studies on smears and histologic sections from 45 cases of adenoid cystic carcinoma. *Acta Cytol.* 13:59, 1969.

36. Farber S.M. Clinical appraisal of pulmonary cytology. *Jama* 175:345, 1961.

37. Farber S.M., Benioff M.A., Frost J.K. Rosenthal M., Tobias G. Cytologic studies of sputum and bronchial secretions in primary carcinoma of the lung. *Dis. Chest* 14:633, 1948.

38. Farber S.M., McGrath A.K. Jr., Benioff M.A., Rosenthal M. Evaluation of cytologic diagnosis of lung cancer. *Jama* 144:1, 1950.

39. Farber S.M., Pharr S.L. The practicing physician and pulmonary cytology. *Lancet* 77:111, 1957.

40. Farber S.M., Rosenthal M., Alston E.F., Benioff M.A., McGrath A.K. Jr. Cytologic Diagnosis of Lung Cancer. Thomas, Springfield, III, 1950.

41. Farber S.M., Wood D.A., Pharr S.L., Pierson B. Significant cytologic findings in non-malignant pulmonary disease. *Dis. Chest* 31:1, 1957.

42. Fennessy J. J., Fry W.A., Manalo-Estrella P., Frias-Hidvegi, D.V. The bronchial brushing technique for obtaining cytologic specimens from peripheral lung lesions. *Acta Cytol.* 14:25, 1970.

43. Fernec J. *Brossage endobronchique. Technique, intérêt du diagnostic cytologique.* Thèse, Rennes (France), 1971.

44. Fod D.K., Fidler H.K., Lock D.R. Dysplastic lesions of the bronchial tree. *Cancer* 14:1226, 1961.

45. Foot N.C. The identification of types of pulmonary cancer in cytologic smears. *Amer. J. Path.* 28:963, 1952.

46. Foot N.C. Cytologic diagnosis in suspected pulmonary cancer. Critical analysis of smears from 1000 persons. *Amer. J. Clin. Path.* 25:223, 1955.

47. Frable W.J. The relationship of pulmonary cytology to survival in lung cancer. *Acta Cytol.* 12:52, 1968.

48. Francis D., Borgeskov S. Progress in preoperative diagnosis of pulmonary lesions. *Acta Cytol.* 19:23, 1975.

49. Fried B.M. Primary carcinoma of the lungs. III. Histogenesis and metaplasia of bronchial epithelium. *Arch. Path.* 8:46, 1929.

50. Friedberg E.C. Giant cell carcinoma of the lungs. *Cancer* 18:259, 1965.

51. Frost J.K., Gupta P.K., Erozan Y.S., Carter D., Hollander D.H., Leven M.L., Ball W.C. Pulmonary cytologic alterations in toxic environment inhalation. *Hum. Path.* 4:521, 1973.

52. Fullmer C.D., Parrish C.M. Pulmonary cytology: a diagnostic method for occult carcinoma. *Acta Cytol.* 13:645, 1969.

53. Funch R.R., von Haam E. MAC in buccal smears. *Acta Cytol.* 15:76, 1975.

54. Funkhouser J.W., Meininger D.E. Cytologic aspects of bronchial brushing in a community hospital. *Acta Cytol.* 16:51, 1972.

55. Gardner A.F. An investigation of the use of exfoliative cytology in the diagnosis of malignant lesion of the oral cavity. The cytologic diagnosis of oral carcinoma. *Acta Cytol.* 8:436, 1964.

56. Garret M. Cellular atypias in sputum and bronchial secretions associated with tuberculosis and bronchiectasis. *Amer. J. Clin. Path.* 34:237, 1960.

57. Giacobbe E., Caruso P.L. Cytologic changes in the oral cavity of reverse smokers. *Acta Cytol.* 17:49, 1973.

58. Greenberg, et al. Bronchiolo-alveolar carcinoma cell of origin. *Amer. J. Clin. Path.* 63:153, 1975.

59. Groos P., Gralley J.J., de Treville R.T.P. "Asbestos bodies." Their none-specificity. *Amer. Hyg. Assoc, J.* 28:541, 1967.

60. Grunze H. *Klinische Zytologie der Thoraxkrankheiten.* Enke-Verlag, Stuttgart, 1955.

61. Grunze H. Long term cytology of squamous metaplasia of bronchial mucosa. Proceedings of the Sixth Annual Meeting of the Inter-Society Cytology Council. New York, 1958.

62. Grunze H. A critical review and evaluation of cytodiagnosis in chest diseases. *Acta Cytol.* 4:175, 1960.

63. Gupta P.K., Verma K. Calcified (psammoma) bodies in alveolar cell carcinoma of the lung. *Acta Cytol.* 16:59, 1972.

64. Haley L.D., Arch R. Use of milliporemembrane filter in the diagnostic tuberculosis laboratory. *Amer. J. Clin. Path.* 27:117, 1957..

65. Hamilton J.D., Brown T.C., McDonald F.W. Morphological changes in smokers' lungs. *Canad. Med. Ass. J.* 77:177, 1957.

66. Hamperl H. Pneumocystis infection and cytomegaly of the lungs in the newborn and adult. *Amer. J. Path.* 32:I, 1956.

67. Hartman P. *Die Cytologie des Bronchiasekretes.* Thieme, Stuttgart, 1955.

68. Hattori S., Matsuda M., Nishihara H., Horai T. Early diagnosis of small peripheral lung cancer.

Cytologic diagnosis of very fresh cancer cells obtained by the TV-brushing technique. *Acta Cytol.* 15:460, 1971.

69. Hayata Y., Oho K., Masatoshi I., Goya Y., Hayashi T. Percutaneous pulmonary puncture for cytologic diagnosis—its diagnostic value for small peripheral pulmonary carcinoma. *Acta Cytol.* 17:469, 1973.

70. Hellstrom H.R., Fisher E.R. Giant cell adenocarcinoma of the lung. *Cancer* 19:1337, 1966.

71. Henderson B.E., Jing J.S.H., Buell P., Gardner, M.B. Risk factors associated with nasopharyngeal carcinoma. *New Engl. J. Med.* 295:1101, 1976.

72. Herbut P.A. Correlation of cytological with pathological findings in tumors of the lung. *Proceedings of the Symposium on Exfoliative Cytology 1951, New York.* Amer. Cancer Soc. 1953, p.50.

73. Herbut P.A., Clerf L.H. Bronchogenic carcinoma: diagnosis by cytologic study of bronchoscopically removed secretions. *Jama* 130:1006, 1946.

74. Hilding A.C. On cigarette smoking, bronchial carcinoma and ciliary action: III. Accumulation of cigarette tar upon artificially produced deciliated islands in the respiratory epithelium. *Ann. Otol.* 65:116, 1956.

75. Hughes A. *A History of Cytology.* Abelard-Schuman, London, 1959.

76. Ide G., Suntzeff V., Cowdry E.V. A comparison on the histopathology fo tracheal and bronchial epithelium of smokers and non-smokers. *Cancer* 12:473, 1959.

77. Jarvi O.H., Hormia M.S., Autio J.V.K., Kangas S.J., Tilvis P.K. Cytologic diagnosis of pulmonary carcinoma in two hospitals. *Acta Cytol.* 11:477, 1967.

78. Johnston W.W., Frable W.J. The cytopathology of the respiratory tract. *Amer. J. Path.* 84:372, 1976.

79. Kanhouwa S.B., Matthews M.J. Reliability of cytologic typing of lung cancer. *Acta Cytol.* 20:229, 1976.

80. Kawecka M. Cytological evaluation of sputum in patients with bronchiectasis and the possibilities of erroneous diagnosis of cancer. *Acta Union Int. Contra Cancrum* 15:469, 1959.

81. Kenney M. Webber C.A. Diagnosis of strongyloidiasis in Papanicolaou-stained sputum smears. *Acta Cytol.* 18:270, 1974.

82. Kern W.H. Cytology of hyperplastic and neoplastic lesions of terminal bronchioles and alveoli. *Acta Cytol.* 9:372, 1965.

83. Kernec J., De Labarthe B., Lefreche J.N., Ramee M.P., Cormier M., Danrigal A. Le brossage bronchique distal. Evolution des techniques. Exploitation des prélèvements. Résultats. *Revue Cytol. Clin.* 7 (1):33, 1974.

84. Kierszenbaum A.L. Bronchial metaplasia observations on its histology and cytology. *Acta Cytol.* 9:365, 1965.

85. Kinsella Jr. D.L. Bronchial cell atypias : a report of a preliminary study correlating cytology with histology. *Cancer* 12:463, 1959.

86. Kirsh,M.M., Orvald T., Naylor B., Kahn D.R., Sloan H. Diagnostic accuracy of exfoliative pulmonary cytology. *Ann. Thor. Surg.* 9:335, 1970.

87. Knudtson K.P. The pathologic effects of smoking tobacco on the trachea and bronchial mucosa. *Amer. J. Clin. Path.* 33:310, 1960.

88. Knudston K.P. Mucolytic action of hyaluronidase on sputum for the cytological diagnosis of lung cancer. *Acta Cytol.* 7:59, 1963.

89. Koss L.G. Cellular changes simulating bronchogenic carcinoma. *Acta Union Int. Contra Cancrum* 14:501, 1958.

90. Koss L.G., Melamed M.R., Goodner J.T. Pulmonary cytology of brief survey of diagnostic results from July 1st 1952 until December 31st 1960. *Acta Cytol.* 8:104, 1964.

91. Koss L.G., Richardson H.L. Some pitfalls of cytological diagnosis of lung cancer. *Cancer* 8:937, 1955.

92. Kovnat D.M., Shankar Rath G., Anderson W.M., et al. Bronchial brushing through the flexible bronchoscope in the diagnosis of peripheral pulmonary lesions. *Dis. Chest* 67:179, 1975.

93. Kyriakos M., Rockoff, S.D. Brush biopsy of bronchial carcinoid. A source of cytologic error. *Acta Cytol.* 16:261, 1972.

94. Lange E., Hoeg K. Cytology typing of lung cancer. *Acta Cytol.* 16:327, 1972.

95. Lazo B.G., Feiner L.L., Seriff N.S. A study of routine cytologic screening of sputum for cancer in 800 men consecutively admitted to a tuberculosis service. *Dis. Chest* 65:646, 1974.

96. Leduc M., Riviere-Theret T.M., Duchatelle P., Voisin C. Possibilités et limites du dépistage du cancer bronchique par examen cytologique de l'expectoration. *Revue Cytol. Clin.* 4:31, 1971.

97. Lerner M.A., Rosbach H., Frank H.A., Fleischner F.G. Radiologic localisation and management of cytologically discovered bronchial carcinoma. *New Engl. J. Med.* 264:480, 1961.

98. Levij I.S. A case of primary cavitary Hodgkin's disease fo the lungs, diagnosed cytologically. *Acta Cytol.* 16:546, 1972.

99. Liu W. *An Introduction to Respiratory Cytology.* Thomas, Springfield, Ill., 1964.

100. Liu W. Concentration and fractionation of cytologic elements in sputum. *Acta Cytol.* 10:368, 1966.

101. McCalum S.M. Ova of the lung fluke *Paragonimus kellicotti* in fluid from a cyst. *Acta Cytol.* 19:279, 1975.

102. Mc Donald J.R. Exfoliative cytology in genitourinary and pulmonary diseases. *Amer. J. Clin. Path.* 24:684, 1954.

103. McDonald J.R. Pulmonary cytology. *Amer. J. Surg.* 89:462, 1955.

104. Makowska W., Zawisza E. Cytologic Evaluation of the nasal epithelium in patients with hay fever. *Acta Cytol.* 19:564, 1975.

105. Masin F., Masin M. Frequencies of alveolar cells in concentrated sputum specimens related by cytologic classes. *Acta Cytol.* 10:362, 1966.

106. Mavrommatis F.S. La détection cytologique du carcinome bronchique au stade de début. *J. Suisse Med.* 92:1094, 1962.

107. Melamed M.R., Cliffton E.E., Mercier C., Koss L.G. The megakaryocyte blood count. *Amer J. Med. Sci.* 252:301, 1966.

108. Melamed M.R., Koss L.G., Cliffton E.E. Roentgenologically occult lung cancer diagnosed by cytology: report of 12 cases. *Cancer* 16:1537, 1963.

109. Meyer J.A., Bechtold E., Jones D.B. Positive sputum cytologic test for five years before specific detection of bronchial carcinoma. *J. Thorac. Cardiovasc. Surg.* 57:318, 1969.

110. Naib Z.M. Pitfalls in the cytologic diagnosis of oat cell carcinoma of the lung. *Acta Cytol.* 8:34, 1964.

111. Naib Z.M. Exfoliative cytology in fungus diseases of the lung. *Acta Cytol.* 6:413, 1962.

112. Naib Z.M. Giant cell carcinoma of the lung: cytological study of the exfoliated cells in sputa and bronchial washings. *Dis. Chest* 40:69, 1961.

113. Naib A.M., Stewart J.A., Dowdle W.R., Casey H.L., Marne W.M., Nahmias A.J. Cytological features of viral respiratory tract infections. *Acta Cytol.* 12:162, 1968.

114. Nasiell M. The general appearance of the bronchial epithelium in bronchial carcinoma: a histopathological study with some cytological viewpoints. *Acta Cytol.* 7:97, 1963.

115. Nasiell M. Abnormal columnar cell findings in bronchial epithelium: a cytologic and histologic study of lung cancer and non-cancer cases. *Acta Cytol.* 11:397, 1967.

116. Nasiell M. The epithelial picture in the bronchial mucosa in chronic inflammatory and neoplastic lung disease and its relation to smoking. A comparative histologic and sputum-cytologic study. Stockholm 1968.

117. Nasiell M. Comparative histological and sputum cytological studies of the bronchial epithelium in inflammatory and neoplastic lung disease. *Acta Path. Microbiol. Scand.* 72:501, 1968.

118. Nasiell M., Roger V., Nasiell K., Enstad I., Vogel B., Bisther A. Cytologic findings indicating pulmonary tuberculosis. I. The diagnostic significance of epitheloid cells and Langhans' giant cells found in sputum or bronchial secretions. *Acta Cytol.* 16:146, 1972.

119. Niskanen K.O. Observations on metaplasia of the bronchial epithelium and its relation to carcinoma of the lung. *Acta Path. Microbiol. Scand.* Suppl. 80, 1949.

120. Non D.P., Lang W.R., Patchefsky A., Takeda M. Pulmonary blastoma: cytopathologic and histopathologic findings. *Acta Cytol.* 20:381, 1976.

121. Olson R.G., Froeb H.F., Palmer L.A. Sputum cytology after inhalation of heated propylene glycol : clinical correlation. *Jama* 178:668, 1961.

122. Oswald V.C., Hinson K.F.W., Canti G., Miller A.B. The diagnosis of primary lung cancer with special reference to sputum cytology. *Thorax* 26:623, 1971.

123. Papanicolaou G.N. Degenerative changes in ciliated exfoliating from the bronchial epithelium as a cytologic criterion in the diagnosis of diseases of the lung. *New York J. Med.* 56:2647, 1956.

124. Papanicolaou G.N. Historical development of cytology as a tool in clinical medicine and in cancer research. *Acta Union Int. Contra Cancrum* 14:249, 1958.

125. Papanicolaou G.N., Cromwell H.A. Diagnosis of cancer of the lung by the cytologic method. *Dis Chest* 15:412, 1949.

126. Papanicolaou G.N., Koprowska I. Carcinoma in situ of the right lower bronchus. A case report. *Cancer* 4:141, 1951.

127. Pappelis G.A., Pappelis A.J., Courtis W.S. Nuclear dry mass and area variations in human buccal nucosa cells. *Acta Cytol.* 17:37, 1973.

128. Pearson F.G., Thompson D.W., Delarue N.C. Experience with the cytologic detection, localisation and treatment of radiologically undemonstrable bronchial carcinoma. *J. Thor. Cardiovasc. Surg.* 54:371, 1967.

129. Pharr S.L., Farber S.M., King E.B. Cellular concentration of sputum for cytologic examination. *Transactions of the Fifth Annual Meeting of the Intersociety Cytology Council, 1957*, p.65.

130. Pierce C.H., Hirsch J.G. Ciliocytophthoria: relationship to viral respiratory infections in humans. *Proc. Soc. Exp. Med.* 98:489, 1958.

131. Pierce C.H., Knox A.W. Ciliocytophthoria in sputum from patients with adenovirus infections. *Proc. Soc. Exp. Biol. Med.* 104:492, 1960.

132. Plamenac P., Nikulin A., Pikula B. Cytology of the respiratory tract in former smokers. *Acta Cytol.* 16:256, 1972.

133. Plamenac P., Nikulin A., Pikula B. Cytologic changes of the respiratory tract in young adults as a consequence of high levels of air pollution exposure. *Acta Cytol.* 17:241, 1973.

134. Plamenac P., Nikulin A., Pikula B. Cytologic changes of the respiratory epithelium in iron foundry workers. *Acta Cytol.* 18:34, 1974.

135. Plamenac P., Pikula B., Kahvic M., Markovic Z., Selak I., Zeger-Vidoc Z. Incidence of "asbestos" bodies in basal lung smear. A postmortem study in residents of Sarajevo. *Acta Med. Iugosl.* 25:325, 1971.

136. Reddy C.R.R.M., Kameswari V.R. Oral exfoliative cytology in reverse smokers having carcinoma of hard palate. *Acta Cytol.* 18:201, 1974.

137. Reddy C.R.R.M., Kameswari, V.R., Raju M.V.S. Carcinoma of palate on Visakhapatnam area. *Indian J. Cancer* 7:84, 1971.

138. Reddy C.R.R.M., Sekhar C., Raju M.V.S., Reddy S.S., Kameswari V.R. Relation of reverse smoking to carcinoma of the hard palate. *Indian J. Cancer* 8:263, 1971.

139. Repsher L.H., Schroter G., Hammond W.S. Diagnosis of pneumocystis carinii pneumonitis by means of endobronchial brush biopsy. *New Engl. J. Med.* 287:340, 1972.

140. Robbins W.T. Bronchial epithelium in smoking and non-smoking college students. *J. Amer. Coll. Health Ass.* 14:265, 1966.

141. Roger V., Nasiell M., Linden M., Enstad I. Cytologic differential diagnosis of bronchiolo-alveolar carcinoma and bronchogenic adeno-carcinoma. *Acta Cytol.* 20:303, 1976.

142. Roger V., Nasiell M., Nasiell K., Hjerpe A., Enstad I., Bisther A. Cytologic findings indicating pulmonary tuberculosis. II. The occurrence in sputum of epitheloid cells and multinucleated giant cells in pulmonary tuberculosis, chronic none-tuberculous inflammatory lung disease and bronchogenic carcinoma. *Acta Cytol.* 16:538, 1972.

143. Rome D.S. Value of aerosol-produced sputum as screening technique for lung cancer. *Acta Union Int. Contra Cancrum* 15:474, 1959.

144. Rosenblatt M.B., Trinidad S., Lisa J.R., Tchertkoff V. Specific epithelial degeneration (ciliocytophthoria) in inflammatory and malignant respiratory disease. *Dis. Chest* 43:605, 1963.

145. Russel W.O., Neidhardt H.W., Mountain C.A., Griffith K.M. Chang J.P. Cytodiagnosis of lung cancer: a report of a four year laboratory, clinical and statistical study with a review of the literature on lung cancer and pulmonary cytology. *Acta Cytol.* 7:1, 1963.

146. Ryan R.F., McDonald J.R., Clagett O.T. Histopathologic observations on bronchial epithelium with special reference to carcinoma of the lung. *J. Thorac. Surg.* 33:264, 1957.

147. Saccomanno G., Archer V.E., Auerbach O., Saunders R.P., Brennan L.M. Development of carcinoma of the lung as reflected in exfoliated cells. *Cancer* 33:256, 1974.

148. Saccomanno G., Saunders R.P., Archer V.E., Auerbach O., Kuschner M., Beckler P.A. Cancer of the lung: the cytology of sputum prior to the development of carcinoma. *Acta Cytol.* 9:413, 1965.

149. Saccomanno G., Saunders R.P., Ellis H., Archer V.E., Wood B.G., Beckler P.A. Concentration of carcinoma or atypical cells in sputum. *Acta Cytol.* 7:305, 1963.

150. Saccomanno G., Saunders R.P., Klein M.G., Archer V.E., Brennan L. Cytology of the lung in reference to irritant, individual sensitivity and healing. *Acta Cytol.* 14:377, 1970.

151. Sanderud K. Squamous metaplasia of the respiratory tract epithelium. An autopsy study of 214 cases. Relation to tabacco smoking, occupation and residence. *Acta Path. Microbiol. Scand.* 44:47, 1958.

152. Sanderud K. Squamous metaplasia of the respiratory tract epithelium. An autopsy study of 214 cases. 3. Relation of disease. *Acta Path. Microbiol. Scand.* 44:21, 1958.

153. Sanderud K. Squamous metaplasia of the respiratory tract epithelium. An autopsy study of 214 cases. 4. Relation to bronchial carcinoma. *Acta Path. Microbiol. Scand.* 44:329, 1958.

154. Sandler H.C. Cytologic screening for early mouth cancer. *Cancer* 15:1119, 1962.

155. Sandler H.C. Morphological characteristics of malignant cells from mouth lesions. *Acta Cytol.* 9:282, 1965.

156. Sassy-Dobray G. The evaluation of cytology in the early diagnosis of pulmonary carcinoma. *Acta Cytol.* 14:95, 1970.

157. Sassy-Dobray G. Possibilities of early diagnosis of bronchogenic carcinoma. *Acta Cytol.* 19:351, 1975.

158. Schroit A.J., Pinkenson M., Beemer A.M. A technical aid in the cytological diagnosis of exfoliative cells in pulmonary pathology. *Acta Cytol.* 17:118, 1973.

159. Schultz, H., Meurers H. Zur elektron mikroskopischen Diagnose der menschlichen lungenadenomatose aus dem Sputum. *Acta Cytol.* 8:242, 1964.

160. Silverman S. Jr. The cytology of benign oral lesions. *Acta Cytol.* 9:287, 1965.

161. Silverman S. Jr, Ware W.H. Comparisons of histologic, cytologic and clinical findings in intra-oral leukopakia and associated carcinoma. *Oral Surg., Oral Med. and Oral Path.* 13:412, 1960.

162. Smith J.H., Frable W.J. Adenocarcinoma of the lung: cytologic correlation with histologic types. *Acta Cytol.* 18:316, 1974.

163. Spain D.M. The distinction between regenerative and atypical alterations in the bronchial mucosa. *Amer. Rev. Tuberc. Pulm. Dis.* 79:591, 1958.

164. Spencer H. *Pathology of the Lung* 2nd Ed., Pergamon, New York, 1968.

165. Spjut H.J., Fier D.J., Ackermann L.V. Exfoliative cytology and pulmonary cancer. A histopathologic and cytologic correlation. *J. Thorac. Surgery* 30:90, 1955.

166. Suprun H. A comparative filter technique study and the relative efficiency of these sieves as applied in sputum cytology for pulmonary cancer cytodiagnosis. *Acta Cytol.* 18:248, 1974.

167. Takahashi M., Hasimoto K., Osada H. Parenteral administration of chymotrypsin for the early detention of cancer cells in sputum. *Acta Cytol.* 11:61, 1967.

168. Takahashi M., Urabe M. A new cell concentration method for cancer cytology of sputum. *Cancer* 16:199, 1963.

169. Umiker W.O. Cytology in bronchiogenic carcinoma. *Amer. J. Clin. Path.* 22:558, 1952.

170. Umiker W.O. False-negative reports in the cytologic diagnosis of cancer of the lung. *Amer. J. Clin. Path.* 28:37, 1957.

171. Umiker W.O. Diagnosis of bronchogenic carcinoma: an evaluation of pulmonary cytology, bronchoscopy and scalene lymph node biopsy. *Dis Chest* 37:82, 1960.

172. Umiker W.O. A new vista in pulmonary cytology: aerosol induction of sputum in *Dis Chest* 39:512, 1961.

173. Umiker W.O. The current role of exfoliative cytopathology in the routine diagnosis of bronchogenic carcinoma: a five-year study of 152 consecutive unselected cases. *Dis. Chest* 40:154, 1961.

174. Umiker W. Cytology in the radiotherapy of carcinoma of the oral cavity. *Acta Cytol.* 9:296, 1965.

175. Umiker W.O., De Weese M.S., Lawrence G.H. Diagnosis of lung cancer by bronchoscopic biopsy, scalene lymph node biopsy and cytologic smears: a report of 42 histologically proved cases. *Surgery* 41:705, 1957.

176. Umiker W.O., Korst D.R., Cole R.P., Manikas S.G. Collection of sputum for cytologic examination: spontaneous vs artficially produced sputum. *N. Engl. J. Med.* 262:565, 1960.

177. Umiker W.O. Storey C. Bronchogenic carcinoma in situ. Report of a case with positive biopsy, cytological examination and lobectomy. *Cancer* 5:369, 1952.

178. Umiker W., Young L., Waite B. The use of chymotrypsin for the concentration of sputum in the cytologic diagnosis of lung cancer. *Univ. Mich. Med. Bull.* 24:265, 1958.

179. Unterman D.H., Reingold I.M. The occurrence of psammoma bodies in papillary adenocarcinoma of the lung. *Amer. J. Clin. Path.* 57:297, 1972.

180. Valentine E.H. Squamous metaplasia of the bronchus. A study of metaplastic changes occurring in the epithelium of the major bronchi in cancerous and non-cancerous cases. *Cancer* 10:272, 1957.

181. Virchow R. Ueber metaplasia. *Virchow Arch. Path. Anat.* 97:410, 1884.

182. Volaitis J., McGrew E.A., Chomet B., Corell N., Head J. Bronchogenic carcinoma in situ in asymptomatic high risk population of smokers. *J. Thor. Cardiovasc. Surg.* 57:325, 1969.

183. Walshe W.H. *Diseases of the Lungs.* London, 1843

184. Wandall H.H. Study on neoplastic cells in sputum as a contribution to the diagnosis of primary lung cancer. *Acta Chir. Scand.* 91, Suppl. 93, 1944.

185. Warner N.E., McGrew E.A., Nanos S. Cytologic study of the sputum in cytomegalic inclusion disease. *Acta Cytol.* 8:311, 1964.

186. Watson W.L., Cromwell H., Craver L., Papanicolaou G.N. Cytology of bronchial secretions: its role in the diagnosis of cancer. *J. Thoracic Surg.* 18:113, 1949.

187. Weller R.W. Metaplasia of bronchial epithelium. A postmortem study. *Amer. J. Clin. Path.* 53:768, 1953.

188. Welsh R.A. The genesis of the Charcot-Leyden crystal in the eosinophilic leukocyte of man. *Amer. J. Path.* 35:1091, 1959.

189. Williams J.W. Alveolar metaplasia: its relationship to pulmonary fibrosis in industries and development of lung cancer. *Brit. J. Cancer* 11:30, 1957.

190. Wood T.A., De Witt S.H., Chu E.W., Rabson A.S., Graykowski E.A. Anitschkow nuclear changes observed in oral smears. *Acta Cytol.* 19:434, 1975.

191. Woolner L.B., Andersen H.A., Bernatz P.E. "Occult" carcinoma of the bronchus: a study of 15 cases of in situ or early invasive bronchogenic carcinoma. *Dis Chest* 278, 1960.

192. Woolner L.B., Andersen H.A., David E., Fontana R.S., Bernatz P.E. In situ and early invasive bronchogenic carcinoma. *J. Thor. Cardiovasc. Surg.* 60:275, 1970.

193. Woolner L.B., McDonald J.R. Bronchogenic carcinoma: diagnosis by microscopic examination of sputum and bronchial secretions: preliminary report. *Proc. Staff Meetings Mayo Clin.* 22:369, 1947.

194. Woolner L.B., McDonald J.R. Carcinoma cells in sputum and bronchial secretions. A study of 150 consecutive cases in which results were positive. *Surg. Gynec. Obstet.* 88:273, 1949.

195. Woolner L.B., McDonald J.R. Diagnosis of carcinoma of the lung: the value of cytologic study of sputum and bronchial secretions. *Jama* 139:497, 1949.

196. Woolner L.B., McDonald J.R. Cytologic diagnosis of bronchogenic carcinoma. *Amer. J. Clin. Path.* 19:765, 1949.

197. Woolner L.B., McDonald J.R. Cytologic diagnosis of bronchogenic carcinoma. *Dis. Chest* 17:1, 1950.

198. Woolner L.B., McDonald J.R. Cytology of sputum and bronchial secretions: studies on 588 patients with miscellaneous pulmonary lesions. *Ann. Intern. Med.* 33:1164, 1950.

199. World Health Organization. *Histological Typing of Lung Tumors.* WHO, Geneva, 1967.

200. Woyke S., Domagala W., Olszewski W. Alveolar cell carcinoma of the lung: an ultrastructural study of the cancer cells detected in the pleural fluid. *Acta Cytol.* 16:63, 1972.

201. Zajicek J., Eneroth C.M. Cytological diagnosis of salivary gland carcinoma from aspiration smears. *Acta Otolaryng.* 263:183, 1970.

202. Ziskin D.E., Moulton R. A comparison of oral and vaginal epithelial smears. *J. Clin. Endocr.* 8:146, 1948.

9
The Digestive Tract

EMBRYOLOGY

The digestive tube develops from the embryonal part of the yolk sac. The epithelial lining originates from the entoderm; the muscularis layer and the serous layer originate from the mesenchyma. The anterior part or foregut forms the esophagus, the stomach, and a portion of the duodenum; the middle part or midgut forms the other portions of the duodenum, the small bowel, and the right large bowel including a part of the transverse colon. Finally, the last portion of the transverse colon, the descending colon, the pelvic colon, the rectum, and the upper part of the anal canal are developed from the hindgut. The liver and the pancreas originate from entodermal buds arising from the gut (Figure 9.1). The mouth or stomodeum is formed by a depression on the ectoderm facing the anterior part of the foregut (entoderm).

Buccal Cavity

See Chapter 8.

HISTOLOGY OF THE ESOPHAGUS

The esophageal mucosa is comprised of a stratified squamous epithelium, the lamina or tunica propria, and the muscularis mucosae. The tunica contains the esophageal glands, which are surrounded by lymphoid tissue. These serous glands are thinly dispersed throughout the tunica and are particularly scarce posteriorly. The cardia portion of the esophagus contains mucous glands.

Deep to the tunica is found the smooth muscle tissue of the lamina muscularis mucosae, the submucosa, and finally in the upper third, two layers of striated muscle, a circular one and a longitudinal one.

In the inferior portion of the esophagus, the stratified squamous epithelium may be replaced by a glandular mucous epithelium (glandular metaplasia or Barrett's syndrome).

CYTOPATHOLOGY OF THE ESOPHAGUS

Aspirations, lavages, and brushings[60] of a normal esophageal mucosa give isolated or clustered squamous elements of the superficial layers. One also finds particles and ingested bronchial elements. Fungi may be present (Plate 9.1).

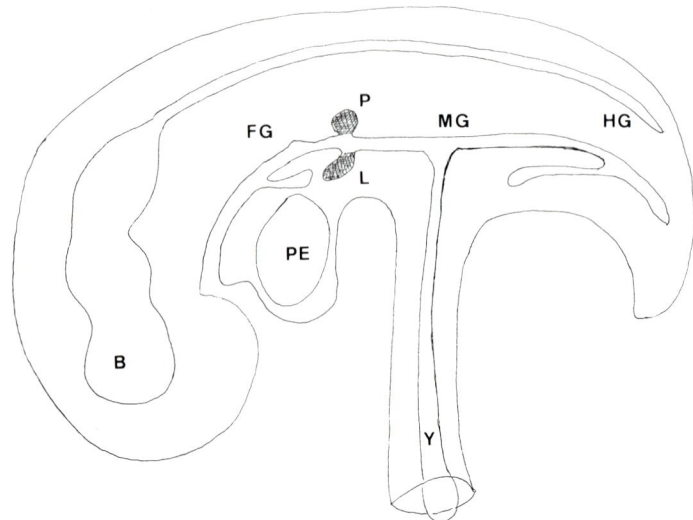

Figure 9.1. *B*, brain; *FG*, foregut; *HG*, hindgut; *L*, liver; *MG*, midgut; *P*, pancreas; *PE*, pericardial cavity; *Y*, yolk sac.

A major problem in esophageal cytology is the search for neoplastic elements since biopsies are often difficult to perform in this area.[18] Most malignant esophageal tumors are epidermoid carcinomas, but investigation may reveal rare examples of in situ[9] and invasive adenocarcinomas. The latter arise from the serous or mucous glands of the tunica or from the glands of the inferior portion of the esophagus which may exhibit glandular metaplasia.

Neoplastic squamous cells exhibit the characteristics of malignancy, typical of this cellular species.[38, 53] It is the predominance of cells of the superficial eosinophilic type or of the parabasal cyanophilic type that, in turn, determines the nature of the tumoral differentiation (Plate 9.2). Nuclear alterations are those described in these cells. Inflammatory cells may be abundant if there is superficial necrosis of the tumor.

Glandular carcinomas desquamate in clusters of small cells whose relatively voluminous nuclei are irregularly shaped and hyperchromatic (Plate 9.3).

Benign esophageal tumors (leiomyoma, neurofibroma, fibroma) are generally found in the submucosa and as such are not of interest to the cytologist.

DIFFERENTIAL DIAGNOSIS OF ESOPHAGEAL LESIONS

Certain benign esophageal lesions can present diagnostic problems. For example, smear samples of chronic erosive esophagitis show altered parabasal or hyperplasic cells, especially when the sample is taken from an area of cellular regeneration. Also, numerous polymorphic histiocytes are found. These histiocytes should not be mistaken for undifferentiated neoplastic cells, whose structural anomalies usually are more pronounced.

Generally speaking, the mode of desquamation facilitates diagnosis, as the cellular organization is better preserved in benign lesions.

Herpetic esophagitis results in cellular anomalies that are often characteristic, such as multinucleation and the presence of eosinophilic nuclear inclusions.

Postradiation lesions of benign cells exhibit classic alterations which do not differ from those discussed in the chapter on genital cytology. Postradiotherapeutic surveillance of esophageal cancers is easily performed with cytologic methods.

GASTRIC CYTOLOGY: GENERALITIES

Cytology must be viewed as representing only one of the several investigative procedures of gastric pathology. It is most valuable when associated with other diagnostic techniques, such as clinical and radiologic examinations and gastroscopy. Cytology can complete or confirm the findings of other techniques and thus constitutes an efficient approach to clinical diagnosis, particularly when results from other examinations prove inconclusive. It, however, cannot be efficiently used in large-scale screening programs for gastric cancer since the human and technical investments required for such an approach would be exorbitant.[1, 15]

As is always the case in clinical cytology, a knowledge of the cellular elements found in normal gastric fluid is an indispensable prerequisite. The very composition of gastric juice presents a serious handicap to the preservation of cell structure. Its high concentration of proteolytic enzymes and its acidity naturally result in cytolysis and cellular necrosis. Thus, a highly practiced technique is essential to obtain adequate results even with the finest method.

Clinical experience has determined the most efficient sampling methods. Smithies et al.[82] have shown that brush cytology of lesions with endoscopic surveillance constitutes the most highly sensitive method, when compared with endoscopy or biopsy performed alone.

HISTOLOGY OF THE GASTRIC MUCOSA

The gastric mucosa is formed by columnar cell glandular structures which open onto the surface epithelium. This surface epithelium is composed of columnar cells, approximately 30μ tall, whose apical cytoplasm is dotted with numerous granulations. These are mucogenic granulations which stain well with mucicarmine. This epithelium desquamates abundantly, and it has been estimated that the cellular population renews itself every 3 days.

The gastric glands include four types of cells: zymogenic cells, parietal cells, mucous cells, and argentaffine cells. While these categories are easily recognized on histologic sections, they are much less apparent in aspirations.

The zymogenic or chief cells are quite fragile and only rapid fixation will conserve their cytoplasmic structure, which is rich in granules containing a pepsin precursor. The ultrastructural study of these chief cells indicates that they represent several cellular types which seem to correspond to different hormonal productions, such as serotonin, gastrin, histamine, glucagon, and catecholamine.

The parietal cells are large polyhedral elements scattered among the chief cells. They have a voluminous and rounded nucleus, and their cytoplasm, which is rich in mitochondria, stains intensely with eosin.

The mucous cells, which are found among the chief cells in the apical portion of the glands, are relatively few in number. They are irregularly shaped due to crowding by neighboring cells.

The argentaffine cells are also relatively rare, and they contain granules which stain with silver salts. They synthesize 5 hydroxy-tryptamine.

GASTRIC CYTOLOGY

The elements found in an aspiration of normal gastric juice are:

- gastric mucosal cells
- squamous cells of esophageal origin, from the mouth or from the upper airways
- histiocytes and leukocytes
- ingested bronchopulmonary elements
- food particles

The quality of a smear depends on the state of preservation of the cells and on the indispensable presence of columnar elements, which assure the examiner of the gastric origin of his sample. The number of these columnar elements depends on the nature of the lesion. For example, atypical cells are more abundant in fungating tumoral forms than in scirrhous forms.

The gastric mucous cells may be found either isolated or in clusters; when their cytoplasm has been well preserved, their columnar aspect is evident. The "honeycomb" alveolar structure is characteristic of cell clusters seen from their apical or basal pole, but when seen laterally, the cells form a palisade. The membrane of the atypical pole of these cells is usually quite visable. The cells measure 20 μ in length and have an elongated or rounded nucleus whose diameter may measure between 10 μ and 15μ, depending on the incidence (Plates 9.4 and 9.5).

The chromatin is finely scattered in small bits in the otherwise clear nuclear background, and the nucleolus is prominent. The nuclear membrane is quite visible, and a few masses of chromatin including the Barr body appear stuck to its internal face. It is the appearance of the chromatin, distributed in distinct clumps over a clear background, that allows discrimination between a glandular and a squamous cell. The cytoplasm is pale and sometimes contains many small vacuoles. It may, however, contain one large vacuole which makes the cell look bloated. Cytolysis is frequent and depends on the state of preservation of the cells. When one of these cells is undergoing degenerative changes, the structures of the chromatin fades and gradually disappears, leaving a partially empty nucleus which can be mistaken for a sign of hypochromatism (bland cell).

A leukocytic and histiocytic infiltration, when abundant, can obscure the epithelial elements. Also, the cellular clusters themselves are often encased in mucus. Squamous cells of esophageal or buccal origin are quite abundant and often constitute the majority of the cellular elements present. These are superficial or intermediate epithelial elements and may be found isolated or in clusters.

Histiocytes may be recognized by their nuclei, which may be horseshoe-shaped or irregular, and their cytoplasm, which is clear and contains fine vacuoles. Histiocytes situated in the submucosa may rarely become loaded with lipids and form clumps of cells whose diameter may attain several millimeters (xanthoma).

Leucocytes and *erythrocytes* are found in variable quantities depending on the pathologic circumstances.

Alimentary debris is sometimes present in appreciable amounts and may cover the cellular elements. In such cases one should abstain from making an immediate diagnosis and request a control sample.

Bacterial flora, fungi, or certain flagellate organisms such as *Trichomonas buccalis* may be present. These organisms are recognized by using Shorr's or Papanicolaou's stain (Plate 9.6).

Deglutition of cells of the bronchiopulmonary tract is not infrequent, and thus one may find columnar ciliated cells. Macrophages also may be found and are recognized by their large size and their cytoplasm, which stains intense green and which often contains carbon particles (Plate 9.7).

ATROPHIC CHRONIC GASTRITIS

Histologically speaking, atrophic gastritis encompasses a series of different lesions appearing with varying intensities. These lesions are:

- atrophy of the mucosa with diminution of the gastric pits
- decrease in the number of parietal and chief cells and increase in the number of mucous cells (mucous metaplasia)
- appearance of an intestinal metaplasia characterized by the presence of intestinal goblet cells, Paneth cells, and numerous, highly developed cells with brush borders histochemical analysis will confirm that these cells, including the mucoid ones, are of intestinal nature
- presence of dilated or cystic glandular structures with flattened cells or conversely the presence of multinucleate cells whose nuclei are hypertrophic and exhibit a high mitotic activity
- presence of submucosal lymphoid infiltration (follicular gastritis)
- edema, lymphocytic, and plasmocytic infiltration and fibrosis of the submucosa

On smears, the inflammatory modifications of chronic gastritis manifest themselves by the presence of altered cells that desquamate abundantly and exhibit nuclear as well as cytoplasmic structural anomalies. The cells tend to assume a cuboidal or columnar form, have rounded nuclei with a clearly visible nuclear envelope, and contain a prominent nucleolus. Some clusters will reveal cells with a pale, large hypochromatic nucleus (bland cell). So-called active columnar cells with large hyperchromatic, regular nuclei have been described and illustrate repair processes. Their recognition is not constant (Plate 9.8).

Intestinal metaplasia is indicated on such smears by the increased number of goblet-shaped mucous cells. Also, histiocytes and leucocytes are numerous. Finally, the presence of cellular necrosis with concomitant leukocytic and histiocytic infiltration may suggest the presence of erosion.

INTESTINAL METAPLASIA

Intestinal metaplasia may be seen in various pathologic conditions, such as gastritis, ulcers, benign or malignant tumors, or pernicious anemia. The presence of clusters of columnar cells whose abundant cytoplasm is denser and whose form is less rounded than normal mucous cells suggests this diagnosis. Sometimes, the cells appear elongated, as if they had been stretched (Plate 9.9).

BENIGN TUMORS OF THE GASTRIC MUCOSA

A diagnosis of polyp is seldom made by cytologic examination. Abundant desquamation of clustered, cytologically normal cells has been described, but a biopsy is necessary for a certain diagnosis.

MALIGNANT TUMORS OF THE STOMACH

Carcinoma of the stomach represents roughly 7 percent of all cancers and generally appears in the prepyloric region. It is basically an affliction of persons over 40 years of age and strikes men two to three times as often as women. Among young adults it is rare but strikes both sexes with equal frequency.

The cause is unknown but several predisposing factors have been ascertained such as the hereditary factor (clinical reports of families in which gastric cancer is frequent) and the ethnic factor (high frequency among the Japanese).[45] Pernicious anemia and the presence of adenomatous polyps of the mucosa are also predisposing factors.

Histology

Gastric cancer may be divided into three macroscopically distinguishable types: fungating, ulcerating, and infiltrating.

Histologically, examination reveals differentiated forms with well-preserved glandular structures, undifferentiated forms exhibiting no particular structural disposition, and intermediate forms. The majority of the cases are poorly differentiated which accounts for the bad clinical prognosis. The cells may be mucosecretory as in the signet-ring cells or nonsecretory.

The infiltrating varieties are characterized by widespread dissemination in the entire gastric wall, either as thin cords or small isolated clumps of cells, whereas the proliferating forms first arise in the gastric lumen. Superficial and in situ lesions are confined to the mucosa. Some cases are multifocal in origin or extend to large areas of the mucosal surface. While sarcomas of the stomach are rare, let us mention leiomyosarcoma and malignant lymphomas. Some rare cases of choriocarcinoma have been described.

Cytology

Cytological methods allow recognition of the different tumors of the gastric wall. The great majority of the malignant tumors of the stomach are adenocarcinomas of the mucosa. These cells desquamate in clusters or as isolated elements and exhibit — in varying degrees — the classic signs of malignancy: anisokaryosis, anisonucleosis, hyperchromatism, nuclear as well as nucleolar hypertrophy, and multinucleation. Abnormal mitotic figures are rarely seen. Bizarre-looking nuclei are seen in anaplastic tumors (Plate 9.10). When the cytoplasm has been well preserved, modifications of staining properties (eosinophilia), anomalies of cytoplasmic size and form, and vacuolation are present. These cells are always larger than normal ones — they measure up to 30μ in diameter.

Disorganization of the cellular arrangement and cellular stratification result in the characteristically thickened and disorderly appearance of the cell clusters (Plates 9.11 and 9.12). If desquamation of the clusters is abundant and cell preservation satisfactory, it is possible to determine the degree of differentiation of the tumor (glandular differentiated forms, signet ring mucosecretory forms, or undifferentiated forms). In some cases, the presence of primitive papillary or acinous structures may suggest the tumor's histologic nature (Plates 9.13 and 9.17). Cellular cords are often embedded in mucus.

Early superficial or in situ lesions exfoliate typical malignant cells that cannot be differentiated from those of more invasive lesions. If cytolytic phenomena are pronounced, the cells are reduced to clusters of bare nuclei whose size, form, and abnormal chromatin distribution may point to malignancy. The presence of hypochromatic nuclei may be the result of poor cell preservation. Inflammatory alterations around the neoplastic area explain the presence of leukocytes, histiocytes, and cellular debris[4] (Plate 9.15).

In infiltrating forms, there is almost no desquamation of neoplastic elements, so unless surface tumoral extensions exist, atypical cells rarely will be found. Therefore, with this type of tumor, the cytologic method is of less value, and a diagnosis must depend more on radiography, clinical examination, and a biopsy of the deeper tissue.

Sarcomas and submucosal tumors are even less frequently diagnosed cytologically. They must invade and ulcerate the mucosa before desquamation in the gastric cavity reveals their presence. Muscular sarcomatous cells appear as elongated cells whose cytoplasmic limits are ill defined and whose nuclei, similarly elongated, are hyperchromatic.

The cells of malignant lymphomas may comprise immature or differentiated lymphocytic elements, histiocytes, and Reed-Sternberg cells. These desquamate principally as isolated elements or in small clusters. Malignant lymphocytic elements are larger than normal lymphocytes and their nuclei often exhibits voluminous nucleoli.

Finally, Langhans' cells have been described in some rare cases of choriocarcinoma. They should be distinguished from multinucleated giant cells described in granulomatous gastritis (gastric Crohn's disease, tuberculosis, etc.).

VALUE OF THE METHOD

In up-to-date laboratories the accuracy of the method is satisfactory. The false-positive rate should not exceed 1 or 2 percent, and the false-negative rate varies from 5 to 30 percent according to the macroscopic type of tumor. Detection is lower for small lesions (under 1 cm in diameter).

Results will also differ according to the method used to collect the cells. As a comparison, the results reported by Shida[81] give an accurate diagnosis 100 percent of the time with the direct brushing technique; 88.8%, with the direct smear technique; and 86%, with the gastric aspiration and lavage method.

GASTRIC CYTOLOGY TECHNIQUES

The proper collection of specimens is essential to obtain high quality gastric cytology; it requires experience and skill as well as the patient's cooperation.

The different techniques used are simple irrigation (with or without use of mucolytic substances), direct smearing, and brushing during gastroscopy.

Gastric Wash Technique

It is recommended that the patient fast for 12 hours prior to undergoing a gastric wash. The tube, which is usually 4 mm in diamether (Levin tube), is passed over the tongue or through the nasopharynx and is swallowed. Passing 55 cm of tubing (counting from the mouth) ensures that the tube has reached the stomach. The patient is asked to roll from left to right on his back and on his stomach, and an abdominal massage is simultaneously performed. Next, 50–100 cc of saline are injected, aspirated, and injected again. This procedure is repeated six times to facilitate cellular desquamation from the mucosal surface. The tube is moved to various distances from the cardia (52 cm) to the pylorus (62 cm). The fluid is finally aspirated, centrifuged at 2500 for 15 minutes and fixed with 95% ethanol. The procedure is repeated, using 300 cc of saline.

Several smears (6 is a minimum) are made from a portion of the centrifugation band and the remaining material is embedded in paraffin for the block technique.

Certain authors recommend the ingestion of 7 mg of chymotrypsin to liquify the gastric mucus and to facilitate cell exfoliation. We do not find that this procedure greatly improves the quality of the samples obtained.

Gastrofiberscopic Techniques

Gastrofiberscopic instruments permit direct visualization of stomach lesions and therefore can be used as valuable adjuncts to the various sampling techniques of gastric cytology. The details of such gastroscopic methods vary with the type of gastrofiberscope used (Kagusai, Kobayaski).

Specimens may be obtained by directly smearing the brush, the suction tube, or the lavage tube on the slide or by washing the instrument in a saline solution which is then centrifuged.

DIFFERENTIAL DIAGNOSIS

Differential diagnosis can be problematic when one is dealing with certain benign lesions, particularly ulcers with or without zones of tissular regeneration, cellular modifications resulting from ionizing radiation, or cyclophosphamides, and atrophic or hypertrophic gastritis (Plate 9.16).

In ulcerated lesions, repair cells and inflammatory cells are abundant. In atrophic gastritis, goblet cells are numerous and mucus is abundant. Malignant lymphomas and anaplastic small-cell carcinomas are often difficult to differentiate. The cells of carcinoma usually reveal a narrow cytoplasmic rim and desquamate in small clusters (Plate 9.14). In lymphomas the cytoplasm is very scanty and cells desquamate as isolated elements. In Hodgkin's disease, the presence of Reed-Sternberg cells is a rare discovery.

When confronted with cellular anomalies, the cytopathologist must consider the possibility of these lesions and advise the clinician. In this case, radiologic, endoscopic, and biopsy correlations are of the utmost importance. Once again, the structure of the chromatin, the nuclear shape, and the hypertrophy of the nucleolus are the more reliable criteria of malignancy.

PRACTICAL UTILIZATION
OF THE CYTOLOGIC METHOD

Both the complexity of the techniques involved and the time expenditure necessary prohibit the use of cytologic methods as means of screening for asymptomatic tumoral lesions of the stomach. Therefore cytology must be reserved for patients who show clinical or radiologic symptoms or for high-risk groups such as those exhibiting pernicious anemia or achlorhydria. The cytologic examination of clinically evident, voluminous tumors is of purely academic interest.

It is important to use the technique for patients whose symptomatology does not suggest a tumor. It sometimes happens that precocious diagnosis permits identification of localized operable cancers whose prognosis is therefore favorable. Such cases, when published at all, are not numerous and are generally the work of specialized laboratories.[78] A more widespread use of the cytologic method, however, should increase the frequency of such cases.

CYTOLOGY OF THE LIVER

See Chapter 11.

CYTOLOGY OF THE PANCREAS

The diagnosis of carcinomas of the pancreas and the biliary tree is a difficult one, and cytology may help to improve their recognition.[14, 40] However, duodenal drainage after stimulation of secretion is a delicate and time-consuming procedure.[29] The use of modern fiberduodenoscopes allows direct vision and sampling of the pancreatic duct.[31] Nevertheless the sensivity of the cytologic method, even in experienced laboratories, does not exceed 60 percent. Cell exfoliation is not constant because of ductal occlusions, and cell preservation is often very poor (Plate 9.16).

Fine-needle puncture of the pancreas during surgery may bring valuable and immediate information, particularly when the surgeon is dealing with a problem of differential diagnosis (pancreatitis versus carcinoma).[3]

Adenocarcinomas exhibit the general criteria of malignancy; well-differentiated forms are more delicate to evaluate. Nucleolar hypertrophy is a constant feature of malignant glandular cells. Calcifications and inflammatory cells suggest a diagnosis of pancreatitis.

CYTOLOGY OF THE COLON

Two methods are available to obtain cellular material: colonic lavage and direct smear from the lesion through the sigmoidoscope.[11,27,46] When the material is well preserved, adenocarcinoma is characterized by the presence of large columnar cells with elongated, hyperchromatic nuclei and large nucleoli.[8,13,20,35] Most of the time, the presence of fecal material and cellular necrosis prevent any valuable cytologic interpretation (Plates 9.19–9.22).

The differential diagnosis of "active" cells of ulcerative colitis from carcinoma may be difficult: these elements probably present the "active" slightly hyperchromatic nucleus of repair cells. Nucleolar hypertrophy is more pronounced in cancer cells.

SUMMARY

Indications for the method:

- the search for malignant lesions
- inflammatory diseases and their causative agent may sometimes be diagnosed or suspected

What should be noted:

- The technical approach is very important in digestive cytology because we are confronted with alterations due to secretions and accumulation of alimentary products and feces.
- The cellular type, the mode of desquamation, the importance of the inflammatory diathesis are the more important structural modifications to be noted.

What the lesions are that may be diagnosed by cytology:

- infections diseases — disorders due to physical agents
- malignant tumors of the esophagus and stomach
- to a lesser extent, neoplastic diseases of duodenum, biliary ducts, liver, pancreas, small bowel, colon, and rectum

BIBLIOGRAPHY

1. Ackermann N.B. An evaluation of gastric cytology: results of a nationwide survey. *J. Chron. Dis.* 20:621, 1967.
2. Anthonisen P., RIIS P. Cytology of colonic secretion in proctosigmoidal disease. *Acta Med. Scand.* 172:375, 1962.
3. Arnesjo B., Stormby N., Akerman M. Cytodiagnosis of pancreatic lesions by means of needle biopsy during operation. *Acta Chir. Scand.* 128:363, 1972.

4. Asakura R., Seto R. Cytology of borderline lesion of stomach cancer. *J. Jap. Soc. Clin. Cytol.* 19:98, 1971

5. Asakura R., Seto R., Katayanagi T., Yokobori T. The morphological study of early gastric cancer cells. *J. Jap. Soc. Clin. Cytol.* 9:204, 1970.

6. Asnaes S. Johansen S.A., OA. Duodenal exfoliative cytology. Duodenal drainage smears after stimulation with secretin. *Acta Pathol. Microbiol. Scand.* Suppl. 212, 1970, p.11

7. Ayre J.E., Oren B.G. A new rapid method for stomach cancer diagnosis: the gastric brush. *Cancer* 6:1177, 1953.

8. Bader G.M., Papanicolaou G.N. Application of cytology in diagnosis of cancer of rectum, sigmoid and descending colon. *Cancer* 5:307, 1952.

9. Belladonna J.A., Hajdu S.I., Bains M.S., Winawer S.J. Adeno-carcinoma in situ of Barrett's esophages diagnosed by endoscopic cytology. *New Engl. J. Med.* 291:895, 1974.

10. Bemvenutti G.A., Hattori K., Cockerham L., Prolla J.C., Kirsner J.B., Reilly R.W. Direct vision brushing cytology of gastrointestinal malignancy; histopathological correlation with biopsy (abstract). *Gastroenterology* 64:696, 1973.

11. Benvenutti G.A., Prolla J.C., Kirsner J.B., Reilly R.W. Direct vision brushing cytology in the diagnosis of colo-rectal malignancy. *Acta Cytol.* 18:477, 1974.

12. Bertalanffy F.D. Evaluation of the acridine-orange fluorescence microscope method for cytodiagnosis of cancer. *Ann. N.Y. Acad. Sci.* 93:715, 1962.

13. Blank W.A., Steinberg A.H. Cytologic diagnosis of malignancies of the lower bowel and rectum. *Amer. J. Surg.* 81:127, 1951.

14. Bowden L., Papanicolaou G.N. The diagnosis of pancreatic cancer by cytologic study of duodenal secretions. *Acta Union Int. Contra Cancum* 16:398, 1960.

15. Brandborg L.L., Taniguchi L., Rubin C.E. Is exfoliative cytology practical for more general use in the diagnosis of gastric cancer? *Cancer* 14:1074, 1961.

16. Brandborj L.L., Wenger J. Cytological examination in gastrointestinal tract disease. *Med. Clin. N. Amer.* 52:1315, 1968.

17. Brits C.J. Liver aspiration cytology. *S. Afr. Med. J.* 48:2207, 1974.

18. Bruinsma A.H. *The value of cytology in the early diagnosis of carcinoma of the esophagus and stomach (making use of the Papanicolaou gastric balloon and its modifications).* Thesis, Utrecht, 1957.

19. Burnett W., McFarlane P.S., Scott S.D., Kay A.W. Carcinoma of the stomach: an evaluation of diagnostic methods including exfoliative cytology. *Brit. Med.* 1:753, 1960.

20. Cameron A.B. A cytologic method of diagnosis of carcinoma of the colon. *Dis. Colon Rectum* 3:230, 1960.

21. Cantrell E.G. Why use gastric cytology? *Gut* 10:763, 1969.

22. Carney C.N. Clinical cytology of the liver. *Acta Cytol.* 19:244, 1975.

23. Clemencon G., Gloor F. Lipid deposits in gastric mucosa. *Endoscopy* 6:192, 1974.

24. Cooper W.A., Papanicolaou G.N. Balloon technique in the cytological diagnosis of gastric cancer. *JAMA* 151:10, 1953.

25. Cowan W.K., Schade R.O.K. Gastric cytology: experience in a district general hospital. *Acta Cytol.* 18:122, 1974.

26. Dablesteen E., Roed-Petersen B., Smith C.J., Pindoborg J.J. The limitations of exfoliative cytology for the detection of epithelial atypia in oral leucoplakia. *Brit. J. Cancer* 25:21, 1971.

27. Deluca V.A., Eisenman L., Moritz M., Feldstein E., Bautista A., Macionus R., Carrillo H., Laborda O. A new technique for colonic cytology. *Acta Cytol.* 18:421, 1974.

28. Dominis M., Celek S., Solter D. Cytology of diffuse liver disorders. *Acta Cytol.* 17:205, 1973.

29. Dreiling D.A., Nieburgs H.E., Janowitz H.D. The combined secretin and cytology test in the diagnosis of pancreatic and biliary tract cancer. *Med. Clin. N. Amer.* 44:801, 1960.

30. Ekelund P., Wasastjerna C. Cytological identification of primary hepatic carcinoma cells. *Acta Med. Scand.* 189:373, 1971.

31. Endo Y., Morii T., Tamura H., Okuda S. Cytodiagnosis of pancreatic malignant tumors by aspiration, under direct vision, using a duodenal fiberscope. *Gastroenterology* 67:944, 1974.

32. Foushee J.H.S., Kalnins Z.A., Dixon F.R., Girsh S. Morehead R.P., O'Brien T.F., Pribor H., Tattary C. Gastric cytology: evaluation of methods and results in 1670 cases. *Acta Cytol.* 13:399, 1969.

33. Fukuda T., Shida S., Takita T., Sawada Y. Cytologic diagnosis of early gastric cancer by the endoscope method with gastrofiberscope. *Acta Cytol.* 11:456, 1967.

34. Galambos J.T. Cytologic examination of benign colonic lesions. *Acta Cytol.* 6:148, 1962.

35. Galambos J.T., Klayman M.I. The clinical value of colonic exfoliative cytology in the diagnosis of cancer beyond the reach of the proctoscope. *Surg. Gynec. Obstet.* 101:673, 1955.

36. Galambos J.T., Massey B.W., Klayman M.I., Kirsner J.B. Exfoliative cytology in chronic ulcerative colitis. *Cancer* 9:152, 1956.

37. Gardner F.N. Observations on the cytology of gastric epithelium in tropical sprue. *J. Lab. Clin. Med.* 47:529, 1956.

38. Gephart T., Graham R.M. The cellular detection of carcinoma of the esophagus. *Surg. Gynec. Obstet.* 108:75, 1959.

39. Gibbs D.D. *Exfoliative Cytology of the Stomach.* New York, Appleton-Century-Crofts. London, Butterworth, 1968.

40. Goldstein H., Ventzke L.E. Value of exfoliative cytology in pancreatic carcinoma. *Gut* 9:316, 1968.

41. Grable E., Zamcheck N., Jankelson O., Shipp F. Nuclear size of cells in normal stomachs, in gastric atrophy and in gastric cancer. *Gastroenterology* 32:1104, 1957.

42. Graham R.M., Rheault M.H. Characteristic cellular changes in epithelial cells in pernicious anemia. *J. Lab. Clin. Med.* 43:235, 1954.

43. Graham R.M., Ulfelder H., Green T.H. The cytologic method as an aid in the diagnosis of gastric carcinoma. *Surg. Gynec. Obstet.* 86:257, 1948.

44. Grossman E., Goldstein M.J., Koss L.G. Winaver S.J., Sherlock P. Cytological examination as an adjunct to liver biopsy in the diagnosis of hepatic metastases. *Gastroenterology.* 62:56, 1972.

45. Haenszel, et al. Stomach cancer in Japan. *J. Natl. Cancer Hist.* 56:265, 1976.

46. Hampton J.M., Bacon H.E., Myers J. A simplified method for the diagnosis of cancer of the colon by exfoliative cytology. *Dis. Colon Rectum* 5:145, 1962.

47. Hatfield A.R.W., et al. Assessment of endoscopic retrograde cholangiopancreatography (ERCP) and pure pancreatic juice cytology in patients with pancreatic disease. *Gut* 17:14, 1976.

48. Heilmann K. Lipid islands in gastric mucosa. *Beitr. Path. Anat.* 149:411, 1973.

49. Henning N., Witte S. Uber eine neue Methode zur Zytodiagnostik der Magenkrankheiten. *Deutsch. Med. Wschr.* 77:1, 1952.

50. Henning N., Witte S. *Atlas of Gastrointestinal Cytodiagnosis.* 2nd Ed. Thieme, Stuttgart, 1970.

51. Hoffler A.S., Rubel L.R. Free-living amoebae identified by cytologic examination of gastrointestinal washings. *Acta Cytol.* 18:59, 1974.

52. Johansen S., Myren J. Fine-needle aspiration biopsy smears in the diagnosis of liver diseases. *Scand. J. Gastroent.* 6:58, 1971.

53. Johnson W.D., Koss L.G., Papanicolaou G., Seybolt J.F. Cytology of esophageal washings. *Cancer* 8:951, 1955.

54. Kalnins Z.A., Rhyne A.L., Dixon F.R., et al. Analysis of cytologic findings in patients with gastric carcinoma. *Acta Cytol.* 11:312, 1967.

55. Kasugai T. Evaluation of gastric lavage cytology under direct vision by the fibergastroscope employing Hank's solution as a washing solution. *Acta Cytol.* 12:345, 1968.

56. Kasugai T., Kobayashi S. Evaluation of biopsy and cytology in the diagnosis of gastric cancer. *Amer. J. Gastroenterol.* 62:199, 1974.

57. Katz S., Sherlock P., Winaver S.J. Rectocolonic exfoliative cytology: a new approach. *Amer. J. Dig. Dis.* 17:1109, 1972.

58. Kernen G.A., Bales C. Cytologic diagnosis of gastric cancer. *Calif. Med.* 108:105, 1968.

59. Kernen G.A., Bales C. Exfoliative cytology in gastric cancer. *Calif. Med.* 108:143, 1968.

60. Kobayashi S., Prolla J.C., Kirsner J.B. Brushing cytology of the esophagus and stomach under direct vision by fiberscopes. *Acta Cytol.* 14:219, 1970.

61. Kobayashi S., Prolla J.C., Winans C.S., Kirsner J.B. Improved endoscopic diagnosis of gastroesophageal malignancy. Combined use of direct vision brushing cytology and biopsy. *JAMA* 212:2086, 1970.

62. Kobayashi S., Yoshii Y., Kasugai T. Selective use of brushing cytology in gastrointestinal structures. *Gastroint. Endosc.* 19:77, 1972.

63. Kobayashi S., Yoshii Y. Kasugai T. Biopsy and cytology in the diagnosis of early gastric cancer. *Endoscopy* 8:53, 1976.

64. Lundquist A. Fine-needle aspiration biopsy for cytodiagnosis of malignant tumor in the liver. *Acta Med. Scand.* 188:465, 1970.

65. Lundquist A. Fine-needle biopsy of the liver. *Acta Med. Scand. Acta Med. Scand.* Suppl. 520, 1971, p.1.

66. McDonald W.C., Brandborg L.L., Taniguchi L., Rubin C.E. Gastric exfoliative cytology: An accurate and practical diagnostic procedure. *Lancet* 2:83, 1963.

67. Martuzzi M., Amadori D., Saragoni A. Le cytodiagnostic du cancer gastrique. *Rev. Cytol. Clin.* 4:9, 1971

68. Miyake T., Yamamoto Y., Ariyoshi J., Takahashi Y., Furukawa H., Suzuki T., Hajiro K., Kuzuya H. The significance of call III cells in gastric cancer cytology. *Jap. Arch. Intern. Med.* 16:1, 1969.

69. Nakamura K., Sugano H., Takagi K. Funchigami A. Histopathologic study on early carcinoma of the stomach. *Gann* 57:613, 1966.

70. Oisen J.H. Duodenal exfoliative cytology. Diagnosis of cancer of duodenum pancreas and biliary tract by exfoliative cytology. *Scand. J. Gastroenterol.* 6 (Suppl. 9): 105, 1971.

71. Prolla J.C., Kirsner J.B. *Handbook and Atlas of Gastrointestinal Exfoliative Cytology.* University of Chicago Press, Chicago, 1972.

72. Prolla J.C., Avier R.G., Kirsner J.B. Morphology of exfoliated cells in benign gastric ulcer. *Acta Cytol.* 15:128, 1971.

73. Prolla J.C., Xavier R.G., Kirsner J.B. Exfoliative cytology in gastric ulcer. Its role in differentiation of benign and malignant ulcers. *Gastroenterology* 63:33, 1972.

74. Raskin H.F., Palmer W.I.., Kirsner J.B. Exfoliative cytology in diagnosis of cancer of the colon. *Dis. Colon Rectum* 2:46, 1959.

75. Raskin H.F., Pleticka S. Exfoliative cytology of the colon. Fifteen years of lost opportunity. *Cancer* 28:127, 1971.

76. Reddy C.R.R.M., Kameswari V.R. Oral exfoliative cytology in reverse smokers having carcinoma of hard palate. *Acta Cytol.* 18:201, 1974.

77. Salhadin A., Gompel C. Le cytodiagnostic du cancer gastrique. *Rev. Cytol. Clin.* 2:7, 1969.

78. Schade R.O.K. *Gastric Cytology* Arnold, London, 1960.

79. Seppala K., Lehtola J., Siurala M. The possible precancerous significance of abnormal cells in gastric cytology specimens: A follow-up study of 545 patients. *Scand. J. Gastroent.* 11:513, 1976.

80. Sherlock P., Kim Y.S., Koss L.G. Cytologic diagnosis of cancer from aspirated material obtained at liver biopsy. *Amer. J. Dig. Dis.* 12:396, 1967.

81. Shida S. *Biopsy Smear Cytology with the Fibergastroscope for Direct Observation.* Gann Monograph II, University of Tokyo Press, 1971, p.223.

82. Smithies A, et al. Value of brush cytology in diagnosis of gastric cancer. *Brit, Med. J.* 4:326, 1975.

83. Stormby N., Akerman M. Aspiration cytology in the diagnosis of granulomatous liver lesions. *Acta Cytol.* 17:200, 1973.

84. Takeda T., Takaso K., Isono S., Sato M., Sato E., Ishioka K. Cytologic studies on so-called atypical epithelium of the protuberant lesions in the stomach. *Acta Cytol.* 19:345, 1975.

85. Wiendl H.J., Schwabe M., Becker G., Kowatsch J. Feulgen-cytophotometric studies of gastric mucosal smears in malignant and benign diseases of the stomach. *Acta Cytol.* 18:222, 1974.

86. Witte S. Gastroscopic cytology. *Endoscopy* 2:88, 1970.

10
Cytology of the Urinary Tract

EMBRYOLOGY

The cloaca, endodermic cul-de-sac, is divided into two parts by the urorectal septum which plunges toward the caudal extremity of the embryo. The anterior part of the cloaca forms the urogenital sinus caudad and the urinary bladder cephalad (Figure 10.1).

The caudal part of the Wolffian duct gives rise to the trigone of the bladder and the urethra; in the male it also forms the epididymis, the vas deferens, and the ejaculatory ducts. In the woman, only vestiges remain: the canal of Gartner, the epoophoron, the organ of Rosenmuller, and the paraoophoron (Figure 10.2).

The urethra in the male originates from the Wolffian duct, the urogenital sinus, and an ectodermic bud. The prostate originates from epithelial buds which develop on the urethra. In the female, the superior two-thirds of the urethra is formed from the mesonephric Wolffian duct and the inferior one-third, from the urogenital sinus; the ureter arises from a bud derived from the Wolffian duct.

The kidneys arise from the mesonephros. The latter is formed by a budding of the Wolffian duct and by a condensation of mesodermal cells surrounding this bud.

It must be remembered that the most anterior part of the urinary system, the pronephros, disappears during the fifth week of embryonic life. Of the middle portion or mesonephros, the testicular excretory ducts, a portion of the epididymis, and some blind canalicular vestiges remain. The urinary tract thus arises primarily from the mesoderm, and to a lesser extent, from the endoderm.

HISTOLOGY

Urethra

The urethral mucosa is a pseudostratified columnar epithelium which exhibits squamous islets (especially at its extremity) and islets of mucous glandular cells (glands of Littré). Lining the mucosa there is a muscular and a serous tunica.

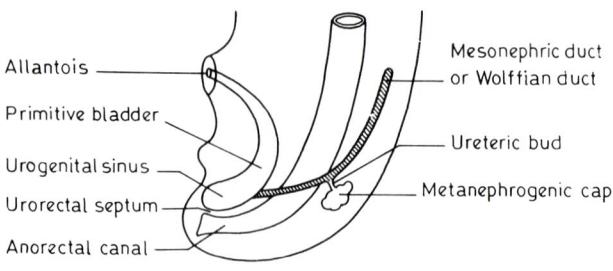

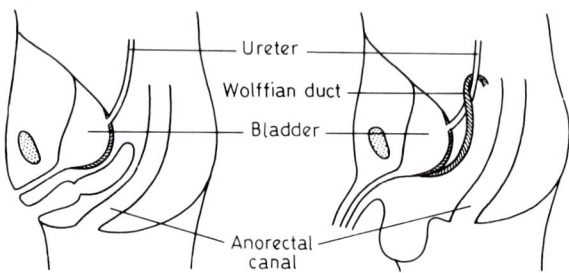

Figure 10.1. Schematic representation of the urinary system (lateral view).

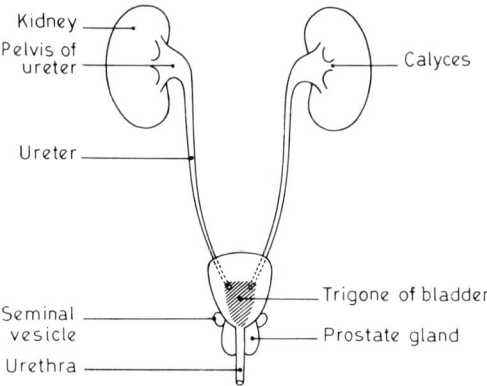

Figure 10.2. Schematic representation of the urinary system (frontal view).

Urinary Bladder

The bladder wall includes a mucous layer, a muscular layer, and a serous tunica (Plate 10.1). The mucosa contains from five to seven layers of cells. The degree of folding of this mucosa and its thickness depend on the volume of urine contained.

The basal layer of the epithelium is composed of rounded cells with voluminous nuclei. Several intermediate layers are formed by large-sized cells with abundant, clear, glycogen-rich cytoplasm. The superficial layer is composed of large cells (umbrella cells) whose diameter varies between 100 and 200μ and which sometimes exhibit marked multinucleation and a dense cytoplasm (Plate 10.5). The ultrastructure of these cells reveals the presence of a superficial membrane about 120 Å thick and composed of three laminae. The superficial lamina is thicker than the deep one; thus the ensemble is called an asymetrical membrane. It constitutes a barrier against the penetration of urine as shown by Hicks with rodents.

In the trigone, a zone set off by the ureteral and urethral orifices, the mucosa contains islets of stratified squamous epithelium. The ureters are lined by a urothelium identical to that of the urinary bladder. Surrounding the urothelium is a layer of smooth muscle and a serosa.

The same epithelium is again found in the renal pelvis and the calyces. The tubuli are lined with a monocellular layer which is columnar or cuboidal depending on the location, and the glomeruli (renal corpuscles) are lined by endothelial cells. The glomerulus, the convoluted uriniferous tubules and the loop of Henle constitute the nephron.

CYTOLOGY OF THE NORMAL URINARY TRACT

The normal urinary tract exhibits only moderate desquamation. Normal urine thus contains only a small number of urothelial and squamous cells. There are no inflammatory cells, erythrocytes, or bacteria.

Urothelial Cells

The urothelial cell is small (about 12 to 20μ in diameter). The nucleus is rounded and may be centrally or peripherally located. The chromatin is finely granular and the nucleolus is visible. The cytoplasm is pale and cyanophilic and has clearly delineated borders. The general form of the cell may be rounded, oblong, or elongated (Plate 10.2). It resembles a squamous parabasal cell. Small cytoplasmic lipid droplets are present in the majority of cells and are a result of necrobiosis.[62]

Squamous Cells

These originate from the urethra, from the prostatic canals, or from the bladder trigone. They are of the superficial, intermediate, or parabasal type and some may be eosinophilic (Plate 10.3).

Renal Tubular Cells

These are small elements that may be found isolated or in clusters. They are columnar, cuboidal, or rounded according to the state of preservation. The round nucleus is surrounded by a cyanophilic cytoplasm. Such tubular cells are very rare in the absence of pathologic conditions of the renal parenchyma.

Prostatic Cells

These are practically absent in normal urine, and to find them one must first perform a prostatic massage. Epithelial prostatic cells are columnar elements with small, round nuclei and clear cytoplasm; they desquamate in small clusters. Acid phosphatase is present in the cytoplasm. Spermatozoa are normally present in prostatic fluid or urine obtained after prostatic massage.

Seminal Vesicle Cells

These are cells of the columnar type; their cytoplasm is well delineated and granular. The nucleus is rounded or oval, the chromatin is uniformly distributed and the nucleolus is prominent. These cells sometimes contain a cytoplasmic pigment and large lobular or folded nuclei, two characteristics that may aid in identification.

HORMONAL EVALUATION FROM URINARY SEDIMENT (UROCYTOGRAM)

The influence of sexual hormones on the morphology of urinary sediment cells has been described by different authors.[34-36, 54-58]

Exfoliated squamous cells originate in the region of the trigone and the urethra. The desquamation is not abundant, and the quantity of collected cells is never greater than a few hundred. We distinguish eosinophilic squamous cells, superficial cyanophilic cells, and intermediate and parabasal cyanophilic cells.

It is possible to establish a hormonal evaluation by using the same morphologic criteria as those used in vaginal cytology: the abundance of cells and the mode of desquamation, the proportion of the different cellular types, the aspect of the cells, and the presence of other elements such as leukocytes, red blood cells, and mucus. A urocytogram is particularly recommended in the following cases:

- a hormonal evaluation of the child or the adolescent
- an examination performed in periods of clinical hemorrhage
- vaginal malformations and hormonal cytodiagnosis in males
- the repetition of hormonal evaluations (without requiring a repeat gynecologic examination)

The disadvantages of the method are the small number of the cells and the possible presence of inflammatory lesions or crystals that may falsify the reading.

Cytologic modifications of the urocytogram during the menstrual cycle are less pronounced than in the vaginal mucosa; we may nonetheless use the urocytogram for characterization of the pre- and postovulation phases. For example, estrogenic deficiency is reflected in the appearance of parabasal cells and estrogenic activity by the presence of superficial eosinophilic cells. For more details, we refer the reader to the works mentioned in the bibliography.

The most widely used and efficient technique is the centrifugation of fresh urine at 800 rpm for 5 minutes. The sediment is washed in Ringer's solution (diluted to 50 percent in water) to separate it from the mucus. The sediment is spread on a slide, fixed with ether-alcohol, and stained with Shorr-hematoxylin or Papanicolaou's stain. A better adherence of cells to the glass is obtained by using completely frosted slides. Air dessication fixation is not advised, since it makes for relatively poor staining.

INFLAMMATORY LESIONS OF THE URINARY TRACT

The frequency of inflammatory lesions of the urinary tract justifies the use of all methods of differential diagnosis available. Among these, cytology plays an important part: a direct examination of centrifugation bands can establish the inflammatory nature of a lesion and in certain cases may even indicate the precise nature of the infection.

Urethritis occurs by direct contamination or is secondary to a renal or bladder infection. Cystitis is secondary to an ascending infection (urethra) or a descending one (ureter and kidney); it is frequently a result of catheterization or urinary bladder endoscopy. In acute infections, the most frequently encountered microorganisms are *Escherichia coli*, *Streptococcus fecalis*, or *Proteus vulgaris*.

Chronic infections are caused by the same microorganisms and are maintained by obstructions such as prostatic hyperplasia or urethral stricture. Certain forms of chronic cystitis must be mentioned because they result in more or less characteristic cytologic images: cystic cystitis, mycotic cystitis (Plate 10.21), follicular cystitis, interstitial cystitis, papillomatous cystitis, tuberculous cystitis, malakoplakia, and bilharzial cystitis (Plate 10.22). Malakoplakia is an inflammatory condition characterized by an accumulation of histiocytic cells with a finely vacuolated cytoplasm located beneath the urothelium. The cytoplasm contains inclusions. The causative agent is unknown.

Acute prostatis is generally a complication of a urinary tract infection. The infection can become chronic with formation of inflammatory granulomas.

Cytology

Infections cause an increase in desquamation of urothelial cells. The nuclei increase in volume and become irregular, and the chromatin is arranged in clumps or is dense and homogeneous; nucleoli may be prominent. The cytoplasm may be vacuolated or exhibit lysis (Plate 10.6). Multinucleation is encountered. Long-lasting chronic irritation of the mucosa by calculi and severe cystitis give rise to marked cellular atypias including dyskaryosis, cytoplasmic vacuolation, and presence of lipid droplets. However, the interpretation of cellular modifications is rendered difficult by the presence of leukocytes, histiocytes, red blood cells, and bacterial flora, and by necrosis.

While such anomalies are encountered in infections they also may arise in various inflammatory irritations, for example, alterations caused by bubble bath detergents.

Viral infections such as herpes, cytomegalic inclusion disease, and adenovirus are characterized by the presence of nuclear or intracytoplasmic inclusions (Plate 10.20) and by multinucleation.

Michaelis-Gutmann bodies, which characterize malakoplakia, can be found in urine specimens when ulceration of the mucosa has occurred. They consist of concentrically laminated bodies containing calcium and iron and surrounded by epithelioid histiocytes. Ultrastructural studies show the lysosomal origin of the Michaelis-Gutmann bodies.[48] The presence of these bodies in urine has been reported.[67]

"Decoy" cells are degenerated epithelial cells whose nuclei are dense, hyperchromatic, sometimes fragmented, or exhibit a coarse chromatin pattern (Plate 10.4). They are erroneously diagnosed as cancer cells by those who are not aware of their existence and significance. They are observed in inflammatory conditions but also coexist with bladder tumors. Computer classification using the TICAS program confirms that these atypical cells represent a distinct cellular category.[51]

Several "inflammatory type" cytologic features have been described in acute renal graft rejection. They consist of the appearance of red blood cells, lymphocytes, cellular debris, granular material, casts, and renal tubular cells often showing anisonucleosis and nucleolar hypertrophy. The rejection can be predicted on the basis of these characteristic features in the urine.[7, 8, 19]

In the presence of atypical transitional cells, one should be aware of the difficulty in making a differential diagnosis between any kind of inflammation and malignancy. The final diagnosis should be made when all clinical, cystoscopic, and cytologic data are available. Urinary cytology is particularly helpful in the diagnosis of nonpapillary lesions with normal or nonspecific cystoscopic findings.

TUMORAL LESIONS

Four types of tumoral lesions of the urinary tract can be revealed by cytology: urothelial tumors, seminal vesicle, and prostatic tumors, tumors of the renal parenchyma, and various metastatic tumors.

Urothelial Tumors

Urothelial tumors originate in the urethra, the urinary bladder, or the ureter. Nonepithelial tumors, which are less frequent, must ulcerate the epithelium before the cells can be found in the urine.

Bladder tumors represent 1 percent of all carcinomas. They gererally appear in persons over 50 and are more common in males than in females. The large majority of bladder tumors (90 percent) are tumors of the papillary type.[50] The papilloma reveals a strictly normal urothelium; this entity is rare and represents less than 5 percent of all bladder tumors. The papillary carcinoma grade I shows a thickened urothelium (Plates 10.7 and 10.8), but the cellular anomalies are very slight.

As with the cervix, one may encounter the whole gamut of abnormalities, ranging from hyperplasia to dysplasia, from in situ carcinoma to invasive carcinoma (Figure 10). These lesions have been described in patients who work in high-risk environments and dc novo in patients with no urologic history.

Histologically, these lesions are characterized by a disorganized cellular disposition of the epithelium. If the lesion is a papilloma, it is characterized by an increase in the number of cellular layers with papillary formations and the appearance of nuclear and cytoplasmic atypias. The number and intensity of such atypias may increase progressively from hyperplasia or dysplasia to intraepithelial carcinoma.

The evolution of these lesions probably follows the same course as equivalent lesions of the cervix. A relationship between these lesions and invasive carcinoma is suggested by three considerations: these images are encountered at the periphery of malignant tumors; they constitute the beginning phase of occupation-associated cancers; and they are incidently reproduced when one experimentally provokes bladder cancer.

The papillomatous form is well known for its recurring or multicentric nature. The papilloma is composed of a fibrovascular axis, covered by a urothelium similar to the normal urothelium. It may be single or multiple. Diffuse papillomatosis spreads over large areas of the mucosa, and it is accompanied by cellular anomalies.

Nonpapillary in situ carcinoma is a much rarer entity that has been discovered in workers exposed to bladder carcinogens and submitted to routine examination of urinary sediment. Atypical cellular elements were observed in the urinary sediment, and the biopsy confirmed the diagnosis.[50]

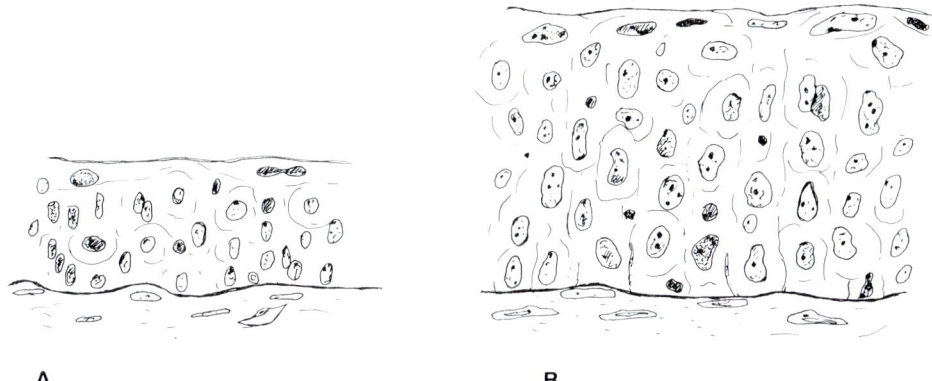

A   B

Figure 10.3. Schematic representation of normal urothelium (*A*) and in situ carcinoma (*B*) (X650).

Rare inverted papillomas growing into the bladder wall and epidermoid papillomas also have been described.

Several systems of classification of urothelial tumors have been proposed; we mention those of WHO[92] and Koss.[50] Stage I comprises the most differentiated tumors in which anomalies are discrete. In stage II, cellular anaplasia becomes more marked; stage III includes undifferentiated tumors (Table 10.1). There is a positive correlation between the degree of anaplasia and the gravity of the clinical diagnosis. The extent and the mode of bladder wall invasion also have a prognostic value. The mucosa, submucosa, and muscular layer may be successively invaded. The invasion occurs either in a mass or, in graver circumstances, in the form of multiple cords or small nodules, or finally, through blood and lymphatic vessels. Figure 10.4 illustrates the different steps of bladder wall tumor invasion.

The urothelial lining exhibits the following signs of malignancy: increased number and loss of polarity of cellular layers; cellular atypias; invasion of the chorion, the muscularis, and the serosa. Neoplastic cords can show foci of squamous and columnar metaplasia. The epidermoid, the glandular, and the undifferentiated types are less frequent and here we need only mention their existence.

The nonepithelial tumors include benign forms: leiomyoma, hemangioma, neurofibroma, and granular cell tumors and malignant forms: rhabdomyosarcoma, lymphoma, fibrosarcoma, leiomyosarcoma, and osteosarcoma. Malignant melanoma also has been reported.

Finally, let us mention the existence of metastatic invasions originating most frequently in the prostate, intestine, and uterus.

Cytology of the Tumors

Intraepithelial Carcinoma

In intraepithelial carcinoma, cellular lesions are discrete and characterized by hyperchromatism and anisonucleosis and by moderate variations of cellular size and form.[2, 66, 68, 82] There are no important necrotic or inflammatory phenomena, and the inflammatory diathesis is identical to that found in intraepithelial carcinoma of the cervix. Cells have a greater tendency to desquamate separately (Plates 10.9 and 10.10).

Table 10.1. Histologic Grading of Uroepithelial Tumors

Grade O	Normal epithelium with no appreciable cellular polymorphism
Grade I	Slightly or moderately thickened epithelium, discrete cellular polymorphism, rare mitotic figures
Grade II	Moderate and irregular hyperplasia of epithelium, loss of normal cell polarity towards the surface, moderate cellular polymorphism, scattered mitotic figures
Grade III	Irregular thickening of epithelium, considerable polymorphism, marked disarrangement of cellular layers, frequent mitotic figures, marked anaplasia of epithelium

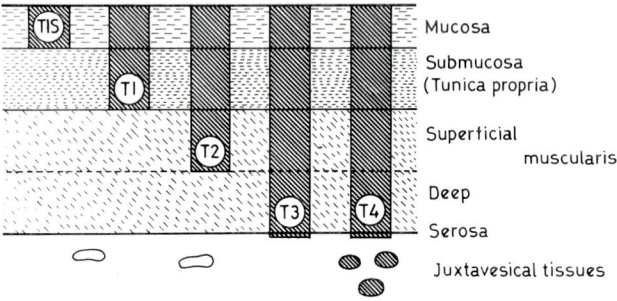

Figure 10.4. Staging of epithelial tumors (U.I.C.C.).

Invasive urothelial cancers are characterized by neoplastic cells whose enlarged, irregular, hyperchromatic nuclei are surrounded by a relatively abundant cyanophilic or eosinophilic cytoplasm.[60] The size and the shape of the malignant cells vary. Desquamation occurs in clumps or more frequently as isolated cells. Medium or large lipid droplets are often present and are a manifestation of cellular necrobiosis. However, their presence does not represent a definitive sign of malignancy since they also are found in normal cells.

Secondary inflammatory phenomena, provoked by ulceration and superficial necrosis of invasive tumors, manifest themselves by the presence of cellular debris, leukocytes, histiocytes, and bacterial flora (Plates 10.11–10.14).

The papillomatous nature of the carcinoma is suggested when one finds clusters of cells that resemble bunches of grapes and appear tridimensional. The more or less differentiated nature of the tumor is reflected in the cellular population of the smear and permits an attempt at classification. The rare epidermoid carcinoma desquamates cells having different degrees of keratinization. The nonkeratinized forms are composed of large cells having a cyanophilic cytoplasm. Keratinized forms have an abundant eosinophilic cytoplasm (Plate 10.15). The glandular form is charcterized by columnar or cuboidal cells.

Seminal Vesicle and Prostatic Tumors

Benign hyperplasia. Benign prostatic hyperplasia is frequent after the age of 50. It consists of glandular hyperplastic nodules disseminated in an abundant fibromuscular stroma. Smears obtained after prostatic massage contain normal-looking prostatic cells. A higher than normal cellular density may suggest the diagnosis (Plates 10.24 and 10.25).

Carcinoma. The reported frequency of prostatic carcinoma varies considerably from author to author. These large differences in frequency (from 14 to 46 percent) are due to variations in clinical and morphologic criteria used to define the tumor. To our own experience, the frequency is around 15 percent.

Prostatic tumoral cytology.   The quality of results depend on the quality of sampling.[77] In order of decreasing reliability are urines collected following prostatic massage,[18, 28] secretions collected after prostatic massage, and finally, catheter-drawn urines.

The cytologic criteria are the same as those usually found in glandular cell types: hyperchromatism and increase in the nuclear and nucleolar volumes with more or less well-preserved cyanophilic cytoplasm and desquamation in small clumps (Plates 10.26 and 10.27). In differentiated forms, the cells remain small in spite of their anomalies. It is only in anaplastic forms that one finds voluminous and monstrous elements.

In prostatic needle aspiration, one may find cells from the seminal vesicles. Most of the cells are columnar with a well-delineated regular cytoplasm. The nucleus is round or oval and of variable size, and the chromatin is uniformly granular. In the absence of spermatozoa and mucus, the recognition of seminal epithelial cells is difficult because these cells do not exhibit any distinctive cytologic characteristics that differentiate them from prostatic elements.

A diagnosis of the seminal origin of the aspiration is easier to make when the highly viscous seminal secretion is abundant and when elongated cells are found. Tumors of the seminal vesicles are very rare and consequently their cytology is not yet well known.

Tumors of the Renal Parenchyma

The recognition of renal tumors by cytology implies that the tumor has invaded the excretory ducts of the calyces or the renal pelvis; the lengthy cellular migration explains both the fact that the cells are poorly preserved and that such cases are not often initially diagnosed by cytology.

The most frequent renal tumor is the clear-cell glandular carcinoma (hypernephroma or Gravitz's tumor). It constitutes about 75 percent of renal tumors and 2 percent of all cancers. It is more frequent in males than in females and occurs in the over-40 age group.

The nuclear DNA content is increased in many malignant tumors and some correlation seems to exist between the degree of this elevation and the clinical malignancy. Cytophotometric DNA analysis of aspirated cells of prostatic carcinoma tend to demonstrate that a pronounced degree of heteroploidy is seen in poorly differentiated and highly malignant carcinoma.[94] On the contrary, low-grade well-differentiated carcinoma shows a diploid DNA content. The same correlations also have been suggested in carcinoma of the cervix and urinary bladder.

Wilms' tumor, or nephroblastoma, is a rare tumor of children appearing most often before the age of 5. Wilms' tumor is rarely encountered in the adult. When it does occur malignant epithelial and sarcomatous elements are associated with it.

Cytology of Renal Tumors

Although neoplastic cells are not frequently found in urine, at one time or another cytopathologists do come across clusters of neoplastic cells whose renal origin is suspected. These epithelial elements desquamate in clusters (Figures 10.30–10.32).

Differentiated tumors exhibit glandular structures whose cells have an abundant, clear cytoplasm and a relatively small nucleus. In undifferentiated forms one finds small clusters whose cells have hyperchromatic nuclei and relatively little cytoplasm.[32, 38, 64, 78, 81]

The essential point here is to recognize the presence of malignant cells even if their renal origin cannot easily be determined cytologically. Subsequent clinical and radiologic examinations will supply the precise location.

Elements desquamated from papillomas of the calyces or hyperplastic cells originating in solitary renal cysts sometimes represent difficult problems of differential diagnosis.

Metastatic Tumors

Uterine and anorectal tumors may invade the bladder wall and desquamate into the urine. Malignant melanoma is easily recognized when melanin-containing malignant cells are present. Melanin should not be mistaken for hemosiderin-laden renal elements.

Differential Diagnosis

Papilloma Versus Carcinoma

The differential diagnosis between benign and malignant cells originating from papillomatous tumors of the bladder is delicate. If we exclude the rare papilloma (grade 0) covered by a strictly normal urothelium, all papillary lesions of the bladder should be classified at least as grade I carcinomas. Since this distinction is difficult even when examining histologic sections, it is presumptuous to try to make this decision only on the basis of cytologic criteria.

The abundance of the cells in the urine is an indication of the existence of a papillomatous lesion and a careful search for cytologic atypias may orientate the diagnosis. From grade 1 to grade 4 carcinoma, the presence of definite atypias makes diagnosis a relatively simple thing.

Nonpapillary Carcinomas

Nonpapillary carcinomas create no problem of diagnosis, since the cellular atypias are marked and evident. Abundant superficial necrosis of the lesion in some cases may create difficulties in recognition of the tumors cells among the numerous inflammatory atypias.

Precancerous Lesions
(Dysplasia, Carcinoma In Situ, Atypical Hyperplasia)

The presence of cancer cells in the urinary sediment without clinical evidence of tumors may suggest the diagnosis of atypical hyperplasia or carcinoma in situ; the absence of an inflammatory diathesis (clear background) is suggestive of noninvasive lesions as in cervical cytology. These cells may be found in routine cytology of asymptomatic patients in high-risk populations such as workers exposed to bladder carcinogens (for example, para-aminodiphenyl).

The early neoplastic transformations should not be confused with iatrogenic alterations, for example, chemotherapy (cyclophosphamide or other alkylating agents) and radiation therapy (Figures 10.16–10.17).

Marked nuclear enlargement as well as nucleolar hypertrophy are the most constant morphologic features of these precancerous lesions.

Summary

The use of the cytologic method is indicated in the following clinical circumstances:

- hematuria
- clinical signs of inflammation manifesting themselves by modifications of the urinary system or its urine output volume
- detection of tumoral lesions in high-risk patients: workers in the chemical and rubber industries (benzidine, naphtylamine), smokers, and persons with chronic infections (e.g., bilharziasis)
- follow-up surveillance of tumoral lesions of the urinary tract
- detection of genital or intestinal metastasis
- prediction of acute renal allograft rejection

It is important to know the clinical history of the patient, particularly if ionizing radiation, chemotherapy, or the administration of certain medicines such as phenacetin have been instituted.

What should be noted:

- Normal urine contains very few urothelial and squamous cells. The presence of any other cellular element has a pathologic significance. Its origin and nature should be carefully determined.

What the lesions are that may be diagnosed by cytology:

- inflammatory lesions, specific or nonspecific
- iatrogenic alterations (for example, postradiation and postchemotherapy modifications)
- benign and malignant tumors

BIBLIOGRAPHY

1. Allegra S.R., Broderick P.S., Corvese N.L. Cytologic and histogenetic observations in well- differentiated transitional cell carcinoma of bladder. *J. Urol.* 107:777, 1972.

2. Allegra S.R., Fanning J.P., Streken J.F., Corvese N.M. Cytologic diagnosis of occult and "in situ" carcinoma of the urinary system. *Acta Cytol.* 10:340, 1966.

3. Atay Z. Cytology of kidney and urinary tract. *Deutsche Medizinische Wochench.* 99:48, 1974.

4. Bergkvist A., Ljungkvist A., Moberger G. Classification of bladder tumors based on the cellular pattern. Preliminary report of clinical-pathological study with a minimum follow-up of eight years. *Acta Chir. Scand.* 130:371, 1965.

5. Bettmann H.K., Meyer C.J. Urocytogrammes et colpocytogrammes au cours du travail de comparaison des urines des nouveaux-nés et du liquide amniotique par la méthode des filtres millipores. *Rev. Cytol. Clin.* 8:12, 1970.

6. Bibbo M., Gill W.B., Harris M.J., Luc C.T., Thomsen S., Wied G.L. Retrograde brushing as a diagnostic procedure of ureteral, renal pelvic and renal calyceal lesions. A preliminary report. *Acta Cytol.* 18:137, 1974.

7. Bossen E.H., Johnston W.W. Exfoliative cytopathologic studies in organ transplantation. The cytologic diagnosis of herpes virus in the urine of renal allograft recipients. *Acta Cytol.* 19:415, 1975.

8. Bossen E.H., Johnston W.W., Amatulli J., Rowlands D.T. Jr. Exfoliative cytopathologic studies in organ transplantation. I. The cytologic diagnosis of cytomegalic inclusion disease in the uterine of renal allograft recipients. *Amer. J. Clin. Path.* 52:340, 1969.

9. Bossen E.H., Johnston W.W., Amatulli J., Rowlands D.R. Jr. Exfoliative cytopathologic studies in organ transplantation. III. The cytologic profile of urine during acute renal allograft rejection. *Acta Cytol.* 14:176, 1970.

10. Bunge R.G. Exfoliative cytology of transitional cell carcinoma. *J. Urol.* 67:740, 1952.

11. Coleman D.V. The cytodiagnosis of human polyomavirus infection. *Acta Ctyol.* 19:93, 1975.

12. Creasman W.T., Lukeman J. Unreliability of urinary cytology in detecting gynecologic malignancy. *Cancer* 30:148, 1972.

13. Cullen T.H., Popham R.C., Voss H.J. Urine cytology and primary carcinoma of the renal pelvis and ureters. *Aust. N.Z. Surg.* 41:230, 1972.

14. Derot M., Fousset A. La cytologie urinaire. Sa valeur diagnostique en néphrologie. *La Vie Med.* 37:269, 1956.

15. De Vooght H.J. Rapid urinary cytology by phase contrast microscopy. A preliminary report. *Urol. Research* 1:113, 1973.

16. De Vooght H.J., Beyer-Boon M.E., Brussee J.M. The value of phase contrast microscopy for urinary cytology, reliability and pitfalls. *Acta Cytol.* 19:542, 1975.

17. De Vooght H.J., Wielenga G. Clinical aspects of urinary cytology. *Acta Cytol.* 16:349, 1972.

18. Droese M., Voeth C. Cytologic features of seminal vesicle epithelium in aspiration biopsy smears of the prostate. *Acta Cytol.* 20:120, 1976.

19. Dukes C.E. The institute of urology scheme for the histologic classification of epithelial tumors of the bladder. In Wallace D.M. (Ed.) *Tumors of the Bladder* Livingston, Edinbugh, 1959, p.105.

20. Eriksson O., Johansson S. Urothelial neoplasms of the upper urinary tract. A correlation between cytologic and histologic findings in 43 patients with urothelial neoplasms of the renal pelvis or ureter. *Acta Cytol.* 20:20, 1976.

21. Eriksson O. Skjeggestad O. Cytologic diagnosis of transitional cell tumors of the urinary bladder using the millipore filter technique. *Acta Path. Microbiol. Scand.* 81:222, 1973.

22. Esposti P.L. Cytologic diagnosis of prostatic tumors with the aid of transrectal aspiration biopsy. *Acta Cytol.* 10:182, 1966.

23. Esposti P.L., Moberger G. Zajicek J. A study of 567 cases of urinary tract disorders including 170 untreated and 182 irradiated bladder tumors. *Acta Cytol.* 14:145, 1970.

24. Esposti P.L., Zajicek J. Grading of transitional cell neoplasms of the urinary bladder from smears of bladder washings: A critical review of 326 tumors. *Acta Cytol.* 16:529, 1972.

25. Foot N.C., Papanicolaou G.N., Holmquist, N.D. Seybolt J.F. Exfoliative cytology of urinary sediments: A review of 2829 cases. *Cancer* 11:127, 1958.

26. Forni A., Ghetti G., Armeli G. Urinary cytology in workers to carcinogenic aromatic amines: A six-year study. *Acta Cytol.* 16:142, 1972.

27. Gardner S., Field A.M., Coleman D.V., Hulme B. New human papovavirus (BK) isolated from urine after renal transplantation. *Lancet* 1:1253, 1971.

28. Garret M., Jassie M. Cytologic examinations of post-prostatic massage specimens as an aid in diagnosis of carcinoma of the prostate. *Acta Cytol.* 20:126, 1976.

29. Gill W.B., Lu C.T., Thomsen D. Retrograde brushing: a new technique for obtaining histologic and cytologic material from ureteral renal pelvic and renal calyceal lesions. *J. Urol.* 109:573, 1973.

30. Grace D.A., Taylor W.N., Taylor J.N., Winter C.C. Carcinoma of the renal pelvis: A 15-year review. *J. Urol.* 96:566. 1968.

31. Hajdu S.I. Exfoliative cytology of primary and metastatic Wilm's tumor. *Acta Cytol.* 15:339, 1971.

32. Hajdu S.I., Savino A., Hajdu E.O., Koss L.G. Cytologic diagnosis of renal cell carcinoma with the aid of fat stain. *Acta Cytol.* 15:31, 1971.

33. Haleem S.A., Sprayregen S., Siegelman S.S. Preoperative diagnosis of renal pelvic carcinoma. *J. Urol.* 108:695, 1972.

34. Haour P. Comparison of radiation cell changes in exfoliated vaginal cells and in exfoliated cells from the urinary tract (urocytogram). *Acta Cytol.* 3:449, 1959.

35. Haour P., Conti C. Le diagnostic cyto-hormonal par l'urocytogramme. *Rev. Cytol. Clin.* 8:26, 1970.

36. Haour P., Delcroix R. Cytologie de l'urèthre chez la femme: aspects hormonaux *Rev. Cytol. Clin.* 7:35, 1974.

37. Harris M.J., Schwinn D.P., Morrow J.W., Gray R.L., Browell B.M. Exfoliative cytology of the urinary bladder of irrigation specimen. *Acta Cytol.* 15:385, 1971.

38. Harrisson J.H., Bostford T.W., Tucker M.R. The use of the urinary sediment in the diagnosis and management of neoplasm of the kidney and bladder. *Surg. Gyn. Obstet.* 92:129, 1951.

39. Haynes M., Trott P.A., Islam A.K.M.S., Hirst G. An ultrastructural study of the urinary bladder in children correlated with histological, bacteriological and clinical findings. *J. Clin. Path.* 28:176, 1975.

40. Hazard J.B., McCormack L.J., Belovich D. Exfoliative cytology of the urine with special reference to neoplasms of the urinary tract: Preliminary report. *J. Urol.* 78:182, 1957.

41. Houghton B.J., Pears M.A. Cell secretion in normal urine. *Brit. Med. J.* 1:622, 1957.

42. Jewett H.J. Cancer of the bladder — diagnosis and staging. *Cancer* 32:1072, 1973.

43. Johansson S., Angervall L., Bengtsson U. Wahlqvist L. Uroepithelial tumors of the renal pelvis associated with abusive of phenacetin containing analgesics. *Cancer* 33:743, 1974.

44. Johnson W.D. Cytopathological correlations in tumors of the urinary bladder. *Cancer* 17:867, 1964.

45. Johnston W.W., Bossen E.H., Amatulli J., Rowlands D.T. Exfoliative cytopathologic studies in organ transplantation. II. Factors in the diagnosis of cytomagalic inclusion disease in urine of renal allograft recipients. *Acta Cytol.* 13:605, 1969.

46. Kalnins Z.A., Rhyne A.L., Morehead R.P., Carter B.J. Comparison of cytologic findings in patients with transitional cell carcinoma and benign urologic disease. *Acta Cytol.* 14:243, 1970.

47. Kern W.H. The cytology of transitional cell carcinoma of the urinary bladder. *Acta Cytol.* 19:420, 1975.

48. Kern W.H., Bales C.E. Quantitative studies of urine cytology. *Amer. J. Clin. Path.* 51:225, 1969.

49. Kern W.H., Bales C.E., Webster W.W. Cytologic evaluation of transitional cell carcinoma of the bladder. *J. Urol.* 100:616, 1967.

50. Koss L.G. *Tumors of the Urinary Bladder. Atlas of Tumor Pathology.* Armed Forces Institute of Pathology, Washington, D.C., 1975.

51. Koss L.G., et al. Computer analysis of atypical urothelial cells. I. Classification by supervised learning algorithms. *Acta Cytol.* 21:247, 1977. II. Classification by unsupervised learning algorithms. *Acta Cytol.* 21:261, 1977.

52. Koss L.G., Melamed M.R., Kelly R.E. Further cytologic and histologic studies of bladder lesions in workers exposed to paraaminodiphenyl. Progress report. *J. Nat. Cancer Inst.* 43:233, 1969.

53. Koss L.G., Tiamson E.M., Robbins M.A. Mapping cancerous and precancerous bladder changes. A study of the urothelium in ten surgically removed bladders. *JAMA* 227:281, 1974.

54. Lencioni L.J. *Urocitograma.* 3rd Ed. Editorial Medica Panamericana, Buenos Aires, 1972.

55. Lencioni L.J. L'urocytogramme. *Diagnostic cyto-hormonal à partir du sédiment urinaire.* Mabine, Ed. Paris 1975.

56. Lencioni L., Amezaga L.M., Badano H. L'urocytogramme et la détermination de l'oestriol et du pregnandiol dans la grossesse chez les diabétiques. *Rev. Cytol Clin.* 8:17, 1970.

57. Lencioni L.J., Martinez Amezaga L.A., Alonso C., Antonio L., Camargo H. Urocytogram and pregnancy. II. Correlation with fetal condition at birth in high risk pregnancies. *Acta Cytol.* 17:125, 1973.

58. Lencioni L.J., Martinez Amezaga L.A., Lo Bianco V.S. Urocytogram and pregnancy. I. Methods and normal values. *Acta Cytol.* 13:279, 1969.

59. Levy E., Jerusalem K. Experience with cytologic examination of urines with the assistance of multiple parallel methods with and without clinical indication of tumor. *Acta Cytol.* 17:121, 1973.

60. Loveless J. The effects of radiation upon the cytology benign and malignant bladder epithelia. *Acta Cytol.* 17:355, 1973.

61. MacMahon B., Cole P., Brown J.B., Aoki K., Lin T.M., Morgan R.W., Woo N.C. Urine oestrogen profiles of Asian and North American women. *Int. J. Cancer* 14:161, 1974.

62. Masin F., Masin M. Sudanophilia in exfoliated urothelial cells. *Acta Cytol.* 20:573, 1976.

63. Masukawa, T., Garancis J.C., Rytel M.W., Mattingly R.F. Herpes genitalis virus isolation from human bladder urine. *Acta Cytol.* 16:416, 1972.

64. Meisels A. Cytology of carcinoma of the kidney. *Acta Cytol.* 7:239, 1963.

65. Melamed M.R., Traganos F., Sharpless T., Darzynkiewicz Z. Urinary cytology automation: preliminary studies with acridine orange stain and flow-through cytolfluorometry. Investigative urology 13:5, 1976.

66. Melamed M.R., Voutsa N.G., Grabstald H. Natural history and clinical behavior of in-situ carcinoma of the human urinary bladder. *Cancer* 17:1533, 1964.

67. Melamed M.R., Wolinska W.H. On the significance of intracytoplasmic inclusions in the urinary sediment. *Amer. J. Path.* 38:711, 1961.

68. Melicow M.M., Hollowell J.W. Intraurothelial cancer; carcinoma in-situ; Bowen's disease of the urinary system. Discussion of 30 cases. *J. Urol.* 68:763, 1952.

69. Melicow M.M. Tumors of the urinary drainage tract: Urothelial tumors. *J. Urol.* 54:186, 1945.

70. Morse N., Melamed M.R. Differential counts of cell populations in urinary sediment smears from patients wtih primary epidermoid carcinoma of the bladder. *Acta Cytol.* 18:312, 1974.

71. Murphy W.M. Herpesvirus in bladder cancer. *Acta Cytol.* 20:207, 1976.

72. Naib Z.M. Exfoliative cytology of renal pelvic lesions. *Cancer* 14:1085, 1961.

73. Newman D.M., et al. Transitional cell carcinoma of the upper urinary tract. *J. Urol.* 98:322, 1967.

74. O'Morchoe et al. Urinary cytologic changes after radiotherapy of renal transplants. *Acta Cytol.* 20:132, 1976.

75. Orell S.R. Transitional cell epithelioma of the bladder: Correlation of cytologic and histologic diagnosis. *Scand. J. Urol. Nephrol.* 3:93, 1969.

76. Papanicolaou G.N. Cytology of the urine sediment in neoplasms of the urinary tract. *J. Urol.* 57:375, 1947.

77. Peters H. The prostatic smear and its clinical usefulness. *J. Urol.* 66:770, 1951.

78. Powder J.R., Naib Z.M., Young Jr. J.D. Cytological examination fo the urine sediment as an aid to diagnosis of epithelial neoplasms of the upper urinary tract. *J. Urol.* 84:666, 1960.

79. Pugh R.C.B. The pathology of cancer of the bladder. An editorial overview. *Cancer* 32:1267, 1973.

80. Reichborn-Kjennerud S., Hoeg K. The value of urine cytology in the diagnosis of recurrent bladder tumors. A Preliminary Report. *Acta Cytol.* 16:269, 1972.

81. Sarnacki C.T., McCormack L.J., Kiser W.J., Hazard J.B., McLaughlin T.C., Belovich D.M. Urinary cytology and the clinical diagnosis of urinary tract malignancy: a clinico-pathologic study of 1400 patients. *J. Urol.* 106:761, 1971.

82. Schade R.O.K. Carcinoma in-situ of the urinary bladder. Histological and cytological observations. *Proc. R. Soc. Med.* 60:109, 1967.

83. Schmid G.H., Hornstein O.P., Mittmann O., Munstermann M. Periodical epithelial exfoliation of the urinary ducts in the male. *Acta Cytol.* 16:352, 1972.

84. Silberblatt J.M. Exfoliative cytology as a screen test for urinary tract malignancy. *Bull. N.Y. Acad. Med.* 29:889, 1953.

85. Smith P., Crozier E.H. Cytology of voided urine in diagnosis of gynecologic malignancy. *Obstet. Gynec.* 41:440, 1973.

86. Suprun H., Bitterman W. A correlative cytohistologic study on the interrelationship between exfoliated

urinary bladder carcinoma cell types and the staging and grading of these tumors. *Acta Cytol.* 19:265, 1975.

87. Theologidis A.D., Jameson R.M., Scott A. The reliability of urinary cytology. *Brit. J. Urol.* 43:598, 1971.

88. Tyler D.E. Urine analysis and urine cytology correlated with the menstrual cycle and age. *Amer. J. Obstet. Gynec.* 90:147, 1964.

89. Tyrkko J. Exfoliative cytology in the diagnosis and follow up of urothelial neoplasms. *Scand J. Urol. Nephrol.* Suppl 19, 1972.

90. Umiker W. Accuracy of cytologic diagnosis of cancer of the urinary tract. *Acta Cytol.* 8:186, 1964.

91. Voutsa N.G., Melamed M.R. Cytology of in situ carcinoma of the human urinary bladder. *Cancer* 16:1307, 1963.

92. WHO *International Classification of Tumors. Histological Typing of Urinary Bladder Tumors.* Ed. F.K. Mostofi, WHO, Geneva, 1973.

93. Wiggishoff C.C., McDonald J.H. Urinary exfoliative cytology in the diagnosis of bladder tumors. *Acta Cytol.* 16:139, 1972.

94. Zetterberg A., Esposti P.L. Cytophotometric DNA-Analysis of aspirated cells from prostatic carcinoma. *Acta Cytol.* 20:46, 1976.

11
Aspiration Cytology

INTRODUCTION

The clinical indications for aspiration cytology are numerous, and in this chapter we shall describe the diagnostic results that can be obtained from a correct use of the method. All organs and tumor masses that are accessible to needle aspiration may constitute an indication for the method. We have dealt separately with breast aspiration and effusion aspiration because of the clinical importance of the subjects. The principles of the technique are identical regardless of the site of the puncture, but the practical manipulations will vary with the anatomic location and the organ involved.

Different types of needles (18–22 gauge) are used with a stylet to avoid contamination by other tissues during the aspiration procedure. Special handles have been manufactured to facilitate the maneuver.

Solid, semisolid, or fluid material may be obtained from lesions. Different techniques of cell concentration will be adopted depending on the consistency of the aspirate.

The major indication for the method is the detection of tumor cells, but we shall mention some other clinical entities whose diagnosis may rely on cytology.

A short enumeration of the different sites of needle aspiration will give an idea of the indications and limitations of the method:

- spleen
- lymph nodes
- salivary glands
- thyroid gland
- lung
- liver
- pancreas
- bone
- accessible tumor masses
- eye
- cerebrospinal fluid (CSF)

TECHNIQUE

The use of a fine needle with a stylet is the most appropriate technique. A syringe will provide an adequate handle as well as the negative pressure necessary to aspirate the material. Special handles that provide a better grip of the syringe and make it easier to guide the needle are available.

Much controversy exists about who should perform the aspiration. Some advocate that the same individual should handle the syringe and read the slide under the microscope. Our opinion is that this "one-man-show" is not an absolute necessity. As long as the person performing the aspiration has had enough practice and gives all the necessary clinical information, it matters little whether it be the clinician or the cytopathologist. One argument in favor of a "two-man-show" is that there is more objectivity when two people are concerned with the same problem.

The aspiration manipulation requires some experience: the lesion must be palpated, the needle correctly inserted, and a slight negative pressure exerted as the needle is gently moved in and out. This ensures that the aspirations will be representative of different parts of the lesion. The syringe is slowly withdrawn and at the same time pressure is released.

If the aspirated material is fluid, it may be centrifuged, or if the quantity is small, it may be expressed onto the slide and smeared. Sometimes small fragments of tissue are present in the fluid; they should be recovered and treated as paraffin-embedded specimens.

If the aspirated material is solid, imprints can be performed by gently squeezing and spreading the material on slides. Remaining material may be embedded in paraffin.

Fixing and Staining

Wet fixing and air drying have their partisans as we have mentioned above. We prefer the wet-fixing techniques, because in most cases preservation of cellular organelles is better. In certain circumstances air-dried smears followed by May-Grunwald Giemsa staining will provide additional information.

Customary procedures of each laboratory and personal preferences play an important role in this choice. Wet fixatives commonly used are alcohol-ether and alcohol. Papanicolaou and hematoxylin-Shorr stains are routinely used.

Cytologic Diagnosis on Surgical Specimens

This technique is a complementary procedure of the frozen section and its indications should be appreciated by the pathologist. It makes for rapid diagnosis, and with adequate stains and fixations, cellular details are very well preserved.

In some cases—determined by the pathologist and under his responsibility—a diagnosis can be given by the examination of the smear alone. But one should keep in mind that this decision can be made only by the cytopathologist who can turn to a frozen section, should cytology prove inconclusive. Used in this way, cytology is an adjuvant, valuable technique.

The various clinical applications are lesions of the breast, the head and neck, the bones, the thorax, the mediastinum, the lungs, the liver, the spleen, the pancreas, the GI tract, the lymphoid tissue, the central nervous system, and the skin.

LYMPH NODES

The cytologic study of lymph nodes represents one of the major fields of application of the needle aspiration technique. The use of imprints on surgically removed lymph nodes is an additional and valuable procedure in surgical pathology (Plate 11.1).

Lymph node aspiration with a fine needle (18 guage) is particularly indicated in the detection of primary and metastatic tumors. It also may provide information in problems of differential diagnosis such as tuberculosis and various other lesions. (The malignant cells will show the conventional criteria of malignancy.)

The cytology of metastatic tumors, particularly in the neck lymph nodes, will indicate the type of carcinoma and sometimes the origin of the lesion. Epidermoid cells are the most easily recognizable: they have dense, irregular, hyperchromatic nuclei surrounded by a more or less keratinized cytoplasm (Plates 11.7 and 11.8).

Adenocarcinomas may originate in various organs: from the GI tract, the respiratory tract, the thyroid, the breast, and the urinary tract.

Comparison with the histology of the primary tumors, when available, will facilitate the determination of the histologic type. For example, well-differentiated tumors of the thyroid may reproduce normal follicular structure in the metastases. These images have been erroneously interpreted as ectopic normal thyroid tissue.

Malignant lymphomas usually provide large quantities of neoplastic cells that can be classified according to the cellular type. Lymphocytic tumors originate in two functional systems: the thymus-dependent system (T cells) and the bursal-equivalent system (B cells). T cells are concerned with the cell-mediated immune response; they have few or no surface immunoglobulins, and they bind with desensitized sheep erythrocytes (rosette formations). B cells deal with the cell's humoral immune response; they possess surface immunoglobulins. These functional characteristics cannot be established by examination of routinely stained preparations, but experimental and clinical data based on specific immunologic techniques give us tentative classifications of these different cellular types. T cells represent approximately 70 percent of the peripheral blood lymphocytes, and they are found in paracortical areas of lymph nodes, in the spleen, and in the gastrointestinal tract. B cells are found in follicular centers of lymph nodes and the spleen, in the bone marrow, and in the gastrointestinal tract.

As the new classifications are not yet universally used, we will present the old ones for the sake of clarity (Table 11.1).

Table 11.1. Hodgkin's Disease

Jackson and Parker Classification[a]
    paragranuloma
    granuloma
    sarcoma

Rye Classification[b]
    lymphocytic predominance
    nodular sclerosis
    mixed cellularity
    lymphocytic depletion

[a]From *N. Engl. J. Med.* 231:35, 1944.
[b]From *Cancer Res.* 26:1063, 1966.

The different forms of both B and T lymphocytes, which were formerly interpreted as representing different steps of cellular differentiation, very probably represent various reversible stages in the metabolic activities of the lymphocytic cell. These stages are morphologically translated as a series of transitions between the small resting lymphocyte and the transformed large cell with its large nucleus, prominent nucleoli, and pyroninophilic (red) cytoplasm.

Lukes and Collins[72] have established a clinicomorphologic correlation between these cellular types and the different lymphomas. Various clinical, experimental, and immunologic data tend to confirm the validity of these correlations. The great majority of the lymphomas are of the B cell type arising from transformed follicular center cells (Figure 11.1).

Other classifications have been proposed,[27, 67, 101] and some difficulty lies in the interpretation of the different terms used in these nomenclatures. It must be noted that the terms "reticulum cell" and "histiocytic cell" should be abandoned in the light of recent experimental data. The so-called reticulum and histiocytic cells are in fact modified lymphocytic cells. In any case, it is recommended that the classification used in cytology should conform to the new histologic nomenclatures (Tables 11.2 and 11.3).

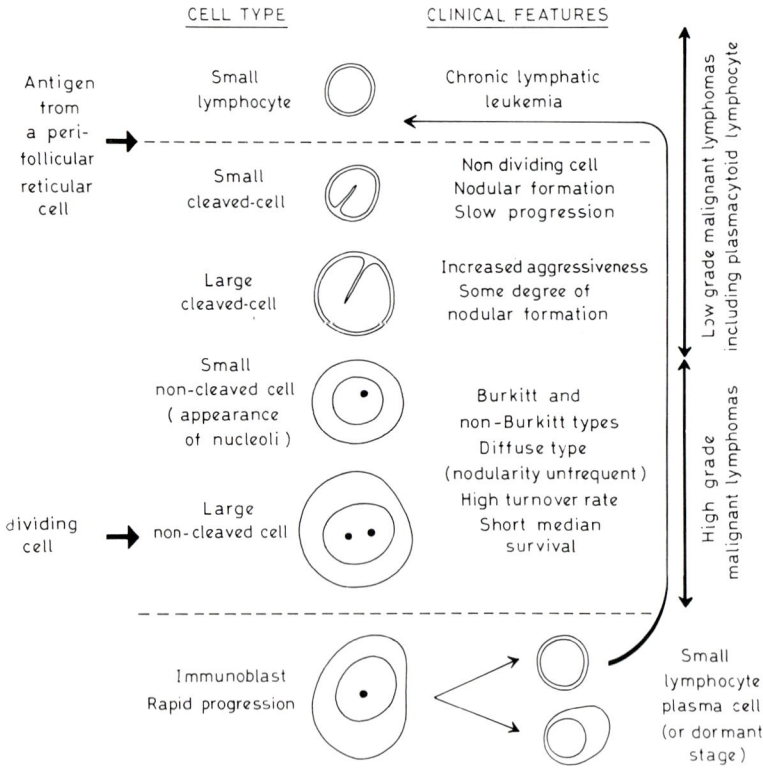

Figure 11.1. Immunologic and morphologic approach of malignant lymphomas. (From Lukes and Collins.[72])

Table 11.2. Non-Hodgkin's Lymphomas (Rappaport Classification)[a]

Malignant lymphoma
Lymphocytic, well-differentiated
    nodular or diffuse
Lymphocytic, poorly differentiated
    nodular or diffuse
Mixed histiocytic-lymphocytic
    nodular or diffuse
Histiocytic
    nodular or diffuse
Undifferentiated

[a]From *Cancer* 9:792, 1956.

Table 11.3. Non-Hodgkin's Lymphomas (Lukes and Collins Classification)[a]

  I. Undefined cell type

 II. T cell types
      convoluted lymphocyte
      immunoblastic sarcoma

III. B cell types
       large cleaved
       small non-cleaved
       large non-cleaved
      immunoblastic sarcoma

IV. Histiocytic type

 V. Unclassifiable

[a]From *Cancer* 34 (Suppl.) : 1448, 1974.

Cytology of Lymphoma

Hodgkin's disease can be suspected in the combined presence of Reed-Sternberg cells, lymphocytes, plasma cells, eosinophils, and cells with lobulated nuclei. Histology will allow more accurate classification of the type of Hodgkin's disease. The large polylobulated nuclei or mirror image nuclei, prominent nucleoli, and often vacuolated cytoplasm characterize Reed-Sternberg cells (Tables 11.1, 11.2 and 11.3). Let us recall that similar cells have been found in infectious mononucleosis, malignant melanoma, and even epithelial neoplasms (Plates 11.2 and 11.3).

In non-Hodgkin's lymphomas, cytologic recognition of the malignant cells is not difficult if one is dealing with immature or transformed cells. It is more delicate with "mature" differentiated or resting lymphocytes which look like benign resting lymphocytes. In this case the monomorphous aspect and cellular richness of the smear are typical of lymphoma. Also, cells from chronic lymphocytic leukemia cannot be distinguished from well-differentiated lymphocytic lymphoma (small resting lymphocytes). Clinical data help to make the distinction (Plates 11.4–11.6).

The cleavage or wrinkling of the nuclei can be better appreciated in paraffin-block preparations than on smears. However, the nuclear size allows the cytologic distinction between large and small cells, and prominent nucleoli characterize the noncleaved cells.

Another important feature of lymphomas is the absence of cohesive cell groups.

Certain histochemical techniques may assist in the diagnosis. Cells of acute lymphocytic and myelogenous leukemias show cytoplasmic diastase-sensitive PAS granules. Nonspecific esterases are present in the cells of paraffin-embedded chronic myelogeneous leukemia.

Splenic puncture is another useful procedure for recognition of malignant lymphomas, myeloid metaplasia, and rare primary splenic neoplasms. This technique is used by hematologists.

Differential Diagnosis

Malignant lymphomas must be differentiated from other small-cell neoplasms. Among these are Wilms' tumor, neuroblastoma,[35] Ewing's sarcoma, embryonal rhabdomyosarcoma, and oat-cell carcinoma of the lung. Cell cohesiveness is absent in lymphomas.[56]

SALIVARY GLANDS

Needle aspiration of salivary gland lesions is a common clinical practice. Any palpable lesion can be punctured with a fine needle (22 gauge) with no major discomfort to the patient. Pathologic conditions of the salivary glands are frequent, and cytologic diagnosis can provide valuable information on nonneoplastic changes such as cysts and inflammations, as well as on benign and malignant neoplasms.

Salivary gland tumors are uncommon: the rate of incidence is around 1.5 per 100,000;[34] and the anatomic site favors their early recognition. Benign tumors are more frequent than malignant growths. Neoplasms of the salivary glands are listed in the WHO classification (Table 11.4). The mixed tumor is the most common benign tumor. Other benign tumors are mentioned in the table.

The more frequent malignant neoplasms are adenoid cystic carcinoma, adenocarcinoma, and epidermoid carcinoma. Mucoepidermoid tumors are locally aggressive but seldom metastasize.

Malignant lymphomas may involve the intraparotid lymphoid tissue. Metastases have been described in the salivary glands: we have personally seen cases of metastatic melanoma and breast carcinoma.

Cytology

In simple cysts the more or less abundant fluid will reveal the presence of degenerated salivary gland cells and typical macrophages with a finely vacuolated cytoplasm.

The mixed tumor (pleomorphic adenoma) is characterized by a mixture of epithelial and mesenchymal structures. The epithelial cells have a round or oval nucleus and a relatively abundant, homogeneous, PAS-positive cytoplasm. Mesen-

Table 11.4. WHO Classification of Histologic Typing of Salivary Glands Tumors

  I. Epithelial tumors
      A. adenomas
         1. pleomorphic adenomas (mixed tumor)
         2. monomorphic adenomas
            adenolymphoma
            oxyphilic adenoma
            other types
      B. mucoepidermoid tumor
      C. acinic cell tumor
      D. carcinomas
         1. adenoid cystic carcinomas
         2. adenocarcinoma
         3. epidermoid carcinoma
         4. undifferentiated carcinoma
         5. carcinoma in pleomorphic adenoma (malignant mixed tumor)
  II. Nonepithelial tumors
  III. Unclassified tumors
  IV. Allied conditions
      A. benign lymphoepithelial lesions
      B. sialosis
      C. oncocytosis

chymal structures often include elongated fibroblastic stromal cells, collagen fibers, and an amorphous, PAS-positive ground substance. This substance may show mucinous hyalin and myxoid or chondroid loci (Plates 11.32 and 11.33).

The absence of cellular anomalies is a valuable criterion which points to a benign growth. However, as we have pointed out, in this domain, the cellular criteria are not always reliable in the evaluation of malignancy. Let us mention two less frequent benign tumors: the presence of lymphocytic cells in large numbers as well as large cuboidal or polygonal oncocytes may suggest a diagnosis of cystadenolymphoma or Warthin's tumor (Plate 11.31).

The oncocytoma or oxyphilic granular cell adenoma occurs only rarely; it is characterized by the presence of large cells with round nuclei and abundant eosinophilic cytoplasm. The oncocyte granules are better observed with Giemsa stain.

Malignant epithelial tumors reveal cellular elements that exhibit the cytologic criteria of malignancy: large or irregular, hyperchromatic nuclei and piling up of elements in disorderly cell aggregates (Plates 11.34 and 11.35).

Great care should be taken when evaluating recurrent mixed tumors: the lesions may reveal highly atypical cells that are benign.

The presence of atypical cells and the absence of stromal elements point to a diagnosis of carcinoma.

THYROID

Fine-needle aspiration of the thyroid is facilitated by the anatomic accessibility of the gland. With the patient in the supine position, the gland is held with one hand, and the patient is advised against swallowing or speaking during the procedure.

Needle aspiration of the thyroid may provide tissue fragments for histologic examination or material for cytologic examination. This depends on the type of needle used: Silverman or Biegeleiser needles provide tissue specimens, and a fine needle gives cytologic material.

Thyroid fine needle puncture is harmless if correctly performed, and no major complication should be feared.[11, 20, 112] Small localized hematomas occasionally have been reported.

The major indication of thyroid cytology is the search for neoplastic cells but the method is applied to all conditions of gland enlargement (goiter). Enlargement may be caused by cellular hyperplasia, inflammatory processes, and distension of follicles by colloid and tumors. The development of goiter may be associated with functional disturbances (hyper- or hypothyroidism). Colloid goiter is a diffuse enlargement of the gland. In nodular goiter, functional and cyclic variations of the gland activity provoke development of cysts, hemorrhages with subsequent fibrosis, and calcifications.

Adenoma represents the development of a single nodule which may exhibit a follicular, trabecular, or papillary structure.

Chronic thyroiditis includes three main types: lymphocytic thyroiditis or Hashimoto's disease, granulomatous or de Quervain's thyroiditis, and ligneous thyroiditis or Riedel's struma. All these pathologic conditions will reveal various combinations of different histologic alterations which are: hyperplastic or hypoplastic modifications of the follicular structure; fibrosis; inflammatory infiltrates including lymphocytes, polymorphonuclear cells, histiocytes, and granulomatous lesions. At the cellular level, hyperactive, normal, or hypoactive modifications as well as inflammatory or degenerative processes will be encountered.

Malignant lesions of the gland include carcinomas, lymphomas, rare sarcomas, and metastatic tumors. The main types of carcinomas are the papillary (50 percent), the follicular (30 percent), and the undifferentiated forms (20 percent). Small-cell carcinoma is in most cases a lymphoma.

Cytology

Aspiration cytology of the thyroid shows follicular cells, various amounts of colloid material, inflammatory cells, and red blood cells (Plate 11.9).

Benign thyroid cells consist of regular, round elements with a pale cyanophilic, homogeneous cytoplasm. Nuclei are round and exhibit a finely dispersed, regular chromatin and a small nucleolus. The cytoplasm may be partially altered by vacuolation, or it is totally destroyed (naked nuclei).

Multinuclear cells and giant cells are present in granulomatous thyroiditis. Large, pale round cells with regular nuclei are Hurthle cells or oncocytic cells. Their cytoplasm may reveal the presence of eosinophilic granulations. Hemorrhage, inflammatory degeneration, and colloid may constitute the major part of the smears and thus make epithelial elements rare, poorly preserved, or even absent (Plate 11.10).

Thyroid neoplastic cells in most cases will show the classical structural modifications of malignancy, and the cytologic diagnosis can be made without any major difficulty. Moreover, malignant aspirations are usually rich in cells. However, a certain number of differentiated carcinomas of follicular or papillary type will show regular elements that are indistinguishable from benign atypical cells of goiter or adenoma. Such cases constitute limitations of the cytologic method (Plates 11.11–11.13).

In conclusion, malignant tumors of the thyroid gland should be suspected in the presence of definite cellular alterations of malignancy or of a large number of differentiated follicular cells that do not exhibit marked nuclear modification. The latter smears may come from well-differentiated carcinomas. Another indication for the method is the detection of metastatic thyroid tumors in lymph nodes.[78]

LUNG

Percutaneous needle biopsy of pulmonary lesions may be successfully performed and is recommended for the diagnosis of peripheral lung masses and to a lesser extent of diffuse lung diseases. The use of adequate needles under the visual guidance of image amplifier fluoroscopy and observation of correct cytologic techniques make the method successful in this domain.[22, 46, 70, 92, 100, 104, 113, 116]

The indications are the detection of primary or metastatic tumors that are inoperable for various reasons, identification of etiologic factors of infections, or unexplained lung diseases.

The main contraindications are pulmonary emphysema or hypertension and systemic bleeding tendency.

Cytology

The material obtained by aspiration with a syringe is smeared on slides and fixed; small fragments and remaining material are fixed in 80 percent alcohol, centrifuged, and submitted to paraffin embedding. Some material is saved for routine bacteria and fungus cultures in a tube of serum broth.

Neoplastic cells will show the characteristic features of malignancy, and sometimes, more specific alterations will permit identification of the tumor type. Examples are squamous cells of epidermoid carcinoma, columnar cells of adenocarcinoma, undifferentiated cells of large or small bronchogenic carcinoma, Reed-Sternberg cells in Hodgkin's disease, lymphocytic cells in lymphomas, and anaplastic sarcomatous cells in osteosarcoma, leiomyosarcoma, or fibrosarcoma (Plate 11.14).

Hyperplastic mesothelial cells as well as reparative or regenerative bronchiolar and alveolar cells may represent difficult problems of differential diagnosis. These cells may reveal hyperchromatic, slightly enlarged nuclei, and hypertrophic nucleoli.

Nonneoplastic lesions for example, granulomatous diseases such as tuberculosis, sarcoidosis, and histoplasmosis have been detected by authors who have a good deal of experience in this field. Etiologic factors have been revealed such as aspergillosis, amebiasis, and specific bacteria. Cases of pneumocystitis carinii and viral pneumonitis have been reported.[68]

Nonneoplastic smears may show inflammatory cells such as leukocytes, lymphocytes, histiocytes, giant cells, and modified bronchiolar and mesothelial cells. The presence of a fragment of tissue that can be submitted to paraffin embedding and histologic techniques helps to make a more accurate diagnosis than the one obtained by examination of isolated cells.

Rare complications encountered during the needle aspiration are pneumothorax and subcutaneous emphysema. Moreover, opponents of the method have claimed that unsatisfactory specimens often are obtained in diffuse lesions, thus creating problems of diagnosis.

LIVER

Fine needle aspiration may be performed to investigate the cytology of the liver paren-chyma; however, the use of special needles that provide combined histologic and cytologic specimens is preferable (Menghini, Silverman, or Terry needles).[24] Washing the needle in saline when a biopsy is performed may provide additional, isolated cells. We at times have detected neoplastic cells in the needle-washing fluid while the tissue fragment did not show metastases.

In our opinion, the fine needle technique has its most practical application in the detection of neoplastic cells. In the series reported by Carney,[15] the cytologic diagnosis of cancer was made in 69 percent of the cases by examining both the biopsy and the cytologic smears. However, false negatives are frequent because in contrast to laparoscopy technique, the puncture is performed blindly.

Inflammatory cellular alterations are much less evident with the fine-needle tech-nique because they are not specific enough. Such alterations are difficult to interpret even in histologic sections. For example, we doubt that the minimal changes of chronic aggressive hepatitis may be recognized on a smear.

Hepatocytes may show signs of degeneration or necrosis. Cytologically, hepatocytes are round, regular, large elements with central, round nuclei. Binucleation is ob-served. The cytoplasm may contain different pigments: lipofuchsin, hemosiderin, and bile pigment (Plate 11.15).

Neoplastic cellular modifications are similar to those described in epithelial cancer cells.

PANCREAS

Information on the cytology of the pancreas may be obtained in two different ways: collection of cells through duodenal aspiration or fine-needle puncture of the pancreas during an operation.

In the hand of the experienced cytopathologist, valuable results have been obtained in the detection and diagnosis of pancreatitis, pancreatic cysts, and adenocarcinoma (see Chapter 9).

BONE TUMOR ASPIRATION

Smears may be prepared from aspiration material obtained by needle biopsy of dif-ferent bone tumors.[19,38,49,80,82,107,118,119] The various bone sarcomas reveal pleomorphic cells exhibiting the typical criteria of malignancy. Experience allows specific recogni-tion of the different types of bone sarcoma. For example, typical malignant osteoblasts of osteosarcoma are large, pleomorphic-bizarre cells. Malignant chondrocytes of chondrosarcoma often are binucleated and the nuclei are densely hyperchromatic. Malignant fibroblasts of fibrosarcoma are spindle-shaped elements with hyper-chromatic, irregular or cigar-shaped nuclei. Tumor cells of Ewing's sarcoma are iden-tified as small elements with round hyperchromatic nuclei and ill-defined cytoplasmic borders (Plates 11.22–11.25).

Plasma cells of multiple myeloma exhibit varying degrees of differentiation. The "spoke wheel" pattern of the nucleus is recognized in most elements.

Multinuclear giant cells are encountered in different neoplastic conditions and therefore are not a specific morphologic criterion of a definite tumor.

TUMOR MASS ASPIRATION

The different primary or metastatic neoplasms that are suitable for the needle aspiration technique are so numerous that we will try to summarize the different cytologic images that may be encountered.[10,18,24,43,78,81,93,105,117]

In a large number of cases, the malignant characteristics of the cells will create no major diagnostic problem, but in some cases even a highly experienced pathologist will have difficulties in characterizing the exact histologic origin of tumors.

Epithelial and sarcomatous tumors, both metastatic and primary, will exhibit specific and nonspecific alterations. If specific changes are present, a more accurate diagnosis can be proposed. If nonspecific changes are dominant, a diagnosis of malignancy without further detail can be made. Therefore, the main advantage of needle aspiration is the rapid diagnosis of any unidentified tumor mass.[44]

The cytologic criteria of diagnosis will vary according to the anatomic site. As we said before, malignancy is usually cytologically diagnosed with acceptable accuracy. If any doubt remains, tissue sections must complement the provisional cytologic diagnosis (Plates 11.16–11.21).

Some cellular structures are very typical of anatomic sites or of lesions and thus may serve as sign posts of diagnosis, for example, osteoclasts, multinucleated histiocytes, oncocytes, the Hürthle cells of the thyroid, striated muscular cells of rhabdomyosarcoma, pigmented cells (melanin, hemosiderin, anthracotic pigment), mast cells, the PAS-positive granules (glycogen) of Ewing's sarcoma, and the plasma cells of myeloma (Plates 11.26 and 11.27).

Cellular cohesiveness is a constant feature of neoplasms. Epithelial neoplastic cells are usually smaller than sarcomatous ones. The main difficulty lies in the distinction between well-differentiated neoplasms with moderate cellular anomalies and hyperplastic, regenerative benign lesions. The possible existence of iatrogenic factors (radiation, chemotherapy) must always be kept in mind when such difficulties arise.

Finally, the constant comparison of cytology and histology is the best procedure to improve the accuracy of one's cytologic judgment.

EYE

Cytology specimens may be obtained from the conjunctiva (scraping) and from the anterior chamber of the eye (needle aspiration).[3, 17, 70, 84, 88, 89, 130] In acute chronic inflammatory diseases (conjunctivitis), the smear is rich in leukocytes; the columnar and goblet cells are increased in number and show inflammatory alterations. Mycotic infections will reveal the presence of spores or hyphae (Monilia, Aspergillus, Actinomyces), and intranuclear or intracytoplasmic inclusions will suggest a viral infection (trachoma, adenovirus infection, herpes).

The conjunctival mucosa is lined by a stratified columnar epithelium. The cornea is composed of a stratified squamous epithelium. The ocular chamber aspiration contains a few lymphocytes and histiocytes but no retinal cells (Plate 11.36).

The most frequently encountered malignant tumors of the eye are squamous cell carcinoma of the cornea or conjunctiva, malignant melanoma of the choroid membrane, and retinoblastoma of the retina. These types of malignant cells have been described previously.

CEREBROSPINAL FLUID CYTOLOGY

The cerebrospinal fluid space consists of two cavities: the ventricular system and the subarachnoid space. The latter is lined by mesenchymal or pial cells, while the ventricular cavities are lined by neuroectodermic elements, the ependymal and the choroid cells.

Normal cerebrospinal fluid may contain a few cellular elements.

Ependymal cells are columnar elements with or without a ciliated border. They are rare in normal conditions.

Choroid plexus cells are cuboidal cells with a central nucleus and a cyanophilic cytoplasm.

Polymorphonuclear leukocytes are only rarely present in normal conditions.

Lymphocytes measure from 10 to 15μ in diameter and have a round, dark nucleus. Lymphocytic cells consist of small lymphocytes ($6-10\mu$ in diameter) and of large activated lymphocytes or lymphoid cells ($10-20\mu$ in diameter). The latter have a dense basophilic cytoplasm as revealed by the methyl-green/pyronin or Giemsa staining procedure. The presence of these activated lymphoid cells is indicative of an antigenic stimulation. Their nucleus is paler and may contain one or two nucleoli.

Plasma cells are never present in normal cerebrospinal fluids; they are observed in various diseases and occasionally in tumors.

Monocytes and macrophages have a common origin in the bone marrow and they are cytogenetically identical. They possess a potential phagocytic activity. Monocytes or monocytoid cells have an eccentric, kidney-shaped, lobular or round nucleus and a rather abundant, basophilic cytoplasm. Nucleoli are not apparent. These cells are $20-30\mu$ in diameter. Activated monocytes are larger cells with a vacuolated cytoplasm; they are very similar to macrophages, but their cytoplasm does not contain visible phagocytic material. Their presence translates irritation of the mesenchymal elements and is indicative of nonspecific meningeal irritation. The storage of particles is apparent only in macrophages. They exhibit various stored materials: lipids, hemosiderin, leukocytes, and erythrocytes.

Inflammatory lesions (meningitis) will show an increase of lymphocytes or polymorphonuclear leukocytes as well as of meningeal cells.

Various nonneoplastic neurologic disorders may be accompanied by the presence of atypical cells in the fluid creating problems of differential diagnosis. Among these disorders are encephalitis, multiple sclerosis, subdural hematoma, and cerebrovascular accidents.

Mesothelial cells exhibit enlarged, lobulated, hyperchromatic nuclei simulating malignant changes. Histiocytic giant cells may be observed.

The detection of cancerous cells is one of the major indications for the method. Cerebral metastases, for example, malignant lymphomas and breast tumors, are more frequent and have a higher tendency to desquamate than do primary intracranial tumors (Plates 11.29 and 11.30).

Primary cerebral tumors will exhibit nuclear polymorphism, hyperchromasia, and cytoplasmic vacuolation.

Cells are isolated or desquamate in clumps. In laboratories that often deal with this type of cytology, more specific diagnosis can be made and different tumors such as medulloblastoma, astrocytoma, glioblastoma multiforme, oligodendroglioma, meningioma, and ependymoma can be recognized cytologically.[48]

Cell concentration may be performed by different methods: centrifugation, filtration, and sedimentation (see Chapter 13). These techniques have their own different qualities, as expressed in various authors' preferences.

The cerebrospinal fluid is centrifuged at 400–800g for 10 minutes at room temperature. Smears are prepared from the centrifugation band and fixed in alcohol. Serum albumin may be added (up to 10 percent of the volume). Smears are stained with Papanicolaou's or Shorr's stain or methyl-green/pyronine. This latter technique results in good details of cellular structure, including the distribution of nucleic acids.

In filtration techniques, using, for example, the cellulose or polycarbonate material, the cells are collected on a filter that is fixed in formalin and stained with the Papanicolaou's, or Shorr's, or hematoxylin and eosin stain. Cell deformation is usually pronounced making the cellular interpretation difficult.

The sedimentation method is based on the spontaneous tendency of cells to settle out of suspension. The cerebrospinal fluid is placed in an open-ended chamber that is applied to a glass slide. A centrally perforated filter paper is placed between the cylinder and the glass slide, and this allows improved sedimentation by suction.

We use the centrifugation technique with the cytocentrifuge. If the fluid is highly cellular, we use centrifugation with smear and paraffin embedding of the cellular band. Rapid and correct fixation is essential to prevent cellular autolysis.

BIBLIOGRAPHY

1. Aaronson A.G., Hajdu S.I., Melamed M.R. Spinal fluid cytology during chemotherapy of leukemia of the central nervous system in children. *Amer. J. Clin. Path.* 63:528, 1975.

2. Ahmed M.N., Feldman M., Seemayer T.A. Cytology of epitheloid sarcoma. *Acta Cytol.* 18:459, 1974.

3. Amsler M., Verrey F. De l'utilité pratique de ponction de la chambre antérieure. *Ophtalmologica* 105:144, 1943.

4. Balhuizen J.C., Bots T.A.M., Schaberg A. The value of cytology in the diagnosis of hypophyseal tumors. *Acta Cytol.* 18:370, 1974.

5. Baringer J.R. A simplified procedure for spinal fluid cytology. *Arch. Neurol.* 22:305, 1970.

6. Bartels P.H., Bahr G.F., Bellamy J.C., Bibbo M., Richards D.L., Wied G.L. A self-learning computer program for cell recognition. *Acta Cytol.* 14:486, 1970.

7. Bartels P.H., Bahr G.F., Bibbo M., Richards D.L., Sonek M.G., Wied G.L. Analysis of variance of the Papanicolaou staining reaction. *Acta Cytol.* 18:522, 1974.

8. Benson P.A. Psammoma bodies found in cervico-vaginal smears. *Acta Cytol.* 17:64, 1973.

9. Bercovici B., Diamant Y., Polishuk W.Z. A simplified evaluation of vaginal cytology in third trimester pregancy complications. *Acta Cytol.* 17:67, 1973.

10. Berg J.W., Robbins G.F. A late look at the safety of aspiration biopsy. *Cancer* 15:826, 1962.

11. Berger J., Yaneva M. Confrontation entre la cytologie et l'histologie en pathologie thyroidienne. *Presse Méd.* 72:2945, 1964.

12. Bibbo M., Bartels P.H., Bahr G.F., Taylor J., Wied G.L. Computer recognition of cell nuclei from the uterine cervix. *Acta Cytol.* 17:340, 1973.

13. Bognel C., Prade M., Weill S., Maillet M., Lemerle J. Etude cytologique du liquide céphalorachidien dans 24 cas de médulloblastomes. *Rev. Cytol. Clin.* 5:79, 1972.

14. Bots G.T.A., Went L.N., Schaberg A. Results of a sedimentation technique for cytology of cerebrospinal fluid. *Acta Cytol.* 8:234, 1964.

15. Carney C.N. Clinical cytology of the liver. *Acta Cytol.* 19:244, 1975.

16. Chang A.H., Ng A.B.P. The cellular manifestations of mycosis fungoides in cerebrospinal fluid. *Acta Cytol.* 19:148, 1976.

17. Charlin C. Erreurs de diagnostic clinique dans certaines tumeurs endoocculaires malignes. *Arch. Ophtal.* (Paris) 33:103, 1973.

18. Chu E., Hoye R. The clinician and the cytopathologist evaluate fine needle aspiration cytology. *Acta Cytol.* 17:413, 1973.

19. Coley B.L., Sharp G.S., Ellis E.B. Diagnosis of bone tumors by aspiration. *Amer. J. Surg.* 13:215, 1931.

20. Cornillot M., Cappelaere P. Les difficultés du diagnostic cytologique des affections thyroidiennes. *Rev. Cytol. Clin.* 3:9, 1970.

21. Cornillot M., Cappelaere P. Le diagnostic cytologique par ponction à l'aiguille fine des lésions tumorales et pseudo-tumorales des glandes salivaires. *Rev. Cytol. Clin.* 4:17, 1971.

22. Dahlgren S., Nordenstrom B. *Transthoracic Needle Biopsy.* Almquist and Wiksell, Stockholm, 1966.

23. Dance E.F., Fullmer C.D. Extrauterine carcinoma cells observed in cervico-vaginal smears. *Acta Cytol.* 14:187, 1970.

24. Deeley T.J. *Needle Biopsy* Butterworths, London, 1974.

25. De Maertelaere Laurent E., Hupin J., Denolin-Reubens R. Procédé nouveau pour l'examen cytologique du liquide céphalo-rachidien. *Acta Clin. Belg.* 24:165, 1969.

26. Den Hartog Jager W.A. Cytopathology of the cerebrospinal fluid with the sedimentation technique after Sayk. *J. Neurol. Sci.* 9:155, 1969.

27. Dorfman R.F. A newly proposed classification of the non-Hodgkin's lymphomas. *Lancet* 1:1295, 1974.

28. Dyken P.R. Cerebrospinal fluid cytology practical clinical usefulness. *Neurology* 25:210, 1975.

29. Ehrmann R.L., Younge P.A., Lerch V.L. The exfoliative cytology and histogenesis of an early primary malignant melanoma of the vagina. *Acta Cytol.* 6:245, 1962.

30. Eneroth C.M., Franzen S., Zajicek J. Aspiration biopsy of salivary gland tumors. A critical review of 910 biopsies. *Acta Cytol.* 11:470, 1967.

31. Eneroth C.M.. Zajicek J. Aspiration biopsy of salivary gland tumors: II. Morphologic studies on smears and histologic sections from onocytic tumors (45 cases of papillary cystadenoma lymphomatosum and 4 cases of oncocytoma). *Acta Cytol.* 9:355, 1965.

32. Eneroth C.M., Zajicek J. Aspiration biopsy of salivary gland tumors: III. Morphologic studies on smears and histologic sections from 368 mixed tumors. *Acta Cytol.* 10:440, 1966.

33. Eneroth C.M., Zajicek J. Aspiration biopsy of salivary gland tumors: IV. Morphologic studies on smears and histologic sections from 45 cases of adenoid cystic carcinoma *Acta Cytol.* 13:59, 1969.

34. Evans W.R., Cruickshank A.H. Epithelial tumors of the salivary glands. Saunders, Philadelphia, 1970.

35. Farr G.H., Hajdu S.I. Exfoliative cytology of metastatic neuroblastoma. *Acta Cytol.* 16:203, 1972.

36. Fentanes De Torres E., Benitez-Bribiesca L. Cytologic detection of vaginal parasitosis. *Acta Cytol.* 17:252, 1973.

37. Fisher E.R., Park E.J., Wechsler H.L. Histologic identification of malignant lymphoma. *Amer. J. Clin. Path.* 65:149, 1976.

38. Forest M., Postel M., Abelanet R., Daudet-Monsac M. La cytologie sur empreintes en pathologie osseuse. *Rev. Cytol. Clin.* 7:23, 1974.

39. Frable W.J. Thin needle aspiration biopsy. A personal experience with 469 cases. *Amer. J. Clin. Path.* 65:168, 1976.

40. Frable W.J., Smith J.H., Perkins J., Foley C. Vaginal cuff cytology. Some difficult diagnostic problems. *Acta Cytol.* 17:125, 1973.

41. Fullmer C.D., Morris R.P. Primary cytodiagnosis of unsuspected mediastinal Hodgkin's disease. *Acta Cytol.* 16:77, 1972.

42. Garelly E. Cyto-diagnostic des tumeurs cutanées. *Rev. Cytol. Clin.* 2:28, 1969.

43. Godwin J.T. Aspiration biopsy: technic and application. *Ann. N.Y. Acad. Sci.* 63:1348, 1956.

44. Godwin J.T. Rapid cytologic diagnosis of surgical specimens. *Acta Cytol.* 20:111, 1976.

45. Graham R.M., Nuovo V.M. Cytology of uterine sarcoma and chorioepithelioma. *Acta Cytol.* 3:555, 1958.

46. Grunze H. Cytologic diagnosis of tumors of the chest. *Acta Cytol.* 17:148, 1973.

47. Guthrie C.G. Gland puncture as a diagnostic measure. *Bull. Johns Hopkins Hosp.* 32:266, 1921.

48. Hajdu S.I. Hajdu E.O. *Cytopathology of Sarcomas and Other Nonepithelial Malignant Tumors.* Saunders, Philadelphia, 1976.

49. Hajdu S.I., Melamed M.R. Needle biopsy of primary malignant bone tumors. *Surg. Gynec. Obstet.* 133:829, 1971.

50. Hajdu S.I., Melamed M.R. The diagnosis value of aspiration smears. *Amer. J. Clin. Path.* 59:350, 1973.

51. Hajdu S.I., Nolan M.A. Exfoliative cytology of malignant germ cell tumors. *Acta Cytol.* 19:255, 1975.

52. Hajdu S.I., Savino A. Cytologic diagnosis of malignant melanoma. *Acta Cytol.* 17:320, 1973.

53. Hansen H.H., Bender R.A., Shelton B.J. The cyto-centrifuge and cerebrospinal fluid cytology. *Acta Cytol.* 18:259, 1974.

54. Harrison V., Peat G. Fetal growth in relation to vaginal cytology. *Acta Cytol.* 18:210, 1974.

55. Hollander D.H., Gupta P.K. Detached ciliary tufts in cervico-vaginal smears. *Acta Cytol.* 18:367, 1974.

56. Janota I. Malignant lymphoma cells in the cerebrospinal fluid. *Lancet* 2:677, 1964.

57. Johnson W.D. The cytologic diagnosis of cancer in serous effusions. *Acta Cytol.* 10:161, 1966.

58. Johnson W.D., Koss L.G., Papanicolaou G.N., Seybolt J.F. Cytology of esophageal washings. *Cancer* 8:951, 1955.

59. Johnston W.W., Ginn F.L., Amatulli J.M. Light and electron microscopic observations on malignant cells in cerebrospinal fluid from metastatic alveolar cell carcinoma. *Acta Cytol.* 15:365, 1971.

60. Kaltenbach F.J., Hillemanns H.G., Fettig O., Hilgarth M. Thrombin cell block technic in gynecologic cytodiagnosis. *Acta Cytol.* 17:128, 1973.

61. Kersey J.H. Gajl-Peczalska K.J. T and B lymphocytes in humans. A review. *Amer. J. Path.* 81, 446, 1975.

62. King E.B., Russel W.M. Needle aspiration biopsy of lung-techniques and cytologic morphology. *Acta Cytol.* 11:319, 1967.

63. Kline T.S. Cytological examination of the cerebrospinal fluid. *Cancer* 15:591, 1962.

64. Kline T.S., Neal H.S. Needle biopsy: A pilot study. *JAMA* 224:1143, 1973.

65. Koivuniemi A., Tyrkko J. Seminal vesicle epithelium in fine-needle aspiration biopsies of the prostate as a pitfall in the cytologic diagnosis of carcinoma. *Acta Cytol.* 20:116, 1976.

66. Kolar O., Zeman W. Spinal fluid cytomorphology description of apparatus, technique and findings. *Arch. Neurol.* 18:44, 1968.

67. Lennert K. Follicular lymphoma. A tumor of the germinal centers. In *Malignant Diseases of the Hematopoietic System,* Gann Monograph in Cancer Research 15. University Tokyo Press, Tokyo 1973, p.217.

68. Lillehei J.P., et al. Pneumocystis carinii pneumonia. *JAMA* 206:596, 1968.

69. Linsk J.A., et al. Diagnosis of intrathoracic tumors by thin needle cytologic aspiration. *Amer. J. Med. Sci.* 263:181, 1972.

70. Liotet S., Rouchy J.P. La cytologie conjonctivale. *Rev. Cytol. Clin.* 6:9, 1973.

71. Lucas P.F. Lymph node smears in diagnosis of lymphadenopathy; review. *Blood* 10:1030, 1955.

72. Lukes R.J., Collins R.D. Immunologic characterization of human malignant lymphomas. *Cancer* 34 (Suppl.): 1448, 1974.

73. Lundquist A. Fine needle aspiration biopsy cyto-diagnosis of malignant tumor in the liver. *Acta Med. Scand.* 188, 465, 1970.

74. McCormack L.J. Exfoliative cytology of cerebro-spinal fluid. *C.A. Bull. Cancer Progr.* 10:165, 1960.

75. McCormack L.J., Coleman A.S. A membrane filter technic for cytology of spinal fluid. *Amer. J. Clin. Path.* 38:191, 1962.

76. McGarry P., Holmquist N.D., Carmel S.A. A post-mortem study of cerebrospinal fluid with histologic correlation. *Acta Cytol.* 13:48, 1969.

77. McMenemey W.H., Cumings J.N. The value of the examination of the cerebrospinal fluid in the diagnosis of intracranial tumors. *J. Clin. Path.* 12:400, 1959.

78. Mallarme J., Jagueux M., Orcel L. Diagnostic cytologique des métastases cervicales des épithéliomas thyroidiens. *Rev. Cytol. Clin.* 5:9, 1972.

79. Marsan C., Edelstein G. Lougnon J. Intérêt du cyto-diagnostic par ponction ganglionnaire chez les malades porteurs de mélanomes malins soumis à une immunothérapie. *Rev. Cytol. Clin.* 8:23, 1975.

80. Marsan C., Sicard A. Cyto-diagnostic des tumeurs des os. Technique des prélèvements. *Arch. Anat. Path.* 11:66, 1963.

81. Martin H.E., Ellis E. Biopsy of needle puncture and aspiration. *Ann. Surg.* 92:169, 1930.

82. Meisels A., Berebichez M. Exfoliative cytology in orthopedics. *Can. Med. Ass. J.* 84:957, 1961.

83. Melamed M.R. The cytological presentation of malignant lymphomas and related diseases in effusions. *Cancer* 16:413, 1963.

84. Michiels J. Dernouchamps J.P. Les modifications de l'humeur aqueuse en cas d'inflammation oculaire. *Arch. Ophtal.* (Paris) 33:145, 1973.

85. Moore R.D., Reagan J.W. A cellular study of lymph-node imprints *Cancer* 6:606, 1953.

86. Moracci E., Berlingieri D. Hormonal evaluation of vaginal smears from artificial vagina. *Acta Cytol.* 17:131, 1973.

87. Mouriquand J., Dargent M. L'empreinte mammaire; étude cytopathologique. *Bull. Cancer* 44:449, 1957.

88. Naib Z.M. Cytology of TRIC agent infection of the eye of newborn infants and their mother's genital tracts. *Acta Cytol.* 14:390, 1970.

89. Naib M. Cytology of ocular lesions. *Acta Cytol.* 16:178, 1972.

90. Naylor B. An exfoliative cytologic study of intra-cranial fluids. *Neurology* 11:560, 1961.

91. Naylor B. The cytologic diagnosis of cerebrospinal fluid. *Acta Cytol.* 8:141, 1964.

92. Nordenstrom B. A new technique for trans-thoracic biopsy of lung changes. *Brit. J. Radiol.* 38:550, 1965.

93. Norquist L.A., Schweid A.I. Preparation of cytologic smears from solid tumors. *Acta Cytol.* 18:459, 1974.

94. Persson P.S. Ettergren L. Cytologic diagnosis of salivary gland tumors by aspiration biopsy. *Acta Cytol.* 17:351, 1973.

95. Pickren J.W., Burke E.M. Adjuvant cytology to frozen sections. *Acta Cytol.* 7:164, 1963.

96. Pierson B. Cytologic diagnosis of central nervous system tumors *Trans. First Intern. Cancer Cytol. Congress Brussels,* 1955.

97. Platt W.R. Exfoliative cell diagnosis of central nervous system lesions. *Arch. Neurol. Psych.* 66:119, 1951.

98. Poulsen R.S. *Automated Prescreening of Cervical Cytology Specimens.* Ph. D. dissertation, Mc Gill University, Montreal, Quebec, March 1973.

99. Pundel J.P. Précis de colposcopie hormonale. Masson, Paris, 1966.

100. Ramzy I. Pulmonary hemartomas: cytologic appearances of fine needle aspiration biopsy. *Acta Cytol.* 20:15, 1976.

101. Rappaport H. Tumors of the hemopoietic system. *Atlas of Tumor Pathology.* Fascicle 8, Armed Forces Institute of Pathology, Washington D.C. 1966.

102. Rich J.R. A membrane filter technique of cerebrospinal fluid cytology. *J. Neuro. Surg.* 37:661, 1972.

103. Sakai Y., Lauslahti K. Comparison and analysis of the results of cytodiagnosis and frozen sections during operation. *Acta Cytol.* 13:359, 1969.

104. Sargent E.N., Turner A.F., Gordonson J., Schwinn C.P. Pashky O. Percutaneous pulmonary needle biopsy. Report of 350 patients. *Amer. J. Roentgenol.* 122:758, 1974.

105. Sayago C. Aspiration and surgical biopsy. *Amer. J. Roentgenol.* 48:78, 1948.

106. Sayk J. Die klinische Bedentung der Liquordiagnostik. *Dtsch. Arztebl.* 20:1476, 1974.

107. Schajowicz F. Aspiration biopsy in bone lesions. *J. Bone Joint. Surg.* 37:465, 1955.

108. Schour L., Chu E.W. Fine needle aspiration in the management of patients with neoplastic diseases. *Acta Cytol.* 18:472, 1974.

109. Schwinn C.P., et al. Cytopathology of percutaneous pulmonary needle aspiration biopsy. *Compendium on Diagnostic Cytology.* Editors' Tutorials of Cytology Chicago 1976, p.517.

110. Sharma S.D., Zeigler O., Trussel R.R. A cytologic study of *Dipetolenema perstans* in cervical smears. *Acta Cytol.* 15:479, 1971.

111. Sherlock P., Kim Y.S., Koss L.G. Cytologic diagnosis of cancer from aspirated material obtained at liver biopsy. *Amer. J. Dig. Dis.* 12:396, 1967.

112. Smejkal V., Soumar J., Smejkalova E. Le diagnostic cytologique par ponctions biopsie des cancers de la thyroïde. *Rev. Cytol. Clin.* 3:9, 1970.

113. Smith W.G. Needle biopsy of lung: with special reference to diffuse lung disease and use of a new needle. *Thorax* 19:68, 1964.

114. Spriggs A.I., Boddington M.M. Leukaemic cells in cerebrospinal fluid. *Brit. J. Haemat.* 5:83, 1959.

115. Stahle A.R., Steukvist B. The imprint method for the cyto-diagnosis of lymphadenopathies and of tumors of the head and neck. *Acta Cytol.* 15:123, 1971.

116. Stein H.L., Evans J.A. Percutaneous trans-thoracic lung biopsy utilizing image amplification. *Radiology* 87:350, 1966.

117. Stewart F.W. The diagnosis of tumors by aspiration. *Amer. J. Path.* 9 (Suppl.): 801, 1933.

118. Stormby N., Akerman M. Cytodiagnosis of bone lesions by means of fine-needle aspiration biopsy. *Acta Cytol.* 17:166, 1973.

119. Stormby N. Aspiration cytology of the pancreas. Compendium on diagnostic cytology. Editor's Tutorials of Cytology. Chicago 1976, p. 574.

120. Strum S.P., Jung K.P., Rappaport H. Observations of cells resembling Sternberg-Reed cells in conditions other than Hodgkin's disease. *Cancer* 26:176, 1970.

121. Suen K.C., Yermakov V., Raudales O. The use of imprint technic for rapid diagnosis in post-mortem examinations. *Amer. J. Clin. Path.* 65:291, 1976.

122. Sultan C. Cytodiagnosis of erythroleukaemia. *Bull. Cancer* Vol. 61, No. 3, July–Sept. 1974.

123. Taylor J., Bartels P.H., Bibbo M., Bahr G.F., Wied G.L. Implememtation of a hierarchical cell classification procedure. *Acta Cytol.* 18:515, 1974.

124. Tindle B.H., Parker J.W., Lukes R.J. "Reed-Sternbery Cells" in infections mononucleosis? *Amer. J. Clin. Path.* 58:607, 1972.

125. Tourtellotte W.W. Cerebrospinal fluid and its reactions in diseases. In. J. Minckler (Ed.): Pathology of the Nervous System. Vol. I. McGraw-Hill, New York, 1968.

126. Trummer M.J., Doohen D.J., Timmes J.J. Open lung biopsy. *Surgery* 53:443, 1963.

127. Ultmann J.E., Koprowska I., Engle R.L. A cytological study of lymph node imprints. *Cancer* 11:507, 1958.

128. Verry F., Defour R. Quelques aspects cytologiques de l'humeur aqueuse dans les uvéités. *Ophtalmologica* 107:98, 1944.

129. Vooijs P.G., Ng A.B.P., Wentz W.B. The detection of vaginal adenosis and clear cell carcinoma. *Acta Cytol.* 17:59, 1973.

130. Walter J.R., Naylor B. A membrane filter method used to diagnose intraocular tumor. *J. Pediat. Ophtal.* 5:36, 1968.

131. Wertlake P.T., Markovits B.A., Stellar S. Cytologic evaluation of the cerebrospinal fluid with clinical and histologic correlation. *Acta Cytol.* 16:224, 1972.

132. Wilsson, The effects of an intrauterine devices on the histologic pattern of the uterine mucosa. *Amer. J. Obstet. Gynec.* 93:802, 1965.

133. Woyke S., Olszewski W., Domagala W., Marzecki Z. Cytodiagnosis of acinic cell carcinoma. Ultrastructural study of material obtained by fine needle aspiration biopsy. *Acta Cytol.* 19:110, 1975.

134. Yamada T., Itou U., Watanabe Y., Ohashi S. Cytologic diagnosis of malignant melanoma. *Acta Cytol.* 16:70, 1972.

135. Zajicek J. *Aspiration Biopsy Cytology. Part I. Cytology of Supradiaphragmatic Organs.* Karger, Basel, 1974.

136. Zajicek J., Eneroth C.M. Cytological diagnosis of salivary gland carcinoma from aspiration biopsy smears. *Acta Otolaryng.* (Stockholm) Suppl. 263. 1970, p.183.

12
Automation

The main purposes of automation are to recognize, to classify, to compare, and to decompose collected information. Automation may be applied to the following aspects of clinical cytology:

- automation of fixation and staining procedures
- automation of cellular identification and interpretation of cellular anomalies
- storage of statistical data

The practical use of such techniques, if available, considerably improves the efficiency of the method and increases the number of specimens analyzed every year, at the same time requiring a smaller number of cytotechnicians.

There are various devices on the market that assure automation of fixation and staining. In our experience, these machines are of some help when the laboratory handles a large number (more than 50,000 slides per year) of routine smears that do not require special techniques.

The most time-consuming procedures in slide preparation are administrative steps (specimen registration and numbering of slides and report sheets), and the covering of slides with the mounting medium and the cover slip. These two steps unfortunately cannot be performed by most automated systems. Only one machine on the market includes automatic covering of slides with diatex solution.

Cytologic diagnosis of cancer and allied dysplasias and precancerous lesions by automated cytology is a more complex matter. A machine that can differentiate malignant cells from normal ones in a reasonable amount of time and with acceptable error rates has not yet been produced. The hope that such a machine will one day become a reality should not prevent public health authorities from providing adequate human and laboratory facilities to deal with the clinical demands in the field of cytology. For the present, an able, well-trained cytotechnician cannot be replaced by a highly sophisticated instrument.

However, automation may have valuable applications, such as cell research, identification of biologic features that cannot be visualized by routine microscopy, permanent registration of cellular data observed by different cytopathologists, teaching, international classification of cellular anomalies, analysis of cellular particularities by different techniques such as histochemistry, enzyme evaluation, and so forth. The accumulation of these data and the computer analysis of the multiple parameters involved should help to improve our knowledge of the cell structure and behavior.

Different techniques are used to gather information on cell structure, which is then fed to the computer. We will only briefly mention these techniques and refer the reader to the bibliography for a more extensive approach to the problem.

In so-called static systems, the morphologic parameters of single cells are registered by the microscope and analyzed by the computer. Various cell features are studied by these methods, including size of cellular components, color of structures obtained by different staining procedures, and cellular contour evaluation. This information is translated into a two-dimensional image, stored in the computer, and compared to the previously acquired data.

The taxonomonic intracellular analytic system (TICAS) is an example of such static systems which permits cell identification by scanning microphotometry (measurement of cell optical density).[30-31] The fluorescence slit-can method[29] entails computer recording and analysis of cell contours and nuclear and cytoplasmic fluorescence. The latter parameter is indicative of cell density.

In so-called dynamic systems, cell analysis is performed by spectrophotometric analysis of a great number of elements. The cells are rapidly moved in single file through an appropriately small capillary tube; their fluorescence and ultraviolet absorption are the most frequently measured parameters, which are translated into electrical pulses, transferred to a signal analyzer, and finally to the computer.

Each of these methods presents technical difficulties that make them impractical for routine work and which can create "false alarms." Incorrect classification of cells and problems of differential diagnosis may be due to cell overlapping, inefficient isolation of inflammatory cells, cellular damage provoked by dispersion techniques, presence of dense objects such as a nucleolus which can be mistaken for the nucleus itself, and loss of the notion of diversity of cell population also due to the use of cell dispersion techniques.

The storage of statistical data both of administrative and medical nature may be achieved in different ways depending on the amount of information to be stored, the frequency of use of such information, and the budget available.

The simplest systems use graphs, tables, sheets, or cards. More elaborate systems of cards will allow sorting of data. Edge-punched cards and feature cards can be stored manually, but their use is limited to a few hundred cases. With edge-punched cards, sorting of the cards is obtained by passing a rod through the appropriate holes. Slotted cards will drop out. Each feature card represents a different bit of information. The punching of the appropriate hole will correspond to a definite number indicating, for example, the patient's identification or any other data. The cards are held up to the light and when different punch holes coincide, the light will shine through.

Machine sorted cards store a large amount of information (up to 960 holes or magnetic dots) and offer an unlimited number of possibilities and programs. Punching and filing are mechanically performed.

The information stored on cards may be submitted to computer analysis and the results are provided in a printed form. More elaborate systems use sophisticated computers. For the purposes of the average laboratory, any system adopted must meet these minimum requirements: numerical identification of specimen and patient, alphabetical classification of patients' names and diseases, diagnosis, and follow-up data.

The choice of the system should be motivated by the needs of the laboratory and not by the desire to appear up to date. As we said before the amount of stored information, the frequency of use of the information, and the ultimate goals of the laboratory (clinical follow-up, epidemiology, teaching and research) will direct the choice of system.

Numerical coding is better because information can be transferred from simple cards to mechanical punch cards and to the computer. Standardized programs such as the SNOP and theWHO nomenclatures are available and obviate the necessity for the individual to write an entire program.

Whatever the system adopted, great care should be taken to provide adequate, reliable, and accurate input information.

BIBLIOGRAPHY

1. Arnould L. Pap ou pas Pap? *Rev. Assoc. Belge Tech. Lab.* 2:5, 1975.

2. Bahr G.F., Bartels P.H., Bibbo M., De Nicolas M., Wied G.L. Evaluation of the Papanicolaou stain for computer assisted cellular pattern recognition. *Acta Cytol.* 17:106, 1973.

3. Bartels P.H., Bahr G.F., Bellamy J.C., Bibbo M., Richards D.L., Wied G.L. A self-learning computer program for cell recognition. *Acta Cytol.* 14:486, 1970.

4. Bartels P.H., Bahr G.F., Bibbo M., Wied G.L. Objective cell image analysis. *J. Histochem. Cytochem.* 20:239, 1972.

5. Bartels P.H., Bahr G.F., Calhoun D., Wied G.L. Cell recognition by neighborhood grouping techniques in TICAS. *Acta Cytol.* 14:313, 1970.

6. Bartels P.H., Bahr G.F., Jeter W.S., Olson G.B., Taylor J., Wied G.L. Evaluation of correlational information in digitized cell images. *J. Histochem. Cytochem.* 22:69, 1974.

7. Bartels P.H., Bhattacharya P.K., Bellamy J.C., Bahr G.F., Bibbo M., Wied G.L. Computer generated, synthetic cell images. *Acta Cytol.* 18:155, 1974.

8. Bartels P.H., Bibbo M., Bahr G.F., Taylor J., Wied G.L. Cervical cytology: descriptive statistics for nuclei of normal atypical types. *Acta Cytol.* 17:449, 1973.

9. Bartels P.H., Bibbo M., Bahr G.F., Wied G.L. Cervical cytology: descriptive statistics for nuclei of normal and atypical cell types. *Acta Cytol.* 17:449, 1973.

10. Bartels P.H., Bibbo M., Taylor J. Wied G.L. Cell recognition from the statistical dependence of gray values in digitized images. *Acta Cytol.* 18:165, 1974.

11. Bartels P.H., Olson G.B., Jeter W.S., Wied G.L. Evaluation of unsupervised learning algorithms in the computer analysis of lymphocytes. *Acta Cytol.* 18:376, 1974.

12. Bartels P.H., Wied G.L. Performance testing for automated prescreening devices in cervical cytology. *J. Histochem. Cytochem.* 22:660, 1974.

13. Bibbo M., Bartels P.H. Bahr G.F. Ng A.B.P., Reagan J.W., Richards D.L., Wied G.L. Data bank for endometrial cells. Operation of the TICAS file project. *Acta Cytol.* 14:574, 1970.

14. Bibbo M., Bartels P.H., Bahr G.F., Taylor J., Wied G.L. Computer recognition of cell nuclei from the uterine cervix. *Acta Cytol.* 17:340, 1973.

15. Cambier M.A., Wheeless L.L., Patten S.F. False alarms: current obstacle to cytopathology automation (abstract). *Acta Cytol.* 20:586, 1976.

16. Collins D.N., Kaufman W. New York State computerized proficiency testing program in exfoliative cytology: development. *Acta Cytol.* 15:34, 1971.

17. Collins D.N., Kaufmann W., Albrecht R. New York State computerized proficiency testing program in exfoliative cytology: evaluation. *Acta Cytol.* 15:468, 1971.

18. Dembitzer H.M., Herz F., Schreiber K., Wolley R.C., Koss L.G. The fine structure of exfoliated cervical and vaginal cells. *Acta Cytol.* 20:243, 1976.

19. Husain O.A.N. Special consideration of multiple parameters. In *Automated Cytology.* A symposium by correspondence. *Acta Cytol.* 15:254, 1971.

20. Jahoda E., Bartels P.H., Bibbo M., Bahr G.F., Holzner H.H., Wied G.L. Computer discrimination of cells in serous effusions. I. Pleural fluids. *Acta Cytol.* 17:95, 1973.

21. Kirland J.A., Parfitt J., Sag T. A computer program for cytology records. *Acta Cytol.* 13:7, 1969.

22. Koss L.G., Bartels P.H., Bibbo M., Freed S.Z., Taylor J., Wied G.L. Computer discriminition between benign and malignant urothelial cells. *Acta Cytol.* 19:378, 1975.

23. Louis C., Poulsen R., Marshall K.G., De Johnston R. The occurrence of isolated dysplastic carcinoma in situ and invasive type cells in cervical smears from patients with invasive squamous cell carcinoma: significance to prescreening using image processing techniques. *Acta Cytol.* 20:158, 1976.

24. Megla G.K. The LARC automatic white blood cell analyzer. *Acta Cytol.* 17:3, 1973.

25. Nadel E.M. Computer analysis of cytophotometric fields by CYDAC and its historical evolution form the cytoanalyzer. *Acta Cytol.* 9:203, 1965.

26. Olson G.B., Wied G.L., Bartels P.H. Differentiation of lymphoid tissue by analysis of digitized images. *Acta Cytol.* 17:89, 1973.

27. Pelzer A., Tibaux G., Chef R. Possibilités de l'automation dans un laboratoire de cytologie clinique. *Rev. Cytol. Clin.* 3:25, 1970.

28. *Systematized Nomenclature of Pathology.* College of American Pathologists, Chicago, 1969.

29. Wheeless L.L., Patten S.F. Split-scan cytofluorometry: basis for an automated cytopathology prescreening system. *Acta Cytol.* 17:391, 1973.

30. Wied G.L., Bahr G.F., Bartels P.H. Automatic analysis of cell images by TICAS. in Wied G.L., Bahr G.F. (Eds), *Automated Cell Identification and Cell Sorting.* Academic Press, New York, 1970, pp.195.

31. Wied G.L., Bartels P.H., Bahr G.G., Oldfield D.G. Taxonomic intracellular analytic system (TICAS) for cell identification. *Acta Cytol.* 12:180, 1968.

32. Wied G.L., Bartels P.H., Bahr G.F., Reagan J.W. TICAS assessment of cells from atypical hyperplasia of the endometrium. *Acta Cytol.* 13:552, 1969.

33. Wied G.L. Bibbo M., Bahr G.F., Bartels P.H. Computer recognition of uterine glandular cells. II. Application of a self-learning recognition program. *Acta Cytol.* 13:662, 1969. I. *Acta Cytol.* 13:611, 1969.

34. Wied G.L., Bibbo M., Bahr G.F., Bartels P.H. Computerized microdissection of cell images. *Acta Cytol.* 14:418, 1970.

35. Zajicek J., Bartels P.H., Bahr G.F., Bibbo M., Wied G.L. Computer analysis of lymphocytes from cases with lymphadenitis and lymphocytic lymphoma. *Acta Cytol.* 16:284, 1972.

13
Techniques
of Clinical Cytology

SAMPLING

The collection, fixation, and staining of specimens are extremely important preliminaries in the cytologic method and as such will be considered at some length.

Because sampling techniques vary with the organ and lesion in question, they are discussed in each chapter as the need arises. Here, we need mention only one principle: whatever the method used, enough cellular material must be obtained to ensure that the specimen is indeed representative of the lesion in question.

FIXATION

Once a cytologic sample is obtained, it must be rapidly fixed. Fixation may be performed immediately on the slide, on a centrifugation band, or on a fluid sample.

Cellular fixation implies the coagulation or insolubilization of cell proteins. This is accomplished by a greater or lesser degree of protein polymerization, a process which arrests the dynamic enzymatic reactions of living cells. Fixation therefore incurs rather dramatic changes; it is nonetheless essential that fixation techniques respect as much as possible the original cellular structure and that of its chemical constituents.[24]

Since the nineteenth century, innumerable fixative agents have been devised and utilized. For example, Gray[26] reports the existence of 700 different mixtures, but even this list is incomplete. Experience and routine have consecrated the use of certain agents whose qualities prove sufficient for many purposes. There is no such thing, however, as an ideal fixative agent, since they all modify—to greater or lesser degrees—the structure of the living cell.[89]

A fixative agent must be relatively nontoxic, volatile, and reasonably priced. For these reasons, cytologists prefer formaldehyde or alcohol-based agents (methanol, ethanol, or propanol). One widely used mixture is that proposed by Papanicolaou: 95 percent ethyl alcohol and ether in equal amounts. For safety reasons, however, many laboratories have discontinued the use of an ether mixture, since it is too inflammable. The 95 percent ethyl alcohol may be replaced by isopropyl alcohol. The addition of acetic acid—up to 20 percent—to alcohol-ether mixtures improves fixation of nuclei but has the disadvantage of modifying the pH.

Carnoy's liquid,* a good nuclear fixative, hemolyzes red blood cells, a fact that can be quite advantageous when one is working with certain hemorrhagic fluids.

Whatever the nature of the specimen, the optimum period of fixation is 15 minutes. Cells may remain in the fixative bath for several days or even several weeks without appreciable denaturation, as long as the slide remains completely immersed.

Fixation by room-temperature desication is frequently practiced in hematology laboratories, but this method should be avoided in other cytologic disciplines; it results in incomplete denaturation of cell proteins, which are therfore still susceptible to partial solubilization in subsequently used aqueous solutions. At any rate, in hematologic procedures, fixation is completed by treatment with the methyl alcohol contained in panoptic stains (Giemsa, May Grunwald).

FIXATION AND CONCENTRATION
OF FLUID SPECIMENS

In fluid specimens such as sputum, urine, mesothelial cavity fluids, cerebrospinal fluid, and bronchial and gastric lavages, cells are diluted and as such must be concentrated. Such concentration may be effected by centrifugation or the use of filtering membranes (Figure 13.1).

CENTRIFUGATION

Centrifugation may be performed on variable quantities of fluid (between a few and several hundred milliliters). When the quantity of fluid is particularly abundant, as in ascites, only a portion of the sample may be used.

If a specimen is fresh, centrifugation may precede fixation. The centrifugation band is then recovered, spread on slides as usual, and immediately fixed. However, if centrifugation cannot be performed rapidly, the fluid specimen should be mixed in

*Carnoy's solution: absolute alcohol, 60.0ml; chloroform, 30.0ml; glacial acetic acid, 10.0ml.

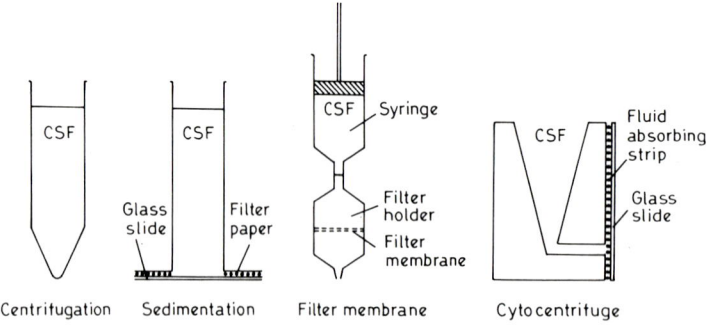

Figure 13.1. Schematic representation of cell centrifugation sedimentation and filtration techniques.

equal volumes with a 50 to 70 percent denatured ethyl alcohol solution or with formaldehyde. Use of a higher percentage alcohol solution is not desirable since this tends to harden the subsequent centrifugation band and makes smearing difficult.

Centrifugation should proceed for 10 minutes at about 1800 rpm (450g) and each tube may contain from 15–50ml.

If a specimen is extremely poor in cells, the resulting band may be so discrete as to be hardly separable. In these cases the cytocentrifuge proves a most useful tool. In this apparatus the centrifugation of cells and their application to slides are effected simultaneously.[60]

A portion of the microscope slide itself constitutes a wall of the centrifugation chamber. The cells are thus directly transferred to the glass, while the fluid is absorbed by a piece of filter paper covering the rest of the slide.

FILTERING MEMBRANES

The object of the filtering membrane technique is to concentrate the cells of fluid specimens on a small surface. The required equipment includes a filter membrane and holder, a filter cup, and a vacuum source. By filtering a small quantity of the specimen (0.5–1.0 ml) one may avoid saturation of the membrane and thus obtain a monocellular layer. This is indeed the greatest advantage of the technique.[9, 21, 23, 34, 37, 47, 59, 75, 79]

The filter membranes are then stained and mounted in a medium of matching refractive index. Prefixation in 50 percent alcohol is recommended to prevent deformation of cells. Modified techniques (dissolution of the filter in acetone or ethylacetate) have been proposed to eliminate the membrane. The cells are transferred by simply pressing the membrane to the slide and then the filter is dissolved.

Two types of filter are available: the cellulose and the polycarbonate types. They differ in their chemical, physical, and optical properties. Cellulose filters are thicker, more porous, and will take up more stain. The birefringence of polycarbonate filters makes the pore outlines more visible and may be distracting to some workers. The pores are 5 or 8μ in diameter according to the type of filter used.

Polycarbonate filters require a slighter negative pressure because the pores are smaller and more regular than cellulose filters. A small quantity (0.5–1.0 ml) of fluid is filtered to prevent saturation of the membrane.

Disadvantages of the method are the cost of the filtering membranes, the rapid cellular saturation of the filter, the distortion of cells during fixation and staining, and the thickening of the filter with subsequent difficulty in focusing the cellular elements.

In summary, the filter membrane method is useful to concentrate cells from fluids containing rare elements such as spinal fluid, clear urine specimens, and synovial fluid. It should be mentioned that fluids containing mucus or other proteins will rapidly clog the membrane. Some authors have advocated the use of the method in pulmonary cytology to reduce screening time.

We use the filtering membranes for cerebrospinal fluid and prefer the cytocentrifuge for all other types of sparse cell population fluids. The latter technique makes for better cell preservation.

STAINING

A stain is capable of coloring one or more substances because it possesses certain molecular groups (chromophoric groups) responsible for the color and other groups responsible for the chemical linkage of the former with the substance to be stained.

In light microscopy, it is absorption by a molecule of certain wave lengths of the visible spectrum that "colors" that molecule. The light reflected by the substance in question is, in effect, lacking the absorbed wave lengths; it is this that creates visible color.

Empirism has fostered the adaptation of the innumerable histologic and cytologic staining techniques in use today. We will limit the present discussion to those stains used in clinical cytology. These methods must meet two basic requirements. First, they must adequately display the structure of the nuclear chromatin, and second, they must fit the tinctorial affinities of the cytoplasm.[8]

The most frequently used procedures of clinical cytology are the Papanicolaou and Harris-Shorr methods, but other techniques, particularly the hematoxylin and Giemsa ones, also are used.[76] This last procedure is utilized after room temperature desication fixation, and it is most often the tool of those cytopathologists who are familiar with techniques used in hematology.

For the imprint smears frequent in surgical pathology, it is convenient to use hematoxylin-eosin or toluidine blue. Both are easy to use and give results comparable to those obtained with the frozen section technique of histology (cryostat).

Papanicolaou's method and the combined methods of Harris and Shorr as proposed by Pundel[56] are both based on the same principles. The nuclear stain common to both methods is Harris hematoxylin. This solution contains hematoxylin, which upon oxidation by mercuric oxide is transformed into hematin. This hematin will stain the nucleus only after mordanting by aluminum (alum of potassium or ammonia).

Mordanting, a process in common use in the textile industry, signifies the formation of a complex between a metal and the stain. It is this complex that attaches to the substance to be stained.

The cytoplasmic stains are orange G, fast green, eosin, and Bismarck brown for Papanicolaou's method; and Biebrich scarlet, orange G, and fast green for Shorr's method.[67]

The results are the following: nuclei stain a blue-violet; acidophilic cells are a pink-orange; and basophilic cells are bright blue or greenish.

The terms "eosinophilic" and "cyanophilic" should be used in preference to acidophilic and basophilic, since the tinctorial affinities of cytoplasm correspond to diverse factors and not merely to acidity or basicity. Among such factors are the quality of fixation, the temperature at which fixation is performed, the nature and concentration of the stain, the ionic strength and the pH of the staining solution. In general terms, basophilia indicates that the structure in question contains nucleic acids, polysaccharides with acid functions, or certain derivatives of the oxidation of lipids. Acidophilia is primarily an indication of keratin, such as in squamous keratinized cells.

Staining Methods of Papanicolaou and Harris-Shorr

Harris Hematoxylin

Hematoxylin crystals	5.0 gm
Ethyl alcohol, absolute	500 ml
Potassium or ammonium alum	100 gm
Mercuric oxide	2.5 gm
Distilled water	1000.0 ml
Glacial acetic acid	2.0–4.0 ml per 100 ml of solution

N.B. Glacial acetic acid is *not* added in Papanicolaou's modification of the technique.

Orange G 6 (OG 6)

Orange G, 0.5% solution in 95% ethyl alcohol	100.0 ml
Phosphotungstic acid	0.015 gm

Eosin Azure 50 (EA 50)

Light green SF, 0.1% solution in 95% ethyl alcohol	45.0 ml
Bismark brown, 0.5% solution in 95% ethyl alcohol	10.0 ml
Eosin, 0.5% solution in 95% ethyl alcohol	45.0 ml
Phosphotungstic acid	0.2 gm
Saturated lithium carbonate solution in distilled water	1 drop

EA 65 is similar to the EA 50 formula, but the light green SF quantity is halved.

Papanicolaou Staining Method

Ethyl alcohol, 70%	1 minute
Distilled water	1 minute
Harris hematoxylin	1–3 minutes
Distilled water	1 minute
Differentiate in Acid alcohol, (1% hydrochloric acid in 95% ethyl alcohol)	1 minute
Tap water	5 minute
Ethyl alcohol, 70%	rinse
Ethyl alcohol, 95%	rinse
Ethyl alcohol, absolute	rinse
Orange G6	2 minutes
Ethyl alcohol, 95%	2 baths
Xylol	1 minute
Mount (balsam, Permount, etc.)	

Harris-Shorr Staining Method

Shorr's Stain

Ethyl alcohol, 50%	100.0 ml
Orange G	0.25 gm

Biebrich scarlet (water soluble)	0.5 gm
Fast green CFC	0.075 gm
Phosphotungstic acid	0.5 gm
Phosphomolybdic acid	0.5 gm
Glacial acetic acid	1.0 ml

Fixation

Ethyl alcohol, 70%	rinse
Distilled water	rinse
Harris' hematoxylin	1–4 minutes
If necessary, differentiate in a 1% aqueous solution of hydrochloric acid	
Tap water	5 minutes
Lithium carbonate saturated aqueous solution	rinse
Ethyl alcohol, 70%	rinse
Ethyl alcohol, 95%	rinse
Ethyl alcohol, absolute	rinse
Xylol	1 minute
Mount (balsam, Permount, etc.)	

Special Staining Methods

A great many staining techniques have been devised to fit particular requirements such as identification of specific enzymes or other substances, or even to improve upon the results of the usual techniques. While the generalities of these special techniques merit some discussion, we refer the reader to more specialized texts for their detailed descriptions.[24,41,48,55]

The Methyl-Green/Pyronin Method

Use: Selective staining of nucleic acids (DNA and RNA).

The use of these two basic stains selectively colors nucleic acids. Methyl green is specific for DNA and, consequently, colors the nucleus green. Pyronin stains the endoplasmic reticulum, RNA, ribosomes, and the nucleolus red.

Methyl-Green/Pyronin Solution

Methyl green*	0.66 gm
Pyronin Y	0.66 gm
Distilled water (hot)	100 ml

N.B. Methyl green must be extracted several times with chloroform to remove the methyl violet. The solution is stable for several months.

Staining Procedure

Distilled water	rinse
Methyl-green pyronin solution	15–20 minutes
Distilled water	rinse quickly
Dehydrate in xylol	rinse
Mount	

The Periodic Acid-Shiff Base (PAS) Method

Use: Detection of simple polysaccharides (glycogen) and mucopolysaccharides (glycoproteins, mucoproteins); acid polysaccharides (mucus, cartilage, thyroid colloid, heparin, basal lamina).

The principle of this method lies in the formation of aldehydes by oxidation of the glycol groups of glycogen or other polysaccharides. These aldehyde groups are then stained via recoloration of bleached basic fuchsin. The basic fuchsin, which is previously rendered colorless with sulfurous acid, resumes its red or purple color via the Schiff reaction. If subsequent enzymatic digestion by ptyalin from saliva causes the stain to disappear, it has been proved that glycogen was indeed present.

Periodic Acid Solution

Periodic acid	1.0 gm
Distilled water	100.0 ml

Shiff Reagent Solution

Basic fuchsin	1.0 gm
Distilled water	200.0 ml

Dissolve fuchsin in boiling distilled water. Cool to 50°C, filter, and add normal hydrochloric acid (20.0 ml). Cool to 20° C and add anhydrous sodium or potassium bisulfite (1.0 gm). Solution must become straw colored. Keep under refrigeration.

Normal Hydrochloric Acid Solution

Hydrochloric acid	83.5 ml
Distilled water	916.5 ml

Reducing Solution

Normal hydrochloric acid	5.0 ml
Sodium or potasium metabisulfite, 10%	5.0 ml
Distilled water	90.0 ml

Staining Procedure

Distilled water	rinse
Periodic acid solution	5 minutes
Distilled water	rinse
Schiff reagent solution	15-20 minutes
Reducing solution	2 minutes in
(three different baths)	each bath
Wash in running water	30 minutes
Harris hematoxylin	10 minutes
Differentiate in 1% hydrochloric acid in 70% ethyl alcohol	
Tap water	rinse
Ammonia water, 3% (to blue)	dip
Tap water	wash 10 minutes

Dehydrate successively in 70%, 90%, and absolute ethyl alcohol
Xylol
Mount

The Lugol Method

Use: Detection of glycogen on smear samples; effected with iodine (Lugol's solution), which yields a mahogany (brown) color.

This technique is of clinical importance in the demonstration of abnormalities of the squamous epithelium of the cervix. The cervix is painted with Lugol solution and then observed with the colposcope. Those areas that do not react to the iodine treatment reveal either an atypical, glycogen-poor epithelium or zones of epithelial erosion.

Lugol's Solution

Potasium iodide	2.0 gm
Iodine	1.0 gm
Distilled water	100.0 ml

The Mucicarmin Method

Mucus is stained red in the Mayer's mucicarmine method.

Mucicarmine Solution

Carmine	1.0 gm
Aluminum chloride, anhydrous	0.5 gm
Distilled water	2.0 ml

Heat carefully while shaking until solution becomes dark red (approximately 2 minutes). Add 50.0 ml of 50% ethyl alcohol. Filter after 24 hours.

Metanil Yellow Solution

Mentanil yellow	0.25 gm
Distilled water	100.0 ml
Acetic acid	0.25 ml

Weigert Iron- Hematoxylin Solution

Hematoxylin crystals	1.0 gm
Ethyl alcohol, 95%	100.0 ml

Use a clean glass bottle and leave in a normally lighted area. Wait a month before use.

Weight Iron-Hematoxylin Solution B

Ferric chloride	2.0 gm
Hydrochloric, acid 24%	1.0 ml
Copper acetate, 4%	1.0 ml
Distilled water	95.0 ml

Ten minutes before use, prepare the working solution: equal parts of solution A and solution B.

Staining Procedure

Weigert's nonhematoxylin	7 minutes
Tap water	10 minutes
Mucicarmine solution	30–60 minutes
	(verify staining
	under micro-
	scope)
Distilled water	rinse quickly
Mentanil yellow solution	a few seconds
Distilled water	rinse quickly
Ethyl alcohol, 95%	rinse
Absolute ethyl alcohol (two baths)	rinse
Xylol	
Mount	

The Alcian Blue Method

A blue-green color is conferred to acid polysaccharides (mucus) by Alcian blue in acid solution.

Alcian Blue Solution

Alcian blue	0.1 gm
Acetic acid, 3% aqueous solution	100.0 ml
Adjust the pH to 2.5. Filter and add one crystal of thymol.	

Nuclear Fast Red Solution (Kernechtrot)

Nuclear fast red	0.1 gm
Aluminum sulfate, 5%	100.0 ml

Heat dissolve nuclear fast red. Filter. Add one crystal of thymol.

Acetic Acid Solution

Glacial acetic acid	3.0 ml
Distilled water	97.0 ml

Staining Procedure

Acetic acid solution	3 minutes
Alcian blue solution	30 minutes
Running water	10 minutes
Distilled water	rinse
Nuclear fast red solution	5 minutes
Running water	1 minute
Dehydrate in 70%, 95% and absolute ethyl alcohol.	
Mount	

The Congo Red Method
Amyloid substances are stained red by an aqueous solution of congo red.

Congo Red Solution

Congo red	1.0 mg
Distilled water	100.0 ml

Lithium Solution

Lithium carbonate	1.3 gm
Distilled water	100.0 ml
Harris hematoxylin (see page 215)	

Staining Procedure

Congo red solution	30–40 minutes
Lithium carbonate	rinse 2 seconds
Lithium carbonate (new bath)	rinse 10 seconds
Tap water	5 minutes
Differentiate in 80% ethyl alcohol until the smear appears slighty red	
Distilled water	30–60 seconds
Harris hematoxylin	1–2 minutes
Differentiate in a 1% aqueous solution of hydrochloric acid	
Tap water	rinse
Dehydrate in 95% and absolute ethyl alcohol	
Xylol	
Mount	

Fontana-Masson Method
Argentaffin granules (melanin-serotonin) are stained black by reduction of ammoniacal silver nitrate.

Silver Nitrate Solution

Silver nitrate	10.0 gm
Distilled water	100.0 ml

Add ammonia drop by drop to 95.0 ml of silver nitrate solution (approximately 2.4 ml) until a dark brown precipitate appears and then disappears.

Gold Chloride Solution

Gold chloride, 1% aqueous	10.0 ml
Distilled water	40.0 ml

Sodium Thiosulfate Solution

Sodium thiosulfate solution	5.0 gm
Distilled water	100.0 ml

Nuclear Fast Red Solution (Kernechtrot)

(See page 219)

Add 5 ml of silver nitrate solution drop by drop until the solution is slightly cloudy. Let stand overnight (stock solution). For *working solution,* mix 25 ml of the stock silver solution with 75 ml of distilled water and filter.

Staining Procedure

Silver nitrate solution	1 hour at 56° C
Distilled water	rinse
Gold chloride solution	10 minutes
Distilled water	rinse
Sodium thiosulfate solution	5 minutes
Distilled water	rinse
Nuclear fast red solution	5 minutes
Distilled water	rinse
Dehydrate in 95% and absolute ethyl alcohol	
Xylol	
Mount	

Kossa's Method
Calcium deposits are stained black by reduction of silver nitrate.

Kossa's Method for Calcium

Silver nitrate	5 gm
Distilled water	100.0 ml
Pyrogallic acid, 10% aqueous solution	
Sodium thiosulfate, 5% aqueous solution	

Staining Procedure

Silver nitrate solution	30–60 minutes, under bright light
Distilled water	rinse briefly
Pyrogallic acid solution	1–3 minutes
Distilled water	rinse briefly
Sodium thiosulfate solution	3–5 minutes
Tap water	wash thoroughly
Nuclear fast red (optional step)	5 minutes
Dehydrate in 70%, 95%, and absolute ethyl alcohol	
Xylol	
Mount	

The Toluidine Blue Method
The heparin granules of mastocytes may be demonstrated by an acid solution of toluidine blue.

Toluidine Blue Solution

Toluidine blue	0.1 gm
Distilled water	100.0 ml

Staining Procedure

Toluidine blue solution	1–2 minutes
Distilled water	rinse

One of the following procedures:

1. Cover slide with distilled water
   Blot around edges of slide
   Put on cover slip
   Seal with fingernail polish
2. Dry the slide completely
   Xylol
   Mount

The Prussion Blue Method (Perls' Method)

Ferric iron is demonstrated in the form of blue granulations of ferric ferrocyanide.

Perls' Solution

Potassium ferrocyanide	10.0 gm
Distilled water	100.0 ml
Hydrochloric acid	20.0 ml
Distilled water	80.0 ml
Nuclear fast red solution (see page 219)	

Staining Procedure

Mix potassium ferrocyanide solution and hydrochloric acid in equal parts	20 minutes
Distilled water	rinse
Nuclear fast red solution	5 minutes
Distilled water	rinse
Dehydrate in 95% and absolute ethyl alcohol	
Xylol	
Mount	

The Oil Red O Method

A solution of oil red O will demonstrate the presence of lipids. (Smears must first be fixed with formaldehyde.)

Oil Red O Solution

Oil red O	2.0 gm
Ethyl alcohol, 70%	50 ml
Acetone	50 ml
Glycerin jelly	
Gelatin	10.0 gm
Distilled water	60.0 ml
Heat until gelatin is dissolved and add:	
Glycerin	70.0 ml
Phenol	1.0 gm

Staining Procedure

Ethyl alcohol, 70%	wash securd
Oil red O solution	5 minutes in a
	closed pan
Ethyl alcohol, 70%	rinse
Distilled water	rinse
Harris hematoxylin	5 minutes
Distilled water .	rinse
Bluing in aqueous solution of ammonia, 3%	
If necessary, differentiate in acid-water solution	1%
Distilled water	rinse
Mount with glycerin jelly	

FLUORESCENCE MICROSCOPY

The Acridine Orange Fluorescence Method

This technique, perfected by Bertalanffy, Masin, and Masin,[10, 81] allows demonstration of nuclei acids via acridine orange staining and fluorescence microscopic examination.[74, 80]

The nuclei appear green and the cytoplasm red (keratinized squamous cells excepted). The higher the concentration in nuclei acids, the greater fluorescence. Trichomonas and fungi appear red.

The advantage of this method resides in the remarkable brilliance of both the nuclear and cytoplasmic staining. This facilitates observation of tinctorial abnormalities and in general makes for rapid reading.

The method is not an ideal routine technique, however, as it presents certain inconveniences. Its manipulations are more delicate and more costly than those of Papanicolaou's technique, and from the morphologic viewpoint, it does not provide any new histologic criteria for the identification of tumor cells. Moreover, problems of differential diagnosis among inflammatory reactions, dysplasia, and cancer are no better solved by this method than by others.

Excellent as an investigative means, its routine accuracy is no better than the classic techniques of Papanicolaou and Harris-Shorr.[88]

HISTOCHEMISTRY

Histochemistry is utilized in certain domains of clinical cytology either to determine the precise nature of a cell or to help understand its function. For example, it is used in the detection of acid and alkaline phosphatases and lacticodehydrogenases in effusions of osseous sarcomas and prostatic carcinomas.

METACHROMASIA

Metachromasia is the color modification of a cationic stain by its fixation and polymerization on electronegative substances of high molecular weight. Metachromasia may be demonstrated with the toluidine blue method (see page 221).

ELECTRON MICROSCOPIC TECHNIQUES

Transmission electron microscopy and scanning electron microscopy have yielded valuable information on the ultrastructural aspects of exfoliated cells.[45]

Transmission electron microscopy reveals the various organelles present in different cell types. They have been described in Chapter 1.

Scanning electron microscopy (SEM) studies reveal the morphology of the cellular surface: microvilli and microridges. The microridges represent small, numerous interlacing convolutions that have been described only in keratin-containing cells. Modifications in the number and shape of microvilli have been described extensively in different types of cells. Figures of scanning electron microscopy illustrate these findings (Figures 13.2 and 13.3).

These electron microscopy techniques have contributed to a better knowledge of cell

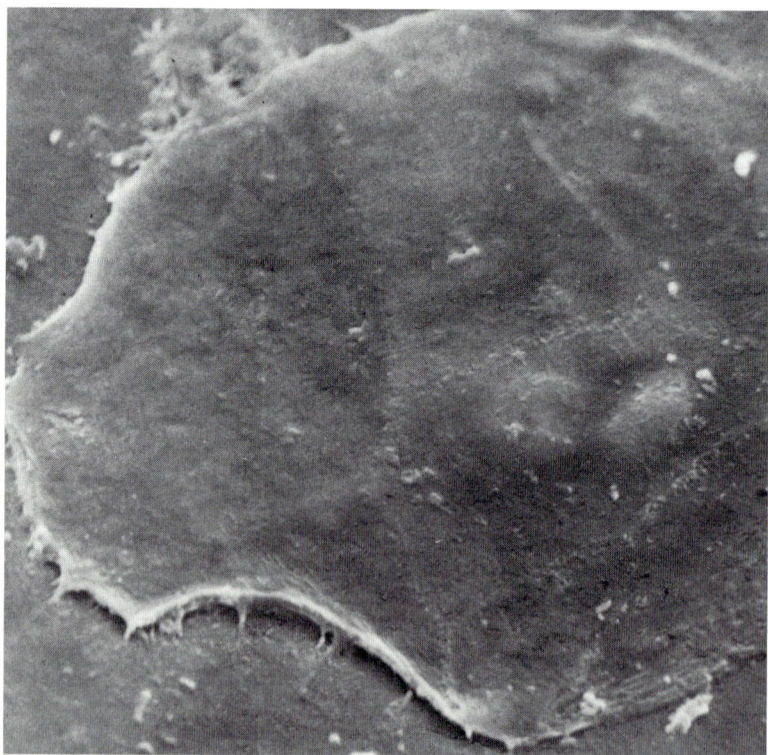

Figure 13.2. Scanning electron microscopy. Normal desquamated superficial squamous cell ($\times$ 1000). (Courtesy of Dr. R. M. Richart, New York.)

Figure 13.3. Scanning electron microscopy. Normal desquamated basal and superficial squamous cells (× 900). (Courtesy of Dr. R. M. Richart, New York.)

structure. However, the sophistication of these methods and the value of observations by light microscopy explain why the former are not used in the routine work of cytology laboratories.

QUALITY CONTROL IN CYTOLOGY

The pursuit of quality in the operation of a cytology laboratory requires the establishment of different systems of controls.[32] Such operational quality is necessary to maintain high professional standards and to obtain the necessary licenses and accreditations from different authorities.

Thus, the following activities must be realized, supervised, and if possible, evaluated: collection of specimens; histologic correlation; educational training; continuing education; proper individual allotment of work; working conditions; fixing, staining, and processing conditions; record keeping; follow-up of cases; permanent supervision of work by the director of the laboratory; and mutual control of cytoscreeners' work.

The rescreening of a certain percentage of negative smears has been recommended, but this technique appears to be impractical. It has been clearly demonstrated that the number of slides that would have to be rescreened to distinguish acceptable from unacceptable error ratios exceeds the technical and financial possibilities of a routine laboratory.[49] The same obstacles arise in evaluating the work of a technician by rescreening.

The introduction of known abnormal smears into the laboratory routine work without the knowledge of the screeners is an acceptable method; it represents the more effective procedure if it is honestly performed. This method requires unstained smears and fictitious requisition forms to be thoroughly effective.

Other possibilities would be the creation of a regional or national reference center with a panel of consulting cytopathologists or the organization of a group of recognized laboratories involved in a rotating system where randomly chosen cases could be regularly sent. The material sent to these centers should be routine slides — and not specially prepared material — to give a true indication of the quality of work done.

False-negative and false-positive smears have been discussed on page 87.

BIBLIOGRAPHY

1. Ahlquist J. Methyl green pyronin staining: effects of fixation; use in routine pathology. *Stain Technol.* 47:17, 1972.

2. Anderson W.A.D. Gunn S.A. The efficacy of a single cytology smear in the detection of cancer of the cervix. *Cancer* 13:92, 1963.

3. Apffel C.A., Baker J.R. Lipid droplets in the cytoplasm of malignant cells. *Cancer* 17:176, 1964.

4. Bahr G.F., Bartels P.H., Bibbo M., De Nicolas M., Wied G.L. Evaluation of the Papanicolaou stain for computer assisted cellular pattern recognition. *Acta Cytol.* 17:106, 1973.

5. Baldet P. Eléments de cytologie ultrastructurale. *Rev. Cytol. Clin.* 1:19, 1972.

6. Baldet P. Eléments de cytologie ultrastructurale (suite). *Rev. Cytol. Clin.* 2:67, 1972.

7. Barr W.T., et al. Cellular contamination during automatic and manual staining of cytological smears. *J. Clin. Path.* 23:604, 1970.

8. Bartels P.H., et al. Analysis of variance of the Papanicolaou staining reaction. *Acta Cytol.* 18:522, 1974.

9. Beaumer M., Nuovo N. Technique de concentration sur filtre millipore. *Arch. Anat. Path.* 15:50, 1967.

10. Bertalanffy F.D. Diagnostic reliability of the acridine orange fluorescence microscope method for cytodiagnosis of cancer. *Cancer Res.* 21:422, 1961.

11. Bibbo M., Bartels P.H., Bahr G.F., Taylor J., Wied G.L. Computer recognition of cell nuclei from the uterine cervix. *Acta Cytol.* 17:340, 1973.

12. Bibbo M., Bartels P.H., Chen M., Harris M.J., Trauttman B., Wied G.L. The numerical composition of cellular samples from the female reproductive tract. I. Carcinoma in situ. *Acta Cytol.* 19:438, 1975.

13. Cellier K.M., Kirkland J.A., Stanley M.A. Statistical analysis of cytogenetic data in cervical neoplasia. *J. Nat. Cancer Inst.* 44:1221, 1970.

14. Conti C., Haour P., Bernhard G. Diagnostic cytologique du cancer par deux techniques associées: fluorescence puis coloration de Papanicolaou. Résultats comparatifs à propos d'une série de 25,000. cas. *Rev. Cytol. Clin.* 4:37, 1971.

15. Creasman W.T., Weed J.C. Screening techniques in endometrial cancer. *Cancer* Vol. 38, Suppl., July 1976.

16. Davis H.J. The irrigation smear. A cytologic method for mass population screening by mail. *Amer. J. Obstet. Gynec.* 84:1017, 1962.

17. De Allende I.L.C., Orias D. *Cytology of the Human Vagina.* Hoeber, New York, 1950.

18. Dixon K.C. Fatty deposition of disorder of the cell. *Quart. J. Exp. Physiol.* 43:139, 1968.

19. Feldman M., Poulsen R., Shepherd L., Marshall K.G. The occurrence of isolated dysplastic and carcinoma in situ type cells in cervical smears from patients with dysplasia and carcinoma in situ: significance to prescreening using image processing techniques. *Acta Cytol.* 17:395, 1973.

20. Fennessy J.J., Fry W.A., Manalo Estrella P., Frias Hidvegi D. The bronchial brushing technique for obtaining cytologic specimens from peripheral lung lesions. *Acta Cytol.* 14:25, 1970.

21. Fields M.J., Martin W.F., Young B.L., Tweedale D.N. Application of the Nedelkoff-Christopherson millipore method to sputum cytology. *Acta Cytol.* 10:220, 1966.

22. Frost J.K. Diagnostic accuracy of "cervical smears." *Obstet. Gynec. Surv.* 24:893, 1969.

23. Miller R.A., Hollander D.H. Cytology filter preparations: factors affecting their quality for study of circulating cancer cells in the blood. *Acta Cytol.* 11:363, 1967.

24. Gabe M. *Techniques histologiques.* Masson, Paris, 1968.

25. Gaudefroy M., Becquet R. Un diagnostic difficile et inattendu: l'oxyurose cervico-vaginale. *Rev. Cytol. Clin.* 2:17, 1969.

26. Gray P. *The microtomists Formulary and Guide.* McGraw-Hill, New York, 1954.

27. Hajdu S.I., Melamed M.R. The diagnostic value of aspiration smears. *Amer. J. Clin. Path.* 59:350, 1973.

28. Hansen H.H., Bander R.A., Shelton B.J. The cytocentrifuge and cerebrospinal fluid cytology. *Acta Cytol.* 18:259, 1974.

29. Hindman W.M. An effective quality control program for the cytology laboratory. *Acta Cytol.* 20:233, 1976.

30. Hoshino K., Ray M., Ward E. Cytophotometric and cytogenetic studies of prolactinsecreting transplantable pituitary tumor cells in C57BL Mice. *Acta Cytol.* 19:337, 1975.

31. Husain O.A.N. Automation in cytology. Internal report, March 1972, London.

32. Husain O.A.N., Butler B., Evans D.M.D., McGregor J.E. Quality control in cervical cytology. *J. Clin. Path.* 27:935, 1974.

33. Ito U., Inaba Y. A simple sedimentation chamber adaptable to the laboratory centrifuge. *Amer. J. Clin. Path.* 58:590, 1972.

34. Juniper K., Chester Cl. Filter membrane technique for cytologic study of exfoliated cells in body fluids. *Cancer* 12:278, 1959.

35. Kaltenbach F.J., Hillemanns H.G., Fettif O., Hilgarth M. Thrombin cell block technic in gynecologic diagnosis. *Acta Cytol.* 17:128, 1973.

36. Kanbour A., Klionsky B., Dym J. The Gravlee jet washer: evaluation of the reflux into the fallopian tubes. *Acta Cytol.* 16:199, 1972.

37. Kington E. Tweeddale D.N. A technique to prepare cytologic filter slides from solid tissues. *Acta Cytol.* 19:155, 1975.

38. Klavins J.V., Flemma R.J. A method for studying the material of the bile ducts. *Acta Cytol.* 8:332, 1964.

39. Kyrkos K., Zacharizdou-Veneti S., Candreviotou S. A comparative study of the Papanicolaou staining method and the fat stain technique in malignant and non-malignant lesions of the urinary tract. *Acta Cytol.* 19:67, 1975.

40. Leif R.C., Gall S., Dunlap L.A., Railey C., Zucker R.M., Leif S.B. Centrifugal cytology. IV. The preparation of fixed stained dispersions of gynecological cells. *Acta Cytol.* 19:159, 1975.

41. Lillie R.D. *Histopathologic Technic and Practical Histochemistry.* 3rd Ed. McGraw-Hill, New York, 1965.

42. Lowhagen T., Nasiell M., Granberg I. Acridine orange fluorescence cytology in detection of cervical carcinoma. *Acta Cytol.* 10:194, 1966.

43. Luna, L.G. (Ed.). *Manual of Histologic Staining Methods of the Armed Forces Institute of Pathology.* McGraw-Hill, New York, 1968.

44. Luse S.A., Reagan J.W. A histocytological study of effusion. *Cancer* 7:1155, 1954.

45. McDowell E.M., Trump B.F. Histologic fixatives suitable for diagnostic light and electron microscopy. *Arch. Path. Lab. Med.* 100:405, 1976.

46. Mainigi K.D. Activities of certain enzymes of glycolytic pathway in normal, chronic cervicitis and malignant human cervix uteri. *Oncology* 26:427, 1972.

47. Mainigi K.D. Studies on the activities of certain enzymes carbohydrate and amino acid metabolism in normal chronic cervicitis and malignant human cervix uteri. *Oncology* 26:438, 1972.

48. Mandard A.M., et al., technique Duigou F. Utilisation des filtres nucléopores à l'étude cytologique des liquides a faible densite cellulaire. *Rev. Cytol. Clin.* 7:41, 1974.

49. Melamed M.R. Presidential address. Twentieth Annual Meeting: American Society of Cytology. *Acta Cytol.* 17:285, 1973.

50. Miller E.M., von Haam E. A comparison of the vaginal aspiration and cervical scraping technics in the screening process for uterine cancer. *Acta Cytol* 5:214, 1956.

51. Nieburgs H.E. *Diagnostic Cell Pathology in Tissue and Smears.* Grune Stratton New York, 1967.

52. Norquist L.A., Scweid A.I. Preparation of cytologic smears from solid tumors. *Acta Cytol.* 15:343, 1971.

53. Okagaki T. Cytology laboratory information system using a mini-computer. *Amer. J. Clin. Path.* 62:797, 1974.

54. Papanicolaou G.N. A new procedure for staining vaginal smears. *Science* 95:438, 1942.

55. Pearse A.G.E. *Histochemistry: Theoretical and Applied.* 2nd Ed. Churchill, London, 1960.

56. Pundel J.P. *Précis de colposcopie hormonale.* Masson, Paris, 1966.

57. Rashad A.L., Evans C.A. Significance of abnormal sites of DNA synthesis in certain lesions of the human uterine cervix. *Amer. J. Obstet. Gynec.* 108:435, 1970.

58. Ravaut G.L., Widal I., Sicard L. A propos du cytodiagnostic du tabes. *Rev. Neurol.* 6:289, 1903.

59. Raynaud A.J., King E.B. A new filter for diagnostic cytology *Acta Cytol.* 11:289, 1967.

60. Rebel A., Simard C., Bertrand G., Francois J. Intérêt de l'utilisation de la cytocentrifugeuse dans le diagnostic des adénopathies et des tumeurs. *Rev. Cytol. Clin.* 7:35, 1974.

61. Richart R.M. The handling of small tissue samples for pathologic examination. *Bulletin of the Sloane Hospital for Women* 9:113, 1963.

62. Richart R.M. Evaluation of the true false negative rate in cytology. *Amer. J. Obstet. Gynec.* 89:723, 1964.

63. Richart R.M., Vaillant H.W. Influence of cell collection techniques upon cytological diagnosis. *Cancer* 18:1474, 1965.

64. Richart R.M., Vaillant H.W. The irrigation smear: false negative rates in a population with cervical neoplasia. *JAMA* 192:199, 1965.

65. Rodney M.B. Data control and quality of the human cytoscreening function. *J. Health and Lab. Sci* 9:215, 1972.

66. Schachter A., Bercovici B. Signification de la cytolyse pendant les phases du cycle menstruel. *Rev. Cytol. Clin.* 5:89, 1972.

67. Shorr, E. New technic for staining vaginal smears. III. Simple differential stain. *Science,* 94:545, 1941.

68. Smith J.W., Townsend D.E., Sparkes R.S. Genetic variants of glucose-6-phosphatase dehydrogenase in the study of carcinoma of the cervix. *Cancer* 28:529, 1971.

69. Soderstrom N. *Fine-Needle Aspiration Biopsy.* Almquist and Wiksell, Stockholm, 1966.

70. Spalsbury C., Brodetsky A.M., Teplitz R.L. Discontinuous ficoll gradient separation of normal and malignant cells: cytologic applicability. *Acta Cytol.* 17:522, 1973.

71. Sprenger E., Moore G.W., Naujoks H., Schlueter G., Sandritter W. DNA content and chromatin pattern analysis on cervical carcinoma in situ. *Acta Cytol.* 17:27, 1973.

72. Spriggs A.I., Boddington M.M. *The Cytology of Effusions.* Heinemann, London, 1968.

73. Stenkvist E.O., Brege K.G. Application of immunofluorescent technique in the cytologic diagnosis of human herpes simplex keratitis. *Acta Cytol.* 19:411, 1975.

74. Stevenson J.L., Von Haam E. The application of acridine orange fluorescence microscopy to experimental carcinoma of the uterine cervix in mice. *Acta Cytol.* 8:402, 1964.

75. Suprun H. A comparative filter technique study and the relative efficiency of these sieves as applied in sputum cytology for pulmonary cancer cytodiagnosis. *Acta Cytol.* 18:248, 1974.

76. Symposium on cytological staining or microscopic techniques other than the original Papanicolaou method. *Acta Cytol.* 2:283, 1958.

77. Tang T.T., McCreadie S.R. A simplified method for the cytologic study of body fluids. *Amer. J. Clin. Path.* 59:113, 1973.

78. Terzand G. The vaginal smear during the luteal phase of the normal menstrual cycle. *Acta Cytol.* 6:24, 1962.

79. Thabet R.J., Knoerschild H.E. Millipore filtration technic for colon washings. *Amer. J. Clin. Path.* 34:185, 1960.

80. Umiker W.L., Pickle L., Waite G. Fluorescence microscopy in exfoliative cytology: evaluation for its application in cancer screening. *Brit. J. Cancer* 13:393, 1959.

81. von Bertalanffy L., Masin F., Masin M. Use of acridine-orange flucrescence technique in exfoliative cytology. *Science* 124:1024, 1956.

82. von Haam E. A comparative study of the accuracy of cancer cell detection by cytologic methods. *Acta Cytol.* 6:508, 1962.

83. Wachtel E.G. *Exfoliative Cytology in Gynaecological Practice.* 2nd Ed. London, Butterworths, 1969.

84. Wachtel E.G. L'examen du liquide amniotique. *Rev. Cytol. Clin.* 1:37, 1972.

85. Way S., Dawson N. Symposium — advantages and disadvantages of various techniques of obtaining material for routine cytological examinations. *Acta Cytol.* 4:247, 1960.

86. Wheeless L.L., Patten S.F., Cambier M.A. Slit-scan cytofluorometry: data base for automated cytopathology. *Acta Cytol.* 19:460, 1975.

87. Wied G.L. Terminology of cytologic reporting of endocrinologic conditions. *Acta Cytol.* 8:383, 1964.

88. Wied G.L., Manglano J.I. A comparative study of the Papanicolaou technic and the acridine-orange fluorescence method. *Acta Cytol.* 6:554, 1962.

89. Wolman M. Problems of fixation in cytology histology and histochemistry. *Intern. Rev. Cytol.* 4:79, 1955.

90. Yataganas X., Mitomo Y., Traganos F., Strife A., Clarkson B. Evaluation of a Feulgen-type reaction in suspension using flow microfluorimetry. *Acta Cytol.* 19:71, 1975.

Index